LAST SLEEP

The Battle of Droop Mountain

November 6,1863

"Poor me, with shock, [Corp. William S.S.] Morris & [2nd Corp. William H.] Hubbard of our old company, together with other true friends & gallant soldiers now 'sleep their last sleep' on Droop Mountain. My loss is great. I mourn sincerely."

–Lt. Col. Andrew R. Barbee
22nd Virginia Infantry

LAST SLEEP

The Battle of Droop Mountain

November 6,1863

by Terry Lowry

Charleston, WV

Charleston, WV

LIBRARY OF CONGRESS
CATALOG CARD NO. 96-70678

ISBN 13: 978-1-942294-49-8
ISBN 10: 1-942294-49-2

10 9 8 7 6 5 4 3

Typography: Leslie Maricelli
Layout: Stan Cohen
Cover Design and Graphics: Egeler Design

Distributed by:

West Virginia Book Company
1125 Central Ave. Charleston, WV 25302
www.wvbookco.com

Acknowledgments

Always in such an endeavor there exist numerous individuals and institutions without whose assistance and understanding such a project would not be possible. First and foremost, a very special thanks to Joan Marker, South Charleston, West Virginia for introducing me to the computer age, typing, aiding with research, creating contemporary maps, and generally keeping me in focus. Unquestionably, her support and care were essential to the completion of this project. Also, much gratitude to Mike Smith, Superintendent, Droop Mountain Battlefield State Park, whose enthusiasm, interest, and cooperation in this project was invaluable. Please forgive me for anybody I may have accidentally omitted.

Michael P. Musick, Military Reference Branch, Military Archives Division, National Archives, Washington, D.C.; Michael T. Meir, Military Reference Librarian, Textual Reference Division, National Archives, Washington, D.C.; Michael E. Pilgrim, Military Reference Branch, Textual References Division, Washington, D.C.; David Wigdor, assistant chief, Manuscripts Division, Library of Congress, Washington, D.C.; John R. Sellers, Historical Specialist, Civil War and Reconstruction, Manuscripts Division, Library of Congress, Washington, D.C.

Christine M. Beauregard, Librarian, Manuscripts and Special Collections, New York State Library, Albany, New York; Elliott C. Meadows, Manuscripts Department and Mary Carey, Reference Librarian, New York Historical Society, New York, New York; Richard Scluato, Technical Assistant, Rare Books and Manuscripts Division, and Andrew H. Lee, Librarian II, New York Public Library, New York, New York; Wendy A. Swik, Military Affairs Librarian, West Point Military Academy, New York.

Sally A. Weikel, Library Technician, State Library of Pennsylvania, Harrisburg, Pennsylvania; Rosine R. Bucher, Librarian, Soldiers and Sailors Memorial Hall, Pittsburgh, Pennsylvania; Hillman Library, University of Pittsburgh, Pittsburgh, Pennsylvania; Michael Winey and Richard Sommers, U.S. Army Military History Institute, Carlisle Barracks, Pennsylvania.

Ervin L. Jordan, Jr., Technical Services Archivist, and Michael Plunkett, Curator of Manuscripts, Archivist, Alderman Library, University of Virginia, Charlottesville, Virginia, for permission to quote from the Micajah Woods Papers; Diane B. Jacob, Archivist, and Patricia F. Wahlrab, Archives Assistant, Preston Library, Virginia Military Institute, Lexington, Virginia; C. Vaughan Stanley, Special Collections Librarian, Washington and Lee University, Lexington, Virginia; Mrs. H. B. Eller, Corresponding Secretary, and Mack H. Sturgill, Smyth County Historical and Museum Society, Inc., Marion, Virginia; Museum of the Confederacy, Richmond, Virginia; Virginia Historical Society, Richmond, Virginia.

Mrs. Ronald D. Shepherd, Curator of Manuscripts, and Mrs. Anne B. Shepherd, Reference Librarian, Cincinnati Historical Society, Cincinnati, Ohio; Ohio Historical Center, Columbus, Ohio; Kenneth Noe, Graduate Assistant, University of Illinois at Urbana-Champaign, Urbana, Illinois, for assistance with the Augustus Moor Papers; Terry Bouton and Marion Hirsch, reference interns, Duke

University, Special Collections Library, Durham, North Carolina.

Regional History Collection, West Virginia University, Morgantown, West Virginia; Cora P. Teel, Manuscripts Librarian, James E. Morrow Library, Marshall University, Huntington, West Virginia; Betty B. Dakan, President, Stonewall Jackson U.D.C., and David R. Houchins, Harrison County Genealogical Society, Clarksburg, West Virginia; the entire staff of the West Virginia State Archives, Cultural Center, Charleston, West Virginia, especially Greg Carroll, Bob Taylor, Jamie Lynch, and Deborah Basham; Ray Swick, Historian, and Larry Reber, Blennerhassett Historical State Park, Parkersburg, West Virginia.

Richard Andre, Charleston, West Virginia; Tim McKinney, Fayetteville, West Virginia, for use of his extensive collection; Bob Driver, Brownsburg, Virginia, for many photos and information; Joe Geiger, Huntington, West Virginia, for information on Col. John H. Oley; Richard E. Blodgett, Jr., North Stonington, Connecticut, for assistance with the William Averell Papers; Vernon Williams, Blue Island, Illinois; Anna B. Atkins, Chesterfield, Virginia, for information on the Mountain House at Droop Mountain; Amber Gail Thompson, Proctorville, Ohio; John Marion Ashcraft, Emmaus, Pennsylvania, for information on the 19th Virginia Cavalry and 31st Virginia Infantry; Richard L. Armstrong, Warm Springs, Virginia, for assistance with information on the 19th and 20th Virginia cavalry regiments; John A. Arbogast, White Sulphur Springs, West Virginia; Dr. Calbert R. Seebert, Spartansburg, South Carolina; Robert J. Trout, Myerstown, Pennsylvania, for material on Lurty's Battery; Brian Kesterson, Washington, West Virginia, for his files on the 3rd West Virginia Mounted Infantry and the Droop Mountain battle; Jacques F. Pryor, College Station, Texas, on his ancestor, Major Robert A. Bailey; Gardner D. Beach, Frankfort, Kentucky, for his extensive notes on Capt. Thomas E. Jackson's Battery; John Scott, Salem, Virginia; A.W. Workman, Osage City, Kansas, for information on James L. Workman, 7th West Virginia Cavalry; Joe H. Ferrell, St. Albans, West Virginia; Frederick Schildkamp, Sinks Grove, West Virginia, for his outstanding research on various regiments, particularly the 22nd Virginia Infantry and 26th Battalion Virginia Infantry; Lanty McNeel, Hillsboro, West Virginia, for his collection and personal knowledge of the Hillsboro area, Robert Cartwright, Dayton, Ohio; and William D. Wintz, St. Albans, West Virginia. Noble Wyatt, Kanawha Valley Civil War Roundtable, Poca, West Virginia; Okey Payne, St. Albans, West Virginia; Richard Payne, St. Albans, West Virginia; Bruce McCune, Charleston, West Virginia; Harold Griffith, Beckley, West Virginia; Jim McNeill, Buckeye, West Virginia; Jennifer O'dell, Mt. Hope, West Virginia; Virginia McLaughlin, Alderson, West Virginia, for artifacts of her ancestor, A.M. McLaughlin; and George and Junior Goode, Droop Mountain, West Virginia.

Gracious thanks to the following land owners for permitting investigation of their property surrounding Hillsboro and the battlefield park: Lanty McNeel; James Burks; Mrs. Leroy Rose; J.T. Thomas; James Burke; and James Fleming.

West Virginia relic hunters: James "Slim" Combs, Charleston; William Price, South Charleston; Jackie Martin, South Charleston; Gary Baker, South Charleston; Gary Bays, South Charleston; Brian Abbott, Parkersburg; Sam Pittman, Charleston; George Musser, Lewisburg; Dean Winebrinner, Charlton Heights; and Tim McKinney.

All photos are credited but I must give particular thanks to Gary Bays, South Charleston, West Virginia; Brian Abbott, Parkersburg, West Virginia; Richard Andre; Tim McKinney; Mike Smith, Droop Mountain Battle State Park; Charles Bracken, Charlton Heights, West Virginia; Joe Geiger, Jr.; Steve Cunningham, Charleston, West Virginia; Mahlon P. Nichols, Bedford, Virginia; Domenick A. Serrano, Bayside, New York; Bill Turner, La Plata, Maryland; Robert Driver, Brownsburg, Virginia; Kenneth D. Swope, archivist, Greenbrier Historical Society, Lewisburg, West Virginia; Bill McNeel, Pocahontas County Historical Society, Marlinton, West Virginia; Mrs. Ernest A. (Margaret) Martin, Rupert, West Virginia; Ronn Palm, Kittanning, Pennsylvania; Richard A. Wolfe, Stafford, Virginia; and Mark Mengeles, Carnifex Ferry Battlefield State Park, Summersville, West Virginia.

Special thanks to those who helped in ways other than research, including: Stan Cohen, Pictorial Histories, for the layout and design; Mike Smith, Droop Mountain Battlefield State Park, whose assistance and cooperation were invaluable; Lisa Scarberry, South Charleston, West Virginia, for proofreading the manuscript; Chris Parsley of Charleston Blueprint, Charleston, West Virginia; Mike Sheets, Co. A, 36th Virginia Infantry Re-enactment Group, for aid with militaria; Dave Tuckwiller, Lewisburg, West Virginia; Dr. Stephen Cassis, Charleston, West Virginia, for making it all possible by saving my eyesight; Judy Bibbee, Circulation Department, Charleston Newspapers for her consideration and patience with my work schedule; Richard Marker, South Charleston, West Virginia, for gracious use of his computer; Denise Smith (D.H.B.) for her outstanding computer abilities; Charlie Tuitt and Vickie Adams, Charleston, West Virginia, for their great support; the Lowry clan: Tim and Skeeter, Chuck and Sandra, and David and Shelley. Most importantly my mother, Ruth Ann Lowry, and my late father, Charles Lowry, without whose sincere belief in me this book could never have been written.

DEDICATION

THIS BOOK IS DEDICATED TO MY MOTHER,
RUTH ANN (SUTTON) LOWRY,
WHOSE LOVE, PATIENCE, UNDERSTANDING AND DEVOTION
MADE THIS BOOK,
AS WELL AS ALL MY WORKS, POSSIBLE.

A Reader's Guide

Readers must keep in mind that on June 20, 1863, West Virginia became the 35th state of the Union. Before that date, West Virginia was yet a part of Virginia, therefore, such designations in this book as (West) Virginia, (West)ern Virginia, and so forth, refer to West Virginia prior to statehood.

For those unfamiliar with Civil War terminology, the Northern military troops are referred to as either Federals, Union, Yankees, United States Army, or by the states they represented, such as West Virginia, Ohio, Pennsylvania, New York, and Illinois. The Southern troops are referred to as either Confederates, rebels, Confederate States Army, secessionists, or the states they represented, such as Virginia. A secessionist, however, does not necessarily represent a soldier.

Some locales are known by a variety of names. Mill Point is sometimes referred to as Cackleytown; Hillsboro (sometimes spelled Hillsborough) was also known as Academy during the war; Marlinton was known as Marlin's and Marling's Bottom; and Keyser was called New Creek.

While every effort was made to present the events in this book in chronological sequence it was not always possible, as many of the events took place at the very same time. This is particularly true of the battle segments, where numerous events transpire on the battlefield simultaneously.

A number of important participants in the Droop Mountain campaign possess the same surname, which can result in confusion. For instance there is a Colonel Jackson and a Captain Jackson, as well as a Confederate and a Federal Lieutenant Colonel Thompson, in addition to a Captain Thompson.

Regarding the photos, many of the officers were photographed after the Droop Mountain fight, therefore the rank designated by their uniform is not necessarily that which they held at the battle. In addition, some are post-war views.

Hillsboro, 1917. PHOTO DOUG CHADWICK

Table of Contents

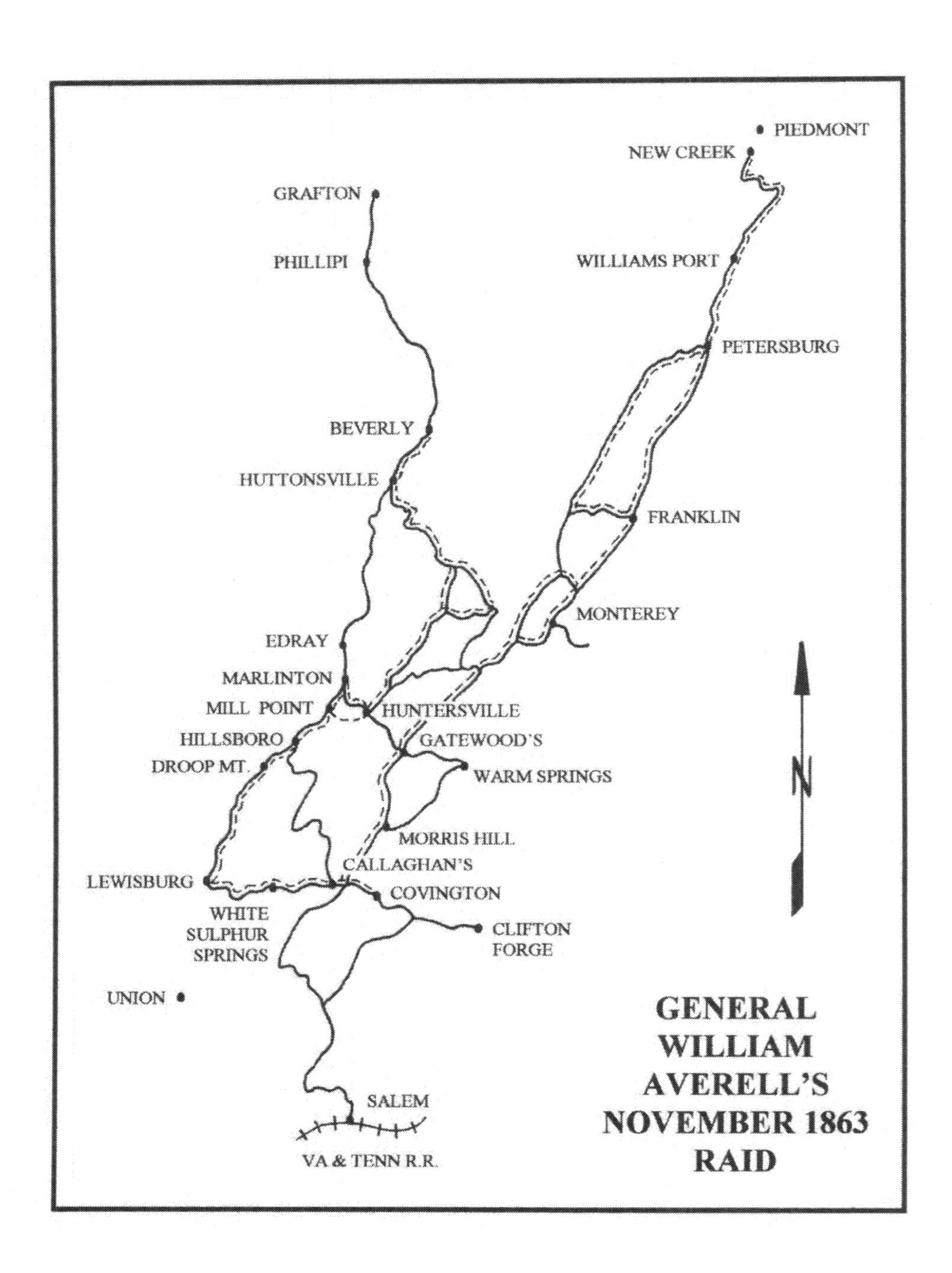
PIEDMONT
NEW CREEK
GRAFTON
PHILLIPI
WILLIAMS PORT
PETERSBURG
BEVERLY
HUTTONSVILLE
FRANKLIN
MONTEREY
EDRAY
MARLINTON
MILL POINT
HUNTERSVILLE
HILLSBORO
GATEWOOD'S
DROOP MT.
WARM SPRINGS
N
MORRIS HILL
CALLAGHAN'S
LEWISBURG
COVINGTON
WHITE
SULPHUR
SPRINGS
CLIFTON
FORGE
UNION
GENERAL
WILLIAM
AVERELL'S
NOVEMBER 1863
RAID
SALEM
VA & TENN R.R.

CHAPTER 1

Foot Soldiers in the Saddle

The battlefield it has no charms
When silence breakes to a clash of arms
The cannons roar—the muskets peal
Proclaims a bloody battlefield

The battles rage with fearful roar
Our comrades [*fall to*] rise no more—

—from The Nicholas Blues, written by Capt. James M. McNeil, Company D (Nicholas Blues), 22nd Virginia Infantry, C.S.A., who was captured at the battle of Droop Mountain November 6, 1863.

Joseph V. Rollins passed away July 3, 1925, at Beard's Fork, West Virginia.. According to family tradition he lingered in and out of consciousness for three days, delirious and often shouting out commands learned as a Confederate soldier in the American Civil War. By the second day the postwar minister's condition drastically deteriorated, and he began to frighten the young children of the family, who were asked to leave. For many years the family had listened to Rollins tell of his wartime exploits, and recognized portions of his deathbed rantings. At his worst it became painfully evident Rollins was reliving the most terrible nightmare of his life—the battle of Droop Mountain. Indeed, his frightening experience at Droop Mountain motivated him to 'desert' to the enemy lines the following month.[1]

Pvt. Joseph V. Rollins, Co. A, 22nd Virginia Infantry, shown here circa 1915. He relived the nightmare of Droop Mountain on his deathbed in 1925. TIM MCKINNEY

Rollins had plenty of reason to harbor fear. Between August and December of 1863 Federal officer Gen. William W. Averell led three daring raids into the Confederate held territory of southern West Virginia and western Virginia, all with the intent of obstructing the vital Virginia and Tennessee railroad. During the second raid, in November, Averell's near 5,000 man force ran up against about 2,000 Confederate soldiers under Gen. John Echols on the summit of Droop Mountain in Pocahontas County, West Virginia, just north of the Greenbrier County line. When the smoke cleared after a near six hour battle, over 78 soldiers lay dead, another 215 wounded, and more than 124 found themselves captured or missing in action, while the Confederate army fled the field in a total rout. Moreover, the Federal victory supposedly ended any further serious Confederate threat in West Virginia, thereby helping to secure West Virginia statehood, a claim yet open to speculation.

Although not very impressive when compared to other, larger Civil War battles, the fight at Droop Mountain has often erroneously been referred to as the largest and most important battle fought on West Virginia soil; the most fratricidal engagement of the war; and the highest elevation contested in West Virginia during the conflict. In disputing the first claim, the 1862 siege and capture of Harpers Ferry during the Antietam campaign involved more men, and the skirmishes that took place in the summer of 1861 were of more importance than Droop Mountain in that they secured northwestern Virginia (northern West Virginia) for the Union. To the claim of being the most fratricidal, Droop Mountain did primarily involve soldiers of both armies who personally knew each other, or were related. Numerous fathers, sons, brothers, and cousins fought against each other on nearly every part of the field. The only battle that could possibly contest such a claim would be Cloyd's Mountain, Virginia in 1864. As to the final boast, Droop Mountain, at an elevation of 3,136 was hardly a match for the actions which took place at Top of Allegheny, at more than 4,200 feet, or even Cheat Summit Fort, some 4,000 plus feet, during Gen. Robert E. Lee's 1861 mountain campaign in West Virginia.

Even the origin of the name Droop Mountain is questionable. It is known the name was used prior to the American Revolution, because about 1775 the court records of Botetourt County, Virginia, excused a Charles Kennison as a juror as he lived "beyond Droop Mountain."[2] One contemporary writer suggested the name possibly came from the crouching, overhanging, or drooping appearance of the mountain from the plains of Greenbrier [County] that inspired early settlers to call it "the drooping mountain."[3] Historian Roy Bird Cook said an "ancient lake "once surrounded Droop Mountain;[4] and Calvin Price added, "When the Levels was a lake Droop Mountain was the dam. The Greenbrier [River] forced a passage through along the extreme eastern side and still plunges through the pass."[5] Yet another journalist wrote, "...after a long level ridge, the mountain droops to the level of the Greenbrier River."[6] But perhaps the best, as well as the most wordy, description of Droop Mountain was given by John Alexander Williams in his history of West Virginia in which he wrote:

> *"Droop Mountain is a flat-topped, steep-sided ridge that cuts across the Greenbrier Valley like the crossbar of a lopsided letter H. The western leg of the H, which actually runs northeast-southwest in line with the general trend of ridges and streams in eastern West Virginia, is formed by the massive knobs of a range formally known as the Yew Mountains but more commonly known by the names of individual peaks and streams. The eastern leg is formed by Allegheny Mountain and its subordinate ridges, which moved off into the sunrise like well-tended roses in a gigantic garden. Here the ridge-and-valley landform of the central Appalachians is easily discernible, with the parallel ridges divided by narrow but fertile valleys and pierced by water gaps that connect the valleys with one another and the streams that carved them with Greenbrier River. The river itself flows south through a canyonlike bed below Droop, but elsewhere it meanders beside rich bottoms such as the one occupied by modern Marlinton...[the Greenbrier Valley's most distinctive features*

are provided by it's] two handsome intermittent basins that occupy the interstices of the H. The pioneers called these basins levels. The Little Levels of Pocahontas County lies below Droop Mountain's sharply rising northeast front, while the Great Levels of Greenbrier County spreads out south and southwest of the mountain. Throughout the nineteenth century these basins were known for the excellence of their gently rolling pastures and fields and for the charming contrast that these features presented against the surrounding backdrop of mountains. Thus while the neatly interconnecting valleys east of Greenbrier River and the river itself have afforded thoroughfares in most periods of history the main road through the Greenbrier Valley has always run between the two Levels, climbing directly over the top of Droop. The flat top of the mountain added to its appeal as a thoroughfare—or as a battlefield."[7]

In early November of 1863, Droop Mountain became such a bloody battlefield, as soldiers, primarily from West Virginia, fought to the death for their beliefs on the mountain's summit.

Mount Up

Events that directly led to the battle of Droop Mountain began during February of 1863. With the Civil War yet raging on all fronts, a delegation from the loyal government of (West) Virginia paid a visit to President Abraham Lincoln in order to request that infantry soldiers in their region be converted to mounted troops. Lincoln interceded with his general-in-chief, Henry Wager Halleck, and the request was granted, although it would be nearly three months before the policy went into effect.[8]

This military strategy may have seemed insignificant at the time but the results would have far reaching effects, as (West) Virginia had always been an enigma for Federal troops. With its rugged, mountainous terrain and irregular natural boundaries, the area consistently baffled Federal commanders, who were unable to cope with the mostly homegrown Confederate troops, whose partisan and guerrilla tactics operated in near perfect tandem with the geography. Despite a few minor combat achievements, and the stationing of various outposts, the Federal command never fully secured the new territory. It was felt the time was right for the "mounted infantry" innovation, which consisted of former foot soldiers who would act as cavalry but could dismount and fight as infantry against an entrenched foe.

In order to strengthen the Federal image in (West) Virginia an election was held in the camps on March 12, 1863 in which the soldiers voted unanimously for the creation of the State of West Virginia, which would officially occur on June 20; and on March 28 the "4th Separate Brigade" was created by General Orders No. 20, Headquarters Middle Department, 8th Army Corps. Given command of this new brigade, comprised mostly of (West) Virginia units, was Gen. Benjamin Stone Roberts, an experienced military man who found himself stationed in an unsympathetic community. Apparently suffering from paranoia, he made war upon the citizenry rather than the soldiers and infuriated the residents as well as his command. He fully displayed his weaknesses during the famed Jones-Imboden Raid, which took place between April 13 and March 22. Following some minor skirmishing he pulled all his outposts in to Clarksburg by April 29 and the Confederate raiders roamed virtually unmolested throughout (West) Virginia.[9]

Although the Confederates eventually retired from the area and the Federals returned to their former posts it was obvious a capable leader was desperately needed, so on May 18 Gen. William W. Averell received orders to replace Roberts. Averell was informed:

Headquarters, 8th Army Corps,
Baltimore, Md., May 18, 1863.

Brig.-Gen'l W.W. Averell, U. S. Volunteers.

General:

* * * * * * * * * *

You will proceed to Weston, in western Virginia, or wherever else you may find Brig.-Gen'l B.S. Roberts, and relieve him of his command of the "Fourth Separate Brigade" of this Army Corps. On assuming command, you will establish your present headquarters at Weston or Buckhannon, or such other point as you may find it best to select south of the Baltimore and Ohio Railroad, drawing your supplies from the depot at Clarksburg. Your command, however, is intended to be, as far as it can be properly made so, a mobile force, and your service will be to keep that region of West Virginia between the railroad and the Kanawha line clear of the enemy, prevent his invasions and supporting and co-operating with Brigadier-General Kelley, commanding on the line of the railroad, and with Brigadier-General Scammon, commanding on the Kanawha and Gauley Rivers. You may be called upon in emergency to follow the enemy, or to cross the mountains east of you, to aid in any movement in the direction of the valley of Virginia. On your left you will find it necessary to guard the passes and approaches by way of the Cheat River Mountain. Keeping these objects in view, it is left to your discretion to station your troops at such points as you may deem most advisable, keeping the body of them, however, together where it may become necessary and best to concentrate, covering your line of supplies.

You will inspect you command and report, at as early a day as possible, its exact condition and wants, with a view to having it supplied and put in the most effective condition. It is designed, as soon as practicable, by reinforcements, if they can be obtained by new organization and by all means of improvement, to convert or exchange the whole or greater part of your troops, so as to make yours a force of cavalry, with light artillery and with little or no infantry.

I am, very respectfully, your ob't servant,
Wm. H. Chesebrough,
Asst. Adjt.-General.[10]

Four days later, on May 22, Averell was in Weston. Unfortunately, Averell would soon find the soldiers he inherited of the "4th Separate Brigade" were in poor condition, comprised mostly of West Virginia "regiments which had been early recruited, mustered into the service, and hurried away to the front. The ruling principal governing them at that time was to know how to make long marches, to endure hardships, and to load and fire *low* in time of battle. The master hand of discipline and drill had not been a part of their military experience."[11]

Making these observations was Major Theodore F. Lang of the 3rd West Virginia Infantry, who was made Acting Assistant Inspector General and Mustering Officer of the brigade. He added the brigade was composed of "loyal, courageous fighters, scattered through half a dozen counties, but who knew little of discipline, or of regimental or brigade maneuvers—scantily supplied with approved arms, equipment, clothing, etc." Lang also noted, "They were inefficient for any reliable defense of the country, and the often hopelessness of any effort to take the offense ..." had recently been demonstrated by the Jones-Imboden Raid. Finally, he mentioned, "That all of the officers and enlisted men alike, were war students with no teachers among them of skilled warfare."[12]

Therefore, the soldiers were elated when a man of such military distinction as Averell assumed command, and he quickly made a "splendid brigade out of the best quality of raw material."[13] Averell wasted no time and stationed the men at three points for "a season of drill in the various arms of the service; they were also re-armed and equipped; 3,000 infantry were changed into cavalry—marched 660 miles, had many skirmishes and fought the battle of Rocky Gap (White Sulphur Springs); all this within 80 days after" Averell took charge.[14]

The Objective

Although West Virginia was officially proclaimed a state on June 20, 1863, and Union forces had won major victories at Gettysburg and Vicksburg during the summer, much of West Virginia was still in Confederate hands. In addition, Gen. Averell had suffered defeat at White Sulphur Springs on August 26-27 in the first of his three raids. Finally, in September Federal troops suffered a major defeat at Chickamauga and were forced westward by men such as Gen. Sam Jones, a West Point graduate and man of considerable military experience, who commanded the Confederate Department of Western Virginia and East Tennessee, with headquarters at Dublin, Virginia. Jones and his contemporaries were quick to utilize the vital Virginia and Tennessee rail line for troop movement.

With this scenario a plan was needed to rid West Virginia of its Confederate presence and to take pressure off the Union forces in Tennessee by drawing Gen. Jones back into western Virginia. Accordingly, General-In-Chief Henry W. Halleck formulated such a strategy, which he presented to the commander of the Department of West Virginia, Brig. Gen. Benjamin Franklin Kelley.

General Kelley, although born in New Hampshire, moved to Wheeling, (West) Virginia at the age of 19. At the outbreak of the war he raised and led the 1st (West) Virginia Regiment Infantry and fought at Philippi, (West) Virginia June 3, 1861, the first land battle of the first campaign of the war. In that engagement he sustained a severe chest wound, recovered, and was promoted to brigadier general May 17, due primarily to his "staunch Union efforts in western Virginia."[15] Kelley was assigned command of the Federal forces in West Virginia, commanding some 32,000 troops that he placed at strategic points throughout the new state.

Kelley's chief responsibility was protection of the Baltimore and Ohio Railroad, which he accomplished by posting strong detachments along a line from his headquarters at Clarksburg to Harpers Ferry in the state's eastern panhandle. Defense of central and southeastern West Virginia was delegated to Gen. Eliakim P. Scammon's 6,000 troops at Charleston in Kanawha County and Gen. Averell's 5,000 men at Beverly in Randolph County, located about 15 miles south of Leadsville (Elkins).

Already the 11th and 12th Corps had been transferred from Washington, D.C. by the B&O rail line to reinforce the Federals in Tennessee. Halleck and Kelley decided that since a portion of the West Virginia Confederates had moved toward Tennessee, the time was right to strike by sending cavalry units from West Virginia to destroy or damage the Virginia and Tennessee Railroad in the vicinity of Lynchburg or Dublin. Gen. William W. Averell was given this assignment, and this is the story of that campaign, the resulting battle of Droop Mountain, and the men who participated—nothing more and nothing less. This is not an essay on economics or politics, just a story of the military in a little known campaign of the Civil War.

7th West Virginia Regiment at Buffalo, West Virginia.

CHAPTER 2

Averell's Yankees

On October 26, 1863, Brigadier General Benjamin Franklin Kelley, U.S. Army, commanding the Department of West Virginia at Cumberland, Maryland, in accordance with instructions from Brigadier General George Washington Cullum, Chief of Staff, Army, Headquarters, Washington, D.C., directed Brigadier General William W. Averell, stationed at Beverly, West Virginia, "to move with his command, [excepting the 10th West Virginia Volunteer Infantry and the companies of the 3rd West Virginia Cavalry under Major Lot Bowen] as soon as possible on Lewisburg, in Greenbrier County, to arrest and capture or drive away the rebel forces stationed in that vicinity, and there, having formed a junction with Brigadier General [Alfred N.] Duffié (commanding a detachment [consisting of two cavalry regiments, two of infantry, and one battery] of Gen. [Eliakim Parker] Scammon's division), to leave the infantry, with orders to hold Lewisburg, and proceed with all the mounted troops [including Duffié's] to the town of Union in Monroe County and thence to the line of the Virginia and Tennessee Railroad, striking it at or near Dublin Station, for the purpose of destroying the bridge over New River [leaving the infantry at Lewisburg until Averell's return]."[1]

Averell was further instructed that if upon reaching Lewisburg he received "satisfactory information that the movement on New River was not practicable, he should then send his infantry, with [Captain John V.] Keepers' battery, back to Beverly, and with the mounted troops of his command and remaining battery, move by the most convenient road into the Valley of the South Branch [of the Potomac River], and by this route return to New Creek [Keyser]," where supplies would be waiting in readiness. Averell was also informed Gen. Duffié "will hold Lewisburg when you leave, or if deemed best he will fall back to Meadow Bluff and hold that place." The troops of Averell and Duffié were told to "take ten day's rations of coffee, hard bread, sugar, and salt." The required supplies of fresh beef and forage were to "be procured from the country" through which the two commands traveled. For such supplies the various owners of these items were to "be given, by the proper officers, proper vouchers, which were to state the facts attending the taking" and to "specify that payment will be made therefor" as loyalty of the individual property owner is determined. Additionally, the orders stated, "should more cattle be found than is necessary for the support of the command[s] while on the march, such surplus cattle will also be taken (and for which similar vouchers will be given), and will be brought forward, with the command[s] to New Creek." Finally, the commanding officers were instructed "stringent measures . . . [would be] taken to prevent interference with private property by the soldiers of the command[s] while on the expedition."[2]

Major Thomas Gibson of the 14th Pennsylvania Cavalry, commanding an independent battalion composed of four West Virginia, one Illinois and one Ohio cavalry companies, received a telegram at Buckhannon, West Virginia, October 28 from Lt. Leopold Markbreit, Acting Assistant Adjutant Gen. at Beverly. Gibson was instructed to move to Beverly and to leave Major Lot Bowen, 3rd West Virginia Cavalry [at Buckhannon] with the companies of 33-year-old Capt. William H. Flesher [Company H] and 22-year-old 1st Lt. George A. Sexton [Company I], both of the 3rd West Virginia Cavalry, along with "all dismounted men and unserviceable men and horses, with all company and garrison equipage and quartermaster's and ordnance stores not allowed by general orders." Gibson gave written instructions to Bowen "relative to picket and guard duty and preservation of drill and discipline."[3]

Major Gibson departed Buckhannon for Beverly on the morning of October 29 and arrived at his destination at 4:30 P.M.[4] On the same day, Capt. Ernst A. Denicke, 68th New York Infantry, commanding a signal corps detachment with Averell, "called in the signal station in advance of Beverly." Corp. George Washington Ordner, Company B, 2nd West Virginia Mounted Infantry, wrote in his diary from Beverly: "Making great preparations for a raid."[5]

Averell's "4th Separate Brigade"

William Woods Averell was born in Cameron, New York November 5, 1832. He received his early education in public schools and later taught school for two winter terms, surveying land and roads during the summer. A few years prior to 1851 he moved to nearby Bath and found employment as a clerk in a drug store. During this period Averell sought appointment to the U.S. Military Academy at West Point, which was secured in 1850 by Congressman David Rumsey. In 1851, at the age of 18, Averell entered West Point, where he thrived on cadet life and "tasted the pleasure of youth despite the disciplined environment." Although never a model cadet, and just an average student in most subjects, his carefree attitude gained him many friends, including future Confederate officer Fitzhugh Lee. He excelled in horsemanship and it was said, "His riding ability would save his life more than once in his military career."

Averell graduated 26th out of a class of 34 in 1855, and was commissioned a brevet 2nd lieutenant of mounted rifles. His first assignment was to the Cavalry School at Jefferson Barracks, Missouri, but he was soon afterward transferred to Carlisle Barracks, Pennsylvania, where he served as adjutant under post commander Col. Charles A. May. He remained there until August of 1857.

His friends gave him the nickname "Swell," derived from Shakespeare and reflecting his ambitious personality. Averell gradually grew bored of garrison duty and was elated when he was transferred with a detachment of recruits to the New Mexico territory in the autumn of 1857. During this time he "enjoyed free Spanish societal traits and had" numerous lovers. In December of 1857 the Kiowa Indians invaded the Rio Grande Valley and Averell confronted and defeated them. In 1858 there was an outbreak of the Navajo tribe and Averell participated in some 25 battles with them until October 8, 1858, when he suffered a serious fracture wound of the left thigh during a night attack on his camp. Reduced to the use of crutches for the following two years, he was sent home on convalescent leave, where he remained with his family. When the Civil War opened with the firing on Ft. Sumter, Averell was yet an invalid on leave but presented himself to the United

Brigadier General William Woods Averell commanded his 4th Separate Brigade, U.S. Army, in the Droop Mountain battle and campaign. NATIONAL ARCHIVES

States War Department for assignment on April 16, 1861.

William W. Averell was immediately "sent to Fort Arbuckle, 300 miles beyond the western border of Arkansas . . . to order commander William H. Emory to destroy all property that could not be transferred to Fort Leavenworth, then leave the post with his troops." Afterwards, Averell returned to New York on June 7, 1861, and mustered volunteers at Elmira. He then served as a courier at 1st Manassas (Bull Run), and two months later was on the staff of the provost marshal at Washington, D.C. In October 1861, he was assigned to the rowdy Kentucky Cavalry Regiment of Col. William H. Young, in the hope that he could instill some discipline. Although Col. Young was a Kentuckian, most of the men in the regiment were from Pennsylvania; therefore, the unit was redesignated the 3rd Pennsylvania Cavalry. The 29-year-old Col. Averell quickly whipped the command into shape and was then assigned to also train the 8th Pennsylvania Cavalry.

For his participation in the 1862 Peninsula Campaign, covering the rear of the retreating Army of the Potomac to Harrison's Landing, Averell found himself appointed acting brigadier general. On September 26, 1862, Averell was promoted to brigadier general with the Army of the Potomac. The Peninsula Campaign took its toll on Averell during the late summer of 1862, coming down with "Chickahominy Fever," that he described as, ". . . a malarial fever and sort of typhoid which exhibits itself in congestive headaches and in a general dislocation of all one's joints." His illness became so severe, for about five weeks his sister had to come from New York to help nurse him back to health. In early 1864, months after the Droop Mountain battle, he would have another attack of "Chickahominy Fever." On March 17, 1863, at Kelly's Ford, Virginia, he engaged and defeated the Confederate forces of his old West Point friend, Fitzhugh Lee. As this was the first time in the war the Federal cavalry had gained a significant victory against Confederate horsemen, Averell immediately gained notoriety. However, less than two months later, during Stoneman's Raid of the Chancellorsville Campaign, Averell sat idle at Rapidan Station "worrying about phantom Confederate cavalrymen." Because of this, as well as his lack of aggressive pursuit at Kelly's Ford, Gen. Joseph Hooker, on May 2, relieved Averell from command of the 2nd Division of the Cavalry Corps, citing "less than satisfactory performances."[6]

Reportedly, Averell was then sent to the backwater theater of action in West Virginia as a sort of punishment. This point is debatable, as West Virginia had always been unable to cope with the Confederate guerrillas and raiders throughout the area, so it must have been felt a leader such as Averell, with his superb riding skills and cavalry tactics, could create a force to contend with such an enemy. In addition, the reputation of the cavalry in the Army of the Potomac had taken a negative turn, as evidenced in a letter written to Averell on November 6 [the very day of the Droop Mountain battle] from his old friend and successor to colonelcy of the 3rd Pennsylvania Cavalry, John B. McIntosh, which read, in part: "Since parting from you we have gone through many changes. It is well for you that you got out of the Army of the Potomac, as your position there must not have been a pleasant one after P. [Alfred Pleasonton] got command of the Cavalry. I understand that most every one is disgusted with his management & well they might be. Under his regime the Cavalry could never win a brilliant name."[7]

Therefore, on May 16, 1863, Averell assumed command of the "Fourth Separate Brigade," 8th Army Corps in West Virginia. He wasted no time and took charge on May 23, as acknowledged in the diary of Sgt. Thomas H. B. Lemley, Company A, 1st West Virginia Cavalry, who wrote: "Gen. [Benjamin Stone] Roberts superseded by Averell."[8] On the following day Lemley wrote: "Review by Averell—warm day."[9] The men of the 2nd West Virginia Mounted Infantry, one of the regiments under Averell's command, said they "always had a great liking for the dashing general and admired his courage and ability," and also stated "he was one of the ablest cavalry officers of the army."[10]

On May 23, when he took command, Averell converted the 2nd, 3rd, and 8th West Virginia infantry regiments to mounted infantry and sent them to a "camp of instruction" near Clarksburg. They received horses in the unusually quick span of two weeks, although saddles, bridles, and miscellaneous cavalry gear did not arrive until a week later. Averell realized the haste involved and reported the 3rd and 8th West Virginia had changed from infantry to "cavalry in . . . forty-eight hours, by mounting the men upon green horses."[11] In addition, Averell felt, the "Horses were not shod, and horses cannot travel over the rugged roads of this country without shoes, without breaking down very soon."[12] This situation, along with a lack of ordnance stores and widely scattered detachments, confronted Averell as he marched and counter-marched his brigade throughout the Williamsport-Martinsburg-Winchester area during most of July, keeping check on the Confederate army retreating from Gettysburg. On August 12 and 14 Averell received orders to drive the Confederate forces out of Pocahontas and Greenbrier counties of West Virginia, eliminate the saltpeter and gunpowder works in Pendleton County, then return to Beverly and either operate "against Staunton or [Gen. John D.] Imboden."[13] Due to a delay in supplies, Averell did not leave until August 18 and eventually was defeated on August 26-27 near White Sulphur Springs by a 1,900 man Confederate force under Col. George S. Patton. Averell, running out of ammunition and with no support from a Federal column from the Kanawha Valley, had to withdraw with heavy casualties. He was safely back in Beverly by August 31, but at Droop Mountain in November there would be "Hell to pay" for most of the rebel forces he had engaged at White Sulphur Springs. Averell would write in his log book that throughout October until November 1 he spent re-arming and drilling the men.[14]

Averell's "4th Separate Brigade," which he would take to Droop Mountain, consisted of the 2nd, 3rd and 8th West Virginia mounted infantry regiments; the 14th Pennsylvania Cavalry; Major Thomas Gibson's Independent Cavalry Battalion [which is believed to have contained four West Virginia cavalry companies; Company C, 16th Illinois Cavalry; and the 3rd Independent Company Ohio Cavalry]; the 10th West Virginia Infantry; the 28th Ohio Infantry; batteries B and G of the 1st West Virginia Light Artillery; and a signal corps detachment of the 68th New York Infantry.

The first regiment comprising Averell's brigade which would fight at Droop Mountain was Lt. Col. Alexander Scott's 2nd West Virginia Mounted Infantry [later 5th West Virginia Cavalry], which was the first West Virginia regiment enlisted for three years service and the first mustered in under Gov. Francis Pierpont, which took place at Wheeling, West Virginia. The regiment was comprised of 10 companies (A through K), the ranks filled with men from the West Virginia communities of Wheeling, Parkersburg, Grafton, and the counties of Wetzel, Taylor, and Ritchie, as well as the Ohio cities of Ironton and Bridgeport. It also included men from the Ohio counties of Monroe and Belmont; Pittsburgh, Pennsylvania; and Greenfield and California, Washington County, Pennsylvania.[15] The regiment boasted "comparatively young men, the average age being about 24 years, a large number of them being but boys of eighteen, while a few had reached the age of forty."[16] They were "young, active, strong and intelligent, the making of a splendid regiment."[17]

The regimental companies met at Beverly, (West) Virginia in late July of 1861 and organized as the 2nd (West) Virginia Infantry. During the first year of the war the regiment guarded the Baltimore & Ohio Railroad at Glover's Gap; they also participated in the activities at Beverly, Elkwater, and Cheat and Allegheny mountains. After serving under Gen. John C. Fremont, the regiment fought in 1862 with Gen. John Pope at Monterey, McDowell, and Cross Keys. During the spring, Company G was detached as an artillery company (Battery G, 1st West Virginia Light Artillery) and was never replaced. Also, West Virginia regiments had to depend upon original members for strength, presenting a difficult situation with dwindling numbers and few recruits. By April 1, 1862, the 2nd West Virginia's total strength was about 900, a figure that would quickly diminish.

During 1863 the regiment saw action at Kelly's Ford, then returned to West Virginia, where they suffered defeat in an action at Beverly on April 24, 1863. The men remained at Beverly until they came under the command of Gen. Averell on May 23, 1863. They were ordered to Grafton to be mounted, thereby becoming the 2nd West Virginia Mounted Infantry. By this time the regiment was known to be "brave to the point of recklessness sometimes. It was said of them that whenever they got in a close place, every man was a general, and that they were almost invincible. They certainly achieved some victories that seemed in the beginning hopeless."[18] At the battle of White Sulphur Springs August 26-27, 1863, the regiment reported 29 casualties, including at least seven killed, 18 wounded, and three captured. Among the mortally wounded and captured was Major F. Patrick McNally. During the Droop Mountain campaign the 2nd West Virginia Mounted Infantry would number 395 men, although only about 200 would participate in the battle.

Lieutenant Colonel Alexander Scott was born about 1821 in Franklin County, Pennsylvania. In 1836 his family moved to Wooster, Ohio while young Alexander went to Pittsburgh to study music. He quit his studies to serve in the Mexican War with Pittsburgh's "Rough and Ready Guards," which was mustered in as Company F, Maryland and District of Columbia volunteers, on October 8, 1847. The company remained on active duty until mustered out July 24, 1848. Scott married at the close of the Mexican War and resided in Nashville, Tennessee; then moved to Mississippi, where he became involved in a furniture business. At the outbreak of the Civil War he commanded the Monroe Rifle Company, which received orders to perform duty at Macon, Georgia and enter the Confederate service. Alexander was given command but declined the offer and left the South in order to return to Pittsburgh, where he aided in recruiting Company F, 2nd (West) Vir-

Lt. Col. Alexander Scott led his 2nd West Virginia Mounted Infantry against the Confederate center at Droop Mountain. FRANK READER'S *HISTORY OF THE 5TH WEST VIRGINIA CAVALRY*

Averell's clerk, Pvt. Francis S. Reader, Co. I, 2nd West Virginia Mounted Infantry. During the Droop Mountain campaign he was clerk on Gen. Averell's staff, which probably greatly aided him years later when he wrote a fine regimental history of the 2nd West Virginia Mounted Infantry (5th West Virginia Cavalry). Reader served as clerk for Gen. Franz Sigel in March and April of 1864 and was captured by Confederate guerrillas June 24, 1864, at Greenbrier Bridge. FRANCIS READER COLLECTION

ginia Infantry. He entered service as captain of the company and was promoted to lieutenant colonel of the regiment either May 20 or July 1, 1862. At the battle of 2nd Bull Run he had a horse shot out from under him and soon "had the confidence of his men and they cheerily followed his leadership, though they knew that it meant danger, and perhaps death." On August 27, 1863, he was wounded in the left arm at White Sulphur Springs, and Col. George R. Latham of the 2nd West Virginia was reportedly slightly wounded in the foot, possibly explaining why Scott led the regiment at Droop Mountain.[19]

Next in Averell's brigade was the 3rd West Virginia Mounted Infantry [later 6th West Virginia Cavalry], led by Lt. Col. Francis Thompson. The regiment was officially organized about July 1, 1861, in the vicinity of Clarksburg, (West) Virginia by town resident David T. Hewes. The 3rd West Virginia was the second regiment raised in (West) Virginia under the call for troops and the men trained at "Camp Hewes" near Clarksburg. The 3rd (West) Virginia Infantry was comprised of ten companies (A through K) and consisted of men from the (West) Virginia counties of Monongalia, Harrison, Preston, Upshur, Taylor, Marshall, Ritchie, and the southern border area of Pennsylvania. During 1861 the regiment was briefly headquartered at Clarksburg, then spent several months guarding outposts from Philippi to Sutton. Throughout the remainder of the year the men of the 3rd (West) Virginia Infantry went to Beverly, then to Elkwater for the winter.

On April 1, 1862, the regiment served in Gen. Robert H. Milroy's brigade and was sent to the Shenandoah

Valley of Virginia to fight at Monterey and McDowell. Later the boys participated in Gen. John C. Fremont's pursuit of "Stonewall" Jackson and fought at Cross Keys, as well as in all the battles in Gen. John Pope's campaign, up to and including 2nd Bull Run. On September 30, 1862, the regiment departed Fort Ethan Allen near Washington, D.C. for Clarksburg, then moved to Point Pleasant and back to Clarksburg, Buckhannon, and various outpost duties. On May 18, 1863, the 3rd (West) Virginia was assigned to the command of Gen. Averell and on June 10 received horses in order to become the 3rd West Virginia Mounted Infantry. On June 17 the regiment received horse equipment and two days later went to Fetterman to drill. The 3rd West Virginia Mounted Infantry participated in Averell's July 1863 movement against the Confederate army retreating from Gettysburg, and fought gallantly at White Sulphur Springs, West Virginia August 26-27, where the regiment reported a total loss of 39, a figure which included at least eight killed, eight wounded, and 13 captured.[20]

Lt. Col. Francis W. Thompson led his 3rd West Virginia Mounted Infantry against the Confederate center and left at Droop Mountain.
THEODORE LANG'S *LOYAL WEST VIRGINIA*

Lieutenant Colonel Frank (Francis) W. Thompson was born January 7, 1828, in Morgantown, (West) Virginia and went to California in 1850, then on to Oregon. He "crossed the plains in 1852, when there was not a house between the Missouri River and the Cascade Mountains in Oregon." In 1855 Thompson served as captain of Company A, 1st Battalion of Oregon Mounted Volunteers in the Yakama and other wars, and from 1856-57 served with a battalion of Oregon and Washington mounted rangers. He studied a number of Indian languages and could speak several quite fluently. Thompson returned to Morgantown and on May 28, 1860 was commissioned a 1st lieutenant in the 76th Militia. Governor John Letcher of Virginia commissioned him a colonel with authority to raise a regiment for southern service but he refused the commission. In 1861, he raised the first company in his county for Federal service, which was possibly the first three year's service regiment in (West) Virginia. He was commissioned a captain of Company

A, 3rd (West) Virginia Infantry either June 5 or July 20, 1861, [depending upon source] at Clarksburg and commissioned lieutenant colonel of the regiment either July 20 or August 6, 1861, although his obituary says July 21, 1862. Thompson was "desparately wounded" at 2nd Bull Run and refused to leave the field until he fell from his horse from the loss of blood. Unconscious, he was taken to a hospital at Washington, D.C. where he was oblivious to his surroundings when he awoke.[21]

Battle colors flag of Lt. Col. Frank Thompson's 6th West Virginia Cavalry (3rd West Virginia Mounted Infantry). Legible battle honors include, in order, Allegheny, McDowell, illegible, Cross Keys, Cedar Mountain, Rappahannock Station, White Sulphur, Waterloo Bridge, [2nd] Bull Run, Rocky Gap [White Sulphur Springs], Droop Mountain, Salem Raid, Cloyd Mountain, Dublin Depot. WEST VIRGINIA STATE ARCHIVES

The 8th West Virginia Mounted Infantry [later 7th West Virginia Cavalry] was organized as the 8th (West) Virginia Infantry at Charleston in the Kanawha Valley of (West) Virginia in the fall of 1861 by Colonel John H. Oley. Comprised primarily of men from the surrounding region, the regiment moved to New Creek (Keyser) in April of 1862 and served in Fremont's Mountain Department. The men were in the advance that pursued "Stonewall" Jackson in the Shenandoah Valley and fought at Harrisonburg and Cross Keys. During Pope's campaign, the 8th (West) Virginia served with Bohlen's Brigade, Sigel's corps. After arriving in Washington City the regiment was transferred to Milroy's brigade and returned to (West) Virginia and the Kanawha Valley. In November of 1862 the command was transferred to Col. Augustus Moor's brigade and in 1863 was assigned to Gen. Averell's "Fourth Separate Brigade." On June 13, 1863, the 8th West Virginia went to a camp of instruction at Bridgeport, (West) Virginia to be mounted and drilled as mounted infantry, officially becoming the 8th West Virginia Mounted Infantry June 20, 1863. According to Averell, "carbines and sabers were obtained from Wheeling for the 8th West Virginia." The regiment served with Averell in his July 1863 campaign against the Confederate army retreating from Gettysburg; they also fought with him at White Sulphur Springs, West Virginia on August 26-27, 1863, where the regiment lost a minimum of 20 men, including six killed, 12 wounded, and two missing or captured.[22]

Colonel John Hunt Oley, "a man of tremendous driving force," was born September 24, 1830 [although his obituary said 1831] at Utica, New York. During his early life he engaged in the mercantile business and later became a telegraph operator, being one of the first who received messages by sound In April of 1861 he enlisted in Company H of the 7th Regiment of New York National Guard, an organization "in which every man was a trained soldier and fit to command." In the summer Gov. Francis Pierpont of the Loyal Government of (West) Virginia telegraphed New York's governor to send him six men from the 7th Regiment of the New York National Guard to drill (West) Virginia troops. Oley was one of those chosen. In the fall he organized the 8th (West) Virginia Infantry, and was commissioned a major in the regiment October 29, 1861. Oley was promoted to lieutenant colonel of the regiment August 24, 1862, and rose to the rank of colonel March 1, 1863. 1st Sgt. George H. Mowrer, Company A, 14th Pennsylvania Cavalry, a personal friend of Oley's, said he was "a fine-appearing officer—intelligent, a good disciplinarian, and in every respect thoroughly qualified to command." Mowrer went on to say Oley "endeared himself to the officers and men of his regiment and had the respect and confidence of his superior officers." One man who would dispute this was Lt. Col. John J. Polsley of the 8th West Virginia, who would make an unsuccessful bid to charge Oley with cowardice in the Droop Mountain fight.[23]

Col. John Hunt Oley of the 8th West Virginia Mounted Infantry. Led his regiment of West Virginians against the Confederate right and center at Droop Mountain.
DROOP MOUNTAIN COMMISSION BOOKLET

The next mounted regiment to comprise Averell's brigade was the 14th Pennsylvania Cavalry (159th Pennsylvania Volunteers), organized by Col. James M. Schoonmaker, who on August 18, 1862, received authority to recruit a cavalry battalion of five companies. Schoonmaker, a citizen of Pittsburgh, Pennsylvania and a lieutenant in the 1st Maryland Cavalry, found the ranks filled so quickly that on August 29 he received authority to expand to twelve companies, making it a regiment. The twelve companies (A through M) were primarily from the Pennsylvania counties of Allegheny, Fayette, Armstrong, Washington, Lawrence, Erie, and Warren, as well as the city of Philadelphia. The regiment first moved to Camp Howe, then to camp Montgomery near Pittsburgh, where they were to be issued horses, arms, and equipments. Before the order was completed the battle of Antietam had taken place, provoking authorities to direct all such materials to Chief Quartermaster of the Army of the Potomac.

8th West Virginia Mounted Infantry officers (later the 7th West Virginia Cavalry). Major Slack led two companies of the regiment against Gen. John D. Imboden at Covington, Virginia, on November 9, 1863. DROOP MOUNTAIN COMMISSION BOOKLET

Mounted and ready, Company G, 7th West Virginia Cavalry [originally the 8th West Virginia Mounted Infantry]. Capt. James S. Cassady of Fayette County, 1st Lt. James D. Fellers, and 2nd Lt. John E. Swaar. Most of the company was recruited in Fayette and Kanawaha counties and the bulk probably fought at Droop Mountain. Date and location of photo unknown.
BOYD STUTLER COLLECTION, WEST VIRGINIA STATE ARCHIVES

Battle colors flag of Col. John H. Oley's 7th West Virginia Cavalry (8th West Virginia Mounted Infantry). Battle honors noted are, in order, Strasburg, Harrisonburg, Cross Keys, Freeman's Ford, Sulphur Springs, Waterloo Bridge, [2nd] Bull Run, Droop Mountain, and the Salem Raid. Note Droop Mountain is incorrectly listed as having taken place in October rather than November of 1863. WEST VIRGINIA STATE ARCHIVES

Field and staff officers of the 7th West Virginia Cavalry [8th West Virginia Mounted Infantry]. The men are believed to be (back row, left to right): Maj. William Gramm, Dr. James H. Rouse, Dr. Lucius L. Comstock, Capt. Jacob M. Rife, Lt. Daniel William Polsley; (second row, left to right): Chaplain Andrew W. Gregg, Lt. Col. John J. Polsley, Col. John H. Oley, Maj. Hedgeman Slack, Lt. John W. Winfield; (front row, left to right); Maj. Edgar B. Blundon, Lt. Thomas H. Burton, Dr. Louis V. Stanford, Lt. John McComb. BOYD STUTLER COLLECTION, WEST VIRGINIA STATE ARCHIVES

On November 24, 1862, the regimental organization was completed and field officers commissioned. The 14th Pennsylvania Cavalry was then sent to Hagerstown, Maryland, where proper materials were obtained and the men were instructed in various troop drills, mounted and dismounted, including platoon, squadron, and evolution of the line. On December 18, 1862, the regiment moved to Harpers Ferry and camped on the Charlestown Pike, serving as the advance post of Gen. Benjamin F. Kelley's command. In early May of 1863 the men were (with the exception of a detachment of dismounted men under Major Shadrack Foley left at Harpers Ferry) sent to Grafton, (West) Virginia, where they were attached to the mounted command of Gen. Averell. During this period the 14th Pennsylvania Cavalry was engaged in protecting Philippi, Beverly and Webster, and on July 4, 1863, assisted Averell in the pursuit of the southern army retreating from Gettysburg, during which time they were rejoined by Major Foley's detail. Reportedly, at the battle of White Sulphur Springs, West Virginia August 26-27, 1863; the 14th Pennsylvania Cavalry "handsomely repulsed three charges" of the enemy and sustained a total loss of 102, which encompassed 10 killed, 42 wounded, and 50 captured or missing. Total strength of the regiment at Droop Mountain was probably around 104.[24]

Colonel James Martinus Schoonmaker was born June 30, 1842, at Allegheny, now a section of Pittsburgh, Pennsylvania, the son of a prominent drug and paint wholesaler. He was barely 19 years of age and a student at Western University of Pennsylvania when the war began. He was an accomplished horseman, enlisted in a local company, and was mustered into the 1st Maryland Cavalry, where he went from a private to 2nd lieutenant in six weeks. With such command he fought at the battle of Cedar Mountain, and in the summer of 1862 was sent to Pittsburgh to recruit a cavalry battalion. He rented a hall in Allegheny County and drew crowds by playing his guitar and singing war songs. He was very popular and respected, prompting the recruits to come in quickly, helping to form the 14th Pennsylvania Cavalry. On September 14, 1862, with Schoonmaker barely 20 years old, he was made colonel of the 14th Pennsylvania Cavalry, "the youngest colonel in the Union army." It was said he would play his guitar and sing to his men in order to keep them from becoming homesick.[25]

Col. James Martinus Schoonmaker of the 14th Pennsylvania Cavalry commanded Federal troops at Mill Point as well as the 14th Pennsylvania Cavalry and Battery B (Keepers'), 1st West Virginia Light Artillery at Droop Mountain. He successfully kept the Confederate right and center occupied as Colonel Moor's column approached the rebel left. U.S. ARMY MILITARY HISTORY INSTITUTE, RONN PALM COLLECTION

Gibson's Independent Cavalry Battalion was led by Major Thomas Gibson of the 14th Pennsylvania Cavalry, and, according to Samuel P. Bates, "was composed of four companies of the Third [West] Virginia [Cavalry] and the Chicago Dragoons, Captain Julius Jaehne." Actually, the composition of the battalion during the Droop Mountain expedition is a bit sketchy, but it is believed to have contained three companies [E, H, and I] of the 3rd West Virginia Cavalry; Company A of the 1st West Virginia Cavalry; Company C, 16th Illinois Cavalry [Chicago Dragoons]; and the 3rd Independent Company Ohio Cavalry. Gibson's Independent Cavalry Battalion is believed to have been organized at Pittsburgh in June and early July of 1863, and led by Major Gibson.[26]

Major Thomas Gibson (Jr.) was born about 1842 in Allegheny County, Pennsylvania, the son of Col. Thomas Gibson, surveyor of the port of Pittsburgh during President James Buchanan's administration, as well as colonel of one of Pennsylvania's militia regiments. His mother's family was influential in the foundry business. Both parents were of Irish descent and the "son inherited the Irish courage and daring." Gibson was a graduate of Western University of Pittsburgh and known as "a gentleman of culture and ability." During the war he was originally commissioned as captain of Company D, (Pittsburgh Fire Zouaves), 2nd West Virginia (Foot Soldiers) Infantry August 2, 1861, to rank from June 14, 1861, and was promoted to Major of the regiment July 7 or 31, 1862. Gibson resigned November 7 or 9, 1862 to accept appointment to Major of the 14th Pennsylvania Cavalry on November 24, 1862, re-enlisting October 8 or December 24, 1862.[27]

Company A, 1st West Virginia Cavalry, commanded by Capt. Harrison H. Hagans, was locally designated the "Kelley Lancers" and was recruited in Monongalia County, (West) Virginia. The company was mustered into U.S. service at Morgantown, (West) Virginia July 18, 1861, for three year's service. The ranks of the 1st West Virginia Cavalry were filled with "superior material, mostly young men from the farms, experienced horsemen and marksmen, who could break and tame the wildest colt, or pierce the head of a squirrel in the top of the tallest hickory with a rifle bullet."[28] Capt. Harrison H. Hagans enlisted in the company as a private at the age of 27 during the regimental organization at Morgantown. He was promoted to 1st lieutenant December 15, 1861, and then to captain of the company March 25, 1862, but resigned July 3, 1862. Hagans was reappointed to captaincy of the company September 3, 1862, upon the promotion of Capt. J. Lowrie McGee to major of the 3rd West Virginia Cavalry. Hagans was again commissioned March 2, 1863, as captain of Company A, 1st West Virginia Cavalry. Early on in 1861 the company was armed with Sharps carbines and Colt army revolvers, which "imparted in the men a new confidence and soon became a terror to bushwhackers," and, according to Major Lot Bowen of the 3rd West Virginia Cavalry, in the spring of 1862 the 1st West Virginia Cavalry was equipped with Spencer rifles.[29]

The three companies [E, H, and I] of the 3rd West Virginia Cavalry which served Averell at White Sulphur Springs [where they lost 2 killed, 1 wounded, and 1 captured or missing] and the Droop Mountain expedition, fell under the command of Major Lot Bowen, who was born in Greene County, Pennsylvania November 17, 1824. In 1850 Bowen moved to Sycamore Dale community [Harrison County, West Virginia] and resided there for the next 40 years. He had been trained in Light Horse Cavalry at the age of 16, served as a delegate from Harrison County to the 1st Wheeling Convention, and in 1863 organized Company E, 3rd West Virginia Cavalry. He was promoted to captain of the company at the age of 37 and later to major of the regiment. It was said Lot Bowen "moved with quick, jerky steps and always seemed in a hurry," and he "displayed qualities of the brave soldier in the vicinity of Webster, Sutton, and Bulltown under General Roberts."[30] Major Bowen would be left in charge of the post at Buckhannon during the Droop Mountain campaign.

Captain Julius Jaehne's Company C, 16th Illinois Cavalry had originally served as Capt. Frederick Schambeck's "Chicago Dragoons." A member of the 1st West Virginia Cavalry noted that, at the outbreak of the war, this company was comprised "of Germans from Chicago, recruited from the shops and among the laboring classes of the city and I doubt if one in twenty had ever mounted a horse before his enlistment. Capt. Schambeck's men were known as good fellows, although but few of them could speak the most broken English" and "they were splendid singers" who often sang their own national airs.[31]

Captain Julius Jaehne, born about 1837, enlisted April 18, 1861, at Chicago, Illinois in Capt. Schambeck's Cavalry Company, Illinois Volunteers, and was mustered in at Bellaire, Ohio July 8, 1861, as 1st sergeant. On November 19, 1861, Jaehne was promoted to brevet 2nd lieutenant, thence to 2nd lieutenant March 25, 1862, to 1st lieutenant December 16, 1862, and finally to captain of Company C, 16th Illinois Cavalry April 17, 1863. In this capacity Jaehne led his men at White Sulphur Springs, losing two wounded; and at Droop Mountain, where it is believed the company numbered less than 90 men.[32]

The 3rd Independent Company Ohio Volunteer Cavalry, led by Capt. Frank Smith, was a three years service company organized at Cincinnati, Ohio by Capt. Philip Pfau, with each man furnishing his own horse and horse equipage. The company was mustered into service July 4, 1861, at Cincinnati and mustered into U.S. service at Camp Chase [Columbus, Ohio] August 16, 1861. The company saw its first action at Princeton,(West) Virginia May 16, 1862, and subsequently at 2nd Bull Run, South Mountain, Antietam, and Beverly, before participation with Gen. Averell's command. Capt. Frank Smith enlisted in the company at the age of 41 as a 1st lieutenant July 4, 1861, and rose to the rank of captain March 10, 1862.[33]

Averell would take two regiments of foot soldiers to Droop Mountain, the 10th West Virginia Volunteer Infantry and the 28th Ohio Volunteer Infantry. The 10th West Virginia Volunteer Infantry was organized by Col. Thomas M. Harris, a practicing physician in Gilmer County, (West) Virginia at the opening of the war. In July of 1861, with the solicitation of Gen. William S. Rosecrans, he visited Governor Francis Pierpont of the loyal government of (West) Virginia at Wheeling and received consent to recruit a regiment, and if successful, to serve as colonel. Harris directed the bulk of his efforts on the northwestern and central counties of present-day West Virginia, as he was familiar with the territory. He traveled throughout twelve counties, beginning in August of 1861 and concluding in May of 1862. Harris recruited from the loyal population of (West) Virginia and gave first priority to suitable line officers, in which he "used great discrimination and made very few mistakes." This policy resulted in the 10th West Virginia being led by "brave, intelligent, and intensely loyal men." The first four or five companies were recruited in the fall of 1861, mostly "hardy mountaineers of the State." As such they were well suited to protect (West) Virginia and were given duty along the border to guard against guerrillas. The 10th West Virginia Volunteer Infantry was officially organized between March 12 and May 18, 1862, at Camp Pickens, Canaan, Glenville, Clarksburg, Sutton, Philippi, and Piedmont, and attached to the Cheat Mountain District. The regiment saw minor duty around Beverly, Bulltown, Sutton, Martinsburg, Winchester, and Summersville until attached to Gen. Averell in May of 1863. The 10th West Virginia became known as the premier West Virginia unit, and was noted for "prowess, courage, intrepidity and general reliability." By the end of the war it was said they were perhaps not excelled by any regiment in the service of West Virginia or any other state. At the Droop Mountain battle the 10th West Virginia would number about 570 men, as Company I was on detached duty at Petersburg in Grant County, West Virginia and, therefore, did not accompany the regiment on the expedition.[34]

Colonel Thomas Maley Harris was born June 17, 1813, [although other dates given include June 10th and 13th, as well as 1817] in Wood (now Ritchie) County, West Virginia. He became a teacher and taught school in local rural communities and later in Clarke and Greene counties, Ohio. He developed an interest in medicine, and in October of 1842, while assistant of the Parkersburg Seminary, he married the principal of the school. Her brother, ironically later became lieutenant colonel of the 10th West Virginia, and was also a physician. In 1843 Harris attended medical lectures in Louisville, Kentucky and then returned to Harrisville, (West) Virginia to practice medicine. In 1856 Harris moved to Glenville in Gilmer County, (West) Virginia where he was a pre-war [as early as 1848] friend of William Lowther ("Mudwall") Jackson, a future Confederate opponent, and was a practicing physician at Glenville as war erupted. He quickly recruited volunteers from the northern counties of present-day West Virginia to form the 10th West Virginia Volunteer Infantry and on March 17, 1862, entered the army as lieutenant colonel of the regiment. He "established a character for energy and faithfulness in obeying orders" and in May of 1862 was commissioned colonel of the 10th West Virginia Volunteer Infantry. In June of 1862 Harris was in command of the forces at Buckhannon, West Virginia fighting bushwhackers, and on July 2, 1863, withstood an attack by his pre-war friend, Col. William L. Jackson, at Beverly, West Virginia. Harris "was always on hand for any duty to which he was assigned and always received honorable mention for his

Col. Thomas Maley Harris of the 10th West Virginia Volunteer Infantry gallantly led his regiment against the Confederate left flank at Droop Mountain, contributing to the subsequent Federal breakthrough. U.S. ARMY MILITARY HISTORY INSTITUTE, MOLLUS COLLECTION

Various Federal participants in the battle from the 8th West Virginia Mounted Infantry and the 10th West Virginia Infantry. Baxter fell mortally wounded while assaulting the Confederate left. Sutton and Bender were both near him when he fell. DROOP MOUNTAIN COMMISSION BOOKLET

intelligent and efficient obedience. He had so schooled his regiment in discipline and tactics that it had early in the service established a character for reliability."[35]

Colonel Augustus Moor's 28th Ohio Volunteer Infantry, the second regiment recruited from the German communities of Ohio, was organized at Camp Dennison, near Cincinnati, between June 10, 1861, and March 1, 1862. The 28th Ohio "was a model of accuracy in drill and neatness in the performance of all camp duties."[36] The regiment saw action in 1861 in (West) Virginia at Carnifex Ferry [where they confronted the rebel 22nd Virginia Infantry, the same opponent they would meet at Droop Mountain], the Sewell Mountain campaign, and Cotton Hill. During 1862 the 28th Ohio fought at Princeton, (West) Virginia and in the Antietam campaign, including South Mountain. By the time of the Droop Mountain battle the regiment had fought and campaigned primarily in the mountains of West Virginia, becoming familiar with Confederate guerrilla tactics, thereby making them a well-suited opponent for the homegrown rebel forces. At Droop Mountain the 28th Ohio Volunteer Infantry would number about 605 effectives, led by 37 year old Lt. Col. Gottfried Becker of the regiment, as Col. Moor would lead the infantry detail of the 10th West Virginia and the 28th Ohio. Becker was commissioned lieutenant colonel of the regiment June 10, 1861.

Colonel Augustus Moor was born in Leipzig, Saxony March 28, 1814, and received military training at the Royal Academy of Forestry in his homeland. He emigrated to America in 1833 and found employment in Philadelphia, Pennsylvania. Moor became a lieutenant colonel in the Washington Guard of Philadelphia and served in the Florida Seminole War of 1836 with a volunteer dragoon company, in the position of lieutenant colonel. He moved to Cincinnati, Ohio, opened a bakery and restaurant, and joined in an Ohio militia unit. He served with distinction in the Mexican War in 1846, where he eventually rose to the rank of colonel of his regiment. After the war, he became a major general of the First Division of the Ohio Militia, resigning his commission two years later. As hostilities erupted in 1861 Moor was placed in command of the 28th Ohio Volunteer Infantry, where he was described as "a German of portly presence and grave demeanor, a gentleman of dignity of character as well as of bearing, and a brave resolute man."[37] He had a reputation as a rigid disciplinarian, with a "square head, with a dark, smooth shaven face, and rather stern expression, and [he] inspired his troops with something like awe, insuring prompt obedience of his commands."[38] When Col. Moor was assigned command of the "Northern Brigade" in January of 1863 he was described as "an intelligent and efficient officer, and gallant soldier, and was well liked by officers and men."[39]

General Averell had two artillery batteries at his disposal at Droop Mountain, batteries B and G of the 1st West Virginia Light Artillery. Battery B (Keepers') 1st West Virginia Light Artillery was organized and mustered in at Ceredo, Wayne County, (West) Virginia October 1, 1861, with Samuel Davey as captain and John V. Keepers, who was approximately 40 years old, as 1st lieutenant. The battery gave impressive service in Gen. Nathaniel P. Banks' 5th Army Corps, Gen. James Shields' Division, and was conspicuous for its outstanding performance at the battle of Kernstown, Virginia March 23, 1862. In January of 1863, Keepers' Battery served in the 8th Army Corps under Gen. Robert C. Schenck, and in Gen. Robert H. Milroy's brigade at Winchester, Virginia. On May 31,

Col. Augustus Moor of the 28th Ohio Volunteer Infantry commanded the Federal flanking party of the 28th Ohio and 10th West Virginia which broke the Confederate left flank at Droop Mountain. U.S. ARMY MILITARY HISTORY INSTITUTE, MOLLUS COLLECTION

The Heroes of Droop Mountain. This rare photographic view is titled, "Headquarters–28th Ohio Vol.–Gauley Bridge, W. Va." The 28th Ohio Volunteer Infantry, along with the 10th West Virginia Infantry, broke the Confederate left flank at Droop. This is the only known view of the 28th Ohio in the field and the officer in the middle front with right hand inside coat appears to be Col. Augustus Moor of the 28th Ohio. The other men are probably members of his staff and the photo was most likely taken in 1861 or 1862. CHARLES BRACKEN, CHARLETON HEIGHTS, WEST VIRGINIA

1863, the battery, with its six guns, was assigned to Gen. Averell and made a courageous display with him at the battle of White Sulphur Springs August 26-27, 1863, losing one killed, three wounded, one captured, and one deserted.[40]

Captain John V. Keepers, born about 1821, married at New Derry, Pennsylvania February, 1843, had 4 children, and entered military service at Athens, Ohio April 19, 1861, when he was appointed a 1st lieutenant in Company A, 18th Ohio Volunteer Infantry (three months service). He was mustered out August 28, 1861 upon the expiration of term of service. On October 1, 1861, Keepers was mustered in as 1st lieutenant of Battery B, 1st West Virginia Light Artillery. Capt. Samuel Davey resigned April 1, 1862, which prompted Keepers' rise to captain of the battery, for which he was commissioned October 18, 1862, to rank from September 20, 1862. Keepers' service record says he actually commanded the battery beginning December 1, 1861 due to neglect by the commanding officer.[41]

Battery G (Ewing's), 1st West Virginia Light Artillery was originally organized as an infantry company out of Pittsburgh, Pennsylvania and known as the "Plummer Guards." The company was one of the first to respond to Pennsylvania's call for volunteers; but, as they feared, Pennsylvania's quota would be quickly met, offered their services to Gov. Francis Pierpont of the loyal government of (West) Virginia, and thus mustered in as Company G, 2nd West Virginia Infantry. In such capacity this company served during 1861 at Winchester, Laurel Hill, Rich Mountain, Beverly, Elkwater; and, on December 13, 1861, a detail of the company participated in the battle of Allegheny Mountain. In the spring of 1862 the company was placed in charge of some old brass six-pounder guns and became proficient artillerists. The authorities were so impressed with their performance as artillerymen, they were transferred to the 1st West Virginia Light Artillery "after which they were recruited to the full battery strength and a splendid equipment of guns was given them."[42] The company took part in all of Gen. Milroy's battles and marches and were with Gen. Fremont and then Gen. Pope up to 2nd Bull Run. Afterwards the command was returned to West Virginia to serve with Gen. Averell. In the fight at White Sulphur Springs August 26-27, 1863, the battery lost one killed and 15 wounded.[43]

Captain Chatham Thomas Ewing, who commanded Battery G, 1st West Virginia Light Artillery, was born January 30, 1839 near New Lisbon, Ohio and in 1852 moved with his family to Pittsburgh, Pennsylvania, where in the summer of 1860 he became a member of the "Pittsburgh Zouaves." At the outbreak of the war Ewing was collecting taxes for Pittsburgh business,' and enlisted May 15, 1861 in Company G, 2nd West Virginia Infantry. He also studied law and was admitted to the bar in the winter of 1861. He was commissioned captain of Company G (later Battery G, 1st West Virginia Light Artillery), 2nd West Virginia Infantry August 2, 1861.[44] Gen. Averell felt Ewing was "an active, brave, and efficient artillery officer."[45] 1st Sgt. George H. Mowrer, Company A, 14th Pennsylvania Cavalry, said Ewing "was a splendid officer and a gentleman of culture. His educational opportunities had been superior, and he evidently made good use of his time."[46] During the battle at White Sulphur Springs August 26-27, 1863, he was seriously wounded and captured, but was released from the rebel authority September 14, 1863, and released on furlough to Pittsburgh. Gen. Averell recalled, "I saw him fall desperately wounded in the thigh or hip. With great

Capt. Chatham Thomas Ewing of Battery G (Ewing's), 1st West Virginia Light Artillery. His guns attempted to strike the Confederate line at Droop Mountain. Having been badly wounded and captured (although he was later released) at the White Sulphur Springs battle in August of 1863, there is some doubt as to his actual presence in the Droop Mountain battle. DROOP MOUNTAIN COMMISSION BOOKLET

fortitude he kept his place, that is, continued for some time to direct the operations of his gunners until the onset of the enemy was repelled," and Averell said he left a surgeon with Ewing.[47] Adam Brown, a member of the battery who lost an arm in the White Sulphur Springs fight, said he was captured along with Ewing and "exchanged about a month later."[48] Averell claimed he exchanged two or three Confederate lieutenants for Ewing and Brown confirmed he was carried in the same ambulance with Ewing to Beverly.[49] This situation lead to speculation as to whether or not Ewing was actually present at Droop Mountain, although his service record shows him present with the battery from October of 1863 to January 1864. Noyes Rand, 22nd Virginia Infantry, acting adjutant general of the Confederate troops at White Sulphur Springs, met Ewing as a prisoner and said he "was a polished man, and highly educated," a view also shared by Sgt. John G. Stephens, Chapman's Battery, who additionally encountered Ewing during his confinement.[50]

If by some odd circumstance Ewing was absent at Droop, and no evidence could be located to validate this charge, then the battery was probably led on the field by Lt. Howard Morton, who "was a brave officer and thoroughly qualified for the position he held."[51]

Rounding out Averell's brigade was a signal corps detachment of the 68th New York Infantry. Organized in New York City between August 1 and 20, 1861, the 68th New York Infantry was recruited by Col. Robert J. Betge. The bulk of the men came

Federal gunner, 1st Lt. Howard Morton, Battery G (Ewing's), 1st West Virginia Light Artillery, was conspicuous in the action at Droop Mountain and was instrumental in advancing his battery against the enemy. FRANCIS READER COLLECTION

from New York City, although a few were from New Jersey, Maryland, and Pennsylvania. A small number were detached March 3, 1863, to serve as a signal corps under the command of Captain Ernst A. Denicke.[52] In an 1864 document, mention is made that this detachment was reduced in the latter part of March [year not given] to 4 officers and 45 enlisted men.[53] During the Droop Mountain campaign, Capt. Denicke reported successfully "using rockets for day signals by removing the parachute and placing in its stead a blank cartridge open at the lower end." He concluded this method was good "in very hilly country, where signaling with flags is not always practicable, this model of signaling by day, I think, is very much to be recommended."[54]

Captain Ernst A. Denicke, a resident of New York, enrolled at the age of twenty-two in the 68th New York Infantry July 22, 1861. He served as a private in Company H August 1, 1861, but seven days later became a 2nd lieutenant in Company E. Denicke transferred to Company G January 1, 1862 and became a 1st lieutenant in Company C July 17, 1862, a position he held until rising to captain of Company E December 18, 1862. Although yet associated with the 68th New York Infantry, he was appointed to the U.S. Volunteer Signal Corps March 3, 1863, and served in the Droop Mountain campaign commanding a signal corps detachment of the regiment.[55] Mark Crayon, a young Greenbrier County boy during the war, recalled, "The signal corps camped in tents and had telescopes and flags. They could make letters or signs with the motion of the flags that the fellows at the next station could understand. They would let a boy look through their glasses when they were not using them."[56]

Also accompanying Averell was Lt. John Rodgers Meigs, Engineer and Acting Aide-de-Camp. Meigs, born in Washington, D. C. in 1842, was the son of Brvt. Major Gen. Montgomery C. Meigs, Quartermaster Gen. of the U. S. Army. He was appointed at large to West Point Military Academy and entered September 7, 1859. While a Cadet on leave of absence he served as Volunteer Aide-de-Camp to Col. Israel Bush Richardson, commanding the 2nd Michigan Volunteers, at the battle of Bull Run July 21, 1861. He graduated first in his class June 11, 1863 and was promoted in the U. S. Army to First Lt., Corps of Engineers. Meigs first served as Assistant Engineer in the construction of the defenses of Baltimore, Maryland "and in opening the communications with Harpers Ferry, before the retreat of Gen. Lee's Rebel Army from Gettysburg, June 15 to July 23, 1863, being engaged in a skirmish at Harpers Ferry, July, 1863." He was Engineer on

Lt. John Rodgers Meigs of the U.S. Engineers. First in his class at West Point in 1863, Meigs served at Droop Mountain as Averell's Engineer and Acting Aide-de-Camp, and as Chief Engineer of the Dept. of West Virginia. He was killed in 1864 by Confederate guerrillas. This is his Cadet photo. U.S. ARMY MILITARY HISTORY INSTITUTE

the staff of Brig. Gen. Benjamin F. Kelley, commanding the Department of West Virginia, from July 23 to November 23, 1863, "being engaged in making surveys, reconnaissances, and various works of defense." In this capacity he was assigned as Engineer and Acting Aide-de-Camp to Averell, and was in the fight at White Sulphur Springs August 26-27, 1863. On November 3, 1863 he was made Chief Engineer of the Department of West Virginia, a position he would hold until August 17, 1864.[57]

The exact number of men in Averell's brigade in the campaign is never clearly stated, with figures ranging from 3000 to 7000, but as Averell reported 3,855 officers and men present for duty on August 31, 1863, and as he was not engaged in any notable activity between that time and the Droop Mountain fight, it must be assumed he had close to this figure, minus men absent for illness, furlough, and detached service. The weaponry of Averell's brigade [besides that already mentioned], according to 1st Sgt. George H. Mowrer, Company A, 14th Pennsylvania Cavalry, "was not first class, only the 14th Pennsylvania [Cavalry] and Gibson's [Independent Cavalry] battalion were armed with carbines. The West Virginia troops had short [.58 caliber] Enfield rifles—muzzle loaders—sabers and Colt's navy revolvers. The rifle is an awkward piece to carry on horseback, and is not a convenient arm for loading while in the saddle; but as the West Virginia troops were to be used as dismounted cavalry, the objection to these arms was not serious."[58] This assessment is confirmed by Corp. George Washington Ordner, Company B, 2nd West Virginia Mounted Infantry, who wrote in his diary October 29, 1863: "The regiment has been supplied with the short Enfield Rifle—Saber Bayonet today."[59] Field recoveries as well as eyewitness accounts conclude the 10th West Virginia Infantry was armed with .58 caliber Enfield rifles and the 28th Ohio Volunteer Infantry used the .69 caliber "big bores."

On October 30 Gen. Kelley sent Gen. Duffié details of the proposed movement with orders to cooperate fully with Averell. Duffié was further informed to "form a junction at or near Lewisburg" . . ."on Saturday, the 7th of November 1863." Kelley also reiterated his strong stance against pillaging of the citizenry.[60]

Duffié's Brigade, Scammon's 3rd Division

Brigadier General Alfred Napoleon Alexander Duffié, nicknamed "Nattie," was born May 1, 1835 in Paris, France, the son of a French count. In 1854 he graduated the military college of St. Cyr and afterwards served as a lieutenant of cavalry in the French army at Algiers and Senegal. Duffié won four decorations in the Crimea and was wounded in action in the battle of Solferino with the Austrians. Due to his wound he took a leave of absence in 1859 and came to the United States, where at the outbreak of the Civil War he resigned his commission in the French Army and offered his services to the Union. Duffié was commissioned a captain of the 2nd New York Cavalry ("Harris Light Cavalry") August 9, 1861 and was promoted to major in October of that year. In July of 1862 he became colonel of the 1st Rhode Island Cavalry and fought in the 2nd Manassas campaign.[61] Historian Stephen Z. Starr would claim Duffié's "main handicaps as a commander of a regiment of American volunteers were an excessive addiction to

Brig. Gen. Alfred Napoleon Duffié led a detachment of Gen. E.P. Scammon's 3rd Division, U.S.A., from Charleston to Lewisburg in an unsuccessful effort to block the Confederate army retreating from Droop Mountain. U.S. ARMY MILITARY HISTORY INSTITUTE, MOLLUS COLLECTION

parades, pomp, and pageantry, and the issuance of congratulatory orders of an overblown floweriness . . .," and he was "greatly given to oratory; his eloquence was marred, but not inhibited by his strong French accent and his lack of familiarity with American idiom."[62]

Duffié distinguished himself in March of 1863 while serving with Gen. William W. Averell at the battle of Kelly's Ford, Virginia. His performance compelled Gen. Joseph Hooker to promote Duffié to brigadier general, to rank from June 23, 1863. He commanded a Division of Cavalry Corps in the Chancellorsville campaign as well as during the early part of the Gettysburg campaign. It was during the latter, at the battles of Aldie (June 17, 1863) and Middleburg (June 19, 1863) in the Loudoun Valley of Virginia that he was accused of mishandling the fighting, and like Averell, relegated to a cavalry command in rural West Virginia as a sort of punishment. He got his orders to report to West Virginia on August 14, 1863 and upon arrival at Charleston in the Kanawha Valley found his command in utter chaos, poorly equipped and trained, detachments widely scattered throughout the area, and, eventually, orders to work in conjunction with a man he did not particularly like – Gen. Averell.[63] Correspondent "T. B. F." of the *Cincinnati Daily Commercial*, who would visit the West Virginia troops, described Duffié as "a man of uncommon energy and ability."[64]

On the expedition to Lewisburg to assist Averell, Duffié would have the 34th Ohio Mounted Infantry, the 2nd West Virginia Volunteer Cavalry, and the 1st Kentucky Independent Battery (Simmonds' Battery). En route he would be joined by the 12th and 91st Ohio volunteer infantry regiments.

The 34th Ohio Mounted Infantry was originally organized as the 34th Ohio Volunteer Infantry at Camp Lucas, Clermont County, Ohio in July and August of 1861. The command was known as the "Piatt Zouaves" in honor of their first colonel, Abraham S. Piatt, and wore "a light blue Zouaves dress." The regiment was sent to the Kanawha Valley of (West) Virginia and saw action in 1861 near Chapmanville in Logan County, (West) Virginia. The 34th Ohio played a prominent role in the 1862 Kanawha Valley campaign, waging combat at Fayetteville, Cotton Hill, Charleston, and Buffalo. The regiment was mounted in May of 1863 and became the 34th Ohio Mounted Infantry, fighting soon afterward at Wytheville, Virginia. In July of 1863 the regiment was assigned to the 3rd Brigade, Scammon's Division, West Virginia, and were headquartered at Charleston in the Kanawha Valley. At this time the 34th Ohio numbered 505 strong. During the Lewisburg expedition the regiment was led by Col. Freeman E. Franklin, previously lieutenant colonel of the regiment, who was made colonel to rank from July 18, 1863; the date Col. John T. Toland was killed in action at Wytheville, Virginia.[65] Correspondent "T. B. F." implied Franklin was a man of "brains and pluck" and a "man who will fight and let fight."[66]

The 2nd West Virginia Volunteer Cavalry was in reality an Ohio unit, as the bulk of the men came from the Ohio counties of Lawrence, Meigs, Vinton, Washington, Morgan, Putnam, and Monroe. Recruitment began in southwest Ohio about August 1, 1861, but when application was made to the Ohio governor he declined their services, as the state of Ohio had been instructed by the War Department to not recruit any additional cavalry. As such, the regiment applied to Gov. Francis Pierpont of (West) Virginia and was accepted into service, becoming the 2nd West Virginia Volunteer Cavalry, instead of the 4th Ohio Cavalry as originally intended.[67] Pvt. John J. Sutton of the regiment would later write: "Many of us have always regretted that we were not allowed to be mustered in as the 4th Ohio Cavalry, where we properly belonged; yet neither during the progress of the war nor since its close, have we had the slightest cause to complain of our treatment at the hands of the little mountain state . . . West Virginia, the 'Child of the Storm.'"[68] The regiment was organized at Parkersburg, (West) Virginia from September to November of 1861, and was officially mustered into the U.S. service as the Second Regiment of Loyal Virginia Cavalry November 8, 1861. Once the regiment was mounted and equipped, it was ordered to Guyandotte, Cabell County, (West) Virginia for drill and discipline. The 2nd West Virginia Cavalry saw its first service in the Sandy River area of eastern Kentucky, followed by extensive campaigning in the mountains of West Virginia. In July of 1863 the regiment numbered 367 effectives, four months prior to the Lewisburg journey. By the time of the Droop Mountain campaign the 2nd West Virginia Cavalry was undoubtedly a hardened West Virginia command. Taking charge of the 2nd West Virginia Cavalry for the Lewisburg jaunt was Major John J. Hoffman.[69]

The 1st Kentucky Independent Battery, or Simmonds' Battery Light Artillery, was commanded by Capt. Seth J. Simmonds and was organized out of Company E, 1st Kentucky Infantry at Camp Clay, in Pendleton, Ohio in 1861. The company was mustered into U.S. service June 3, 1861, and detached as artillery October 31 of the same year. The battery "served with great credit in all the campaigns in western Virginia in the early part of the war" and fought at South Mountain and Antietam in 1862. The battery was distinguished for gallantry, discipline and soldierly conduct throughout the whole enlistment.[70] One section of the battery accompanied Duffié to Lewisburg, and based upon July and November 1863 regimental figures, the battery possibly numbered approximately 98 men during the Droop Mountain expedition.

Duffié said these three commands numbered "in all 970 officers and men and 1,025 horses." En route to Lewisburg, at Tyree's on Sewell Mountain in Fayette County, he would be joined by the 12th and 91st Ohio infantry regiments, under the command of Col. Carr B. White of the 12th Ohio. According to correspondent "Q. P. F." of the *Cincinnati Daily Commercial*, White was "a little man, but a mighty [man]."[71]

Col. Carr B. White of the 12th Ohio Volunteer Infantry. He commanded General Duffié's two infantry regiments, the 12th Ohio and 91st Ohio, in the expedition from Charleston to Lewisburg. U.S. ARMY MILITARY HISTORY INSTITUTE, MOLLUS COLLECTION

The 12th Ohio Volunteer Infantry was organized at Camp Dennison near Cincinnati, Ohio on May 3, 1861. Most of the men came from the Ohio cities of Morrowtown, Wilmington, New Richmond, Xenia, Newark, Lebanon, Middletown, Ripley, Dayton, and Hillsboro. The 12th Ohio saw action in the 1861 Kanawha Valley campaign in (West) Virginia, including Scary Creek and Carnifex Ferry, where Col. John W. Lowe of the 12th Ohio was killed in action. Georgetown, Ohio resident Lt. Col. Carr B. White of the regiment, who served as major until appointed lieutenant colonel June 8 or 28, 1861, immediately assumed colonelcy of the 12th Ohio. Following hard campaigning in the mountains of West Virginia, the regiment fought at Bull Run Bridge, South Mountain, and Antietam. The regiment returned to West Virginia in late 1862 and was quartered at Fayetteville when the Droop Mountain campaign began.[72]

Commanding the 12th Ohio on the expedition to Lewisburg was Lt. Col. Jonathan D. Hines, who had been commissioned Major of the regiment June 28, 1861 and lieutenant colonel September 10, 1861.[73]

The 91st Ohio Volunteer Infantry was comprised of men from the southern Ohio counties of Adams, Scioto, Lawrence, Gallia, Jackson, and Pike. The regiment was organized at Camp Ironton, Ohio August 26, 1862, and was mustered into U.S. service at Guyandotte September 4, 1862. The regiment served in the Kanawha Valley and experienced their baptism of fire at Buffalo, (West) Virginia September 27, 1862. The remainder of 1862 and 1863 was spent in the mountains of West Virginia.[74] Corp. John W. Weed, Company A, 91st Ohio recalled: "The 91st Ohio together with a few other regiments was left in that part of West Va. for eighteen months to prevent future raids and to keep bushwhackers in check. The 18 months time was spent in drilling, scouting and building forts. We had some skirmishes and got to learn what the whistle of a bullet sounded like, but in the 18 months we didn't have a man of the regiment killed by an enemy. We had several wounded and a good many died from sickness."[75]

In command of the 91st Ohio during the Lewisburg trip was Col. John A. Turley of Portsmouth, Ohio, who was commissioned lieutenant colonel of the 22nd Ohio Volunteer Infantry April 23, 1861. He was appointed colonel of the 81st Ohio Volunteer Infantry August 19 or 29, 1861 but resigned December 1, 1861. Turley became colonel of the 91st Ohio Volunteer Infantry August 22, 1862.[76]

CHAPTER 3

Mountaineers In Gray

On October 31, the very day after Gen. Kelley sent Duffié his orders to cooperate with Averell, Gen. Sam Jones, commanding the Army of Western Virginia and East Tennessee, Confederate States Army, reported the brigade of Gen. John Echols had 1,558 present for duty, and an aggregate present and absent of 2,225.[1] Indeed, it would be this command, along with the cavalry brigade of Col. William L. Jackson and a mounted detachment of Gen. Albert Gallatin Jenkins' Brigade that would block the Federals' southern movement. Stationed at Lewisburg, a town at the 2,800 foot level in Greenbrier County, was Echols' Brigade, consisting of the 22nd Virginia Infantry, the 23rd and 26th battalions of Virginia infantry, and Chapman's Battery. Col. Jackson, with his headquarters at Mill Point in Pocahontas County, seven miles north of Droop Mountain, was in charge of the Huntersville Line, an area stretching from Nicholas County to most of Pocahontas and Greenbrier counties. Under Jackson's supervision could be found his own 19th Virginia Cavalry, which had detachments scattered at such places as Edray and Glade Hill as the main body remained with Jackson; the 20th Virginia Cavalry, preparing winter quarters at Marlin's Bottom [Marlinton]; and Lurty's Battery, which also rested at Mill Point with Jackson. Jenkins' Brigade, temporarily under the command of Col. Milton J. Ferguson of the 16th Virginia Cavalry, and recuperating from the Gettysburg campaign in the vicinity of Lewisburg, consisted of the 14th Virginia Cavalry; Ferguson's 16th Virginia Cavalry; and Jackson's Battery. At Droop Mountain these three brigades would fall under the field command of Gen. Echols.

Echols' Brigade

John Echols, the man who would command all Confederate forces at Droop Mountain, was born March 20, 1823, at Lynchburg, Virginia. After graduating with honors at Washington College (now Washington and Lee) in 1840, he entered the Virginia Military Institute in 1843 and served only one term, "standing at the end of the year number nine in a class of twenty-six cadets." Following his resignation he studied law at Harvard, taught school briefly at Harrisonburg, Virginia, and shortly afterward was admitted to the bar in Rockbridge County. He moved to Union, in Monroe County, (West) Virginia, and was elected attorney for the Commonwealth; became a member of the state legislature representing Monroe County, and was a member of the secession convention from that district. From 1858 to 1861 he was a member of V.M.I.'s governing Board of Visitors, and would be declared an honorary graduate of V.M.I. after the war.

Echols was a prominent member of the Virginia Convention when the Ordinance of Secession was adopted and in 1861 offered his services to the Confederacy. He was authorized to muster in the volunteer forces in the vicinity of

Brigadier General John Echols commanded the 1st Brigade, Army of Southwest Virginia, C.S.A., and was field commander of all Confederate forces at Droop Mountain. U.S. ARMY MILITARY HISTORY INSTITUTE, MOLLUS COLLECTION

Staunton. After completion of this task he was assigned as lieutenant colonel of the 27th Virginia Infantry, which he commanded at 1st Bull Run (Manassas). His regiment, along with four other Virginia units, gained fame as the "Stonewall Brigade." Promoted to colonel of the 27th Virginia, Echols took part in the early phases of Gen. "Stonewall" Jackson's 1862 Shenandoah Valley Campaign, until suffering a severe wound at the battle of Kernstown, Virginia March 23, 1862. Upon recovery he was made a brigadier general and on October 16, 1862, succeeded Gen. William W. Loring as commander of the Army of Western Virginia, a position he held briefly. During the summer of 1863 he served at Atlanta on the court of inquiry created to investigate the surrender of Vicksburg. This assignment temporarily took him away from his command, leaving Col. George S. Patton in charge of Echols' (1st) Brigade at the battle of White Sulphur Springs, West Virginia August 26-27, 1863.[2]

Echols was an imposing figure, standing at six-foot four-inches tall, and weighing some 260 pounds, "making him an easy target" for the enemy. Purportedly, this body structure may have caused undue pressure upon his heart, often making him ill, and possibly contributing to some poor military judgments, particularly his reported ignorance of the obscure road on his left flank that the Federal army utilized at Droop Mountain. Historian William C. Davis noted Echols ". . . was a skillful organizer and a good fighter, though sometimes slow in the field. The reason for the latter lay possibly in his great size, for it weakened his heart . . . [creating a] . . . neuralgia of the heart."[3]

Davis also said the amiability of Echols "was remarkable—somehow a grin seemed always lurking behind his round, mustachioed face."[4] It was said Echols "possessed a mind trained in nature's mould, broad of grasp, quick of perception, accurate in discrimination and just in equipoise—a heart full to the over flowing with the milk of human kindness."[5] Another admirer noted he "was one of the most splendid specimens of manhood. He towered above his fellows with a stateliness and grandeur always imposing and added to his magnificently proportioned physique, was a mind peculiarly adapted to the times, and the business chosen in life."[6] George M. Edgar, who would serve under Echols said, "He was not a field man, though an efficient organizer."[7] At a veterans' reunion in 1903 the speaker said Echols "gave to the service the great advantage of a scientific military training, mental endowments of a very high order, a trained and sure-footed judgment, a physical frame of exceptional strength and endurance, and energy that defied the effects of privation and toil, and a calm deliberate courage, above the power of danger, emergency or disaster to daunt or disturb."[8]

The first regiment comprising Echols' Brigade that would clash at Droop Mountain was the 22nd Virginia Infantry, organized about April 25, 1861, and evolved out of Capt. George S. Patton's Kanawha Riflemen. The regiment was originally known as the 1st Kanawha Regiment from May to August of 1861, and consisted of eight infantry companies, one company of cavalry, and one company of artillery. Most of the men were from the (West) Virginia counties of Kanawha, Putnam, Fayette, Jackson, Nicholas, Monroe, Boone and Wyoming. This command participated in the 1861 Kanawha Valley campaign of Gen. Henry A. Wise and fought in most of the battles and skirmishes, including Scary Creek on July 17, 1861. In early August the regiment was organized as the 22nd Virginia Infantry, consisting of ten infantry companies (A through K) and saw active service at Cross Lanes, Carnifex Ferry, Sewell Mountain, and Cotton Hill. The unit had been mustered into service July 1, 1861, (although some members claimed they were officially mustered into the Confederacy August 31, 1861) and finally brought to completion at the reorganization of the Confederate army in (West) Virginia May 1, 1862.

During 1862 the 22nd Virginia Infantry fought at Giles Court House and suffered the highest rate of casualties among Confederate units engaged at the battle of Lewisburg May 23, 1862. The regiment finished out the year in the 1862 Kanawha Valley campaign and opened 1863 with participation in the famed Jones-Imboden Raid. On August 26-27, 1863 the 22nd once again suffered the highest rate of losses among rebel commands at White Sulphur Springs.[9] This provoked Pvt. Henry Devol McFarland, Company H, 22nd Virginia, to write his mother, "It seems as if we always have to bear the brunt of battle in every fight [in which] we are engaged."[10]

The 22nd Virginia Infantry, which also had a number of men from the (West) Virginia counties of Greenbrier and Roane, and the Virginia counties of Craig and Allegheny, "often campaigned in some of the worst possible geographic and environmental conditions, led by some of the most incompetent and incompatible generals the south could produce."[11] As a result, direct leadership of the regiment suffered heavily early in the war. By the time of the Droop Mountain battle the 22nd Virginia had shed enough blood and sweat to truly be considered a veteran West Virginia Confederate regiment. They boasted they had never lost a battle, somehow forgetting Lewisburg, and were alternately called by their comrades as the "Star of the South" and their enemy as a "brag regiment."

According to various members of the regiment as well as one of Jackson's Battery, the 22nd Virginia would enter the Droop Mountain fight 550 strong. One group that would not participate in the battle was Lt. George Cofer and Lt. Robert S. Campbell, "who took 35 men of their own Company D (Nicholas Blues), 22nd Virginia, and went on a scout of Nicholas County, West Virginia starting on November 4. Two days later, on the same day as the Droop Mountain fight, they attacked a small party of Federals at Summersville under the command of Capt. Riley Ramsey, capturing four men, four horses, several guns, and blankets and overcoats. Upon gaining knowledge of the disaster at Droop Mountain the same day, they decided to parole Ramsey's men and remained in Webster County until Averell left Greenbrier County. Then they rejoined the regiment at Camp Beauregard November 17."[12]

Leading the 22nd Virginia was Col. George Smith Patton, who was born June 26, 1833, at Fredericksburg, Virginia. Early in his youth Patton moved to Richmond, where his father practiced law and George received his education. In 1849 he entered the Virginia Military Institute at Lexington, Virginia, graduated in 1852, and spent the following two years teaching in Richmond, where he also studied law. He married and in 1856 moved his new family to Charleston, (West) Virginia, where he formed a law partnership with Thomas Lee Broun. Prior to the war Patton served as Commissioner in Chancery to Kanawha Circuit Court and Kanawha County Court, and he and Broun also served on the board of directors of the Coal River Navigation Company. George formed a militia company in

Colonel George Smith Patton of the 22nd Virginia Infantry led the 1st Brigade (Echols'), Army of Southwestern Virginia, at Droop Mountain. (Damage is in original photo.) NATIONAL ARCHIVES

1856 dubbed the Kanawha Minutemen, and he was designated as captain. Shortly thereafter the name changed to the Kanawha Rifles and in November of 1859 was permanently changed to the Kanawha Riflemen. Patton enlisted May 8, 1861, as captain of Company H (Kanawha Riflemen), 22nd Virginia Infantry, and on July 7, 1861, was commissioned lieutenant colonel of the regiment. He participated in Gen. Henry A. Wise's 1861 Kanawha Valley campaign and was in command of the victorious Confederates at Scary Creek July 17, 1861, where he received a severe bullet wound in the left shoulder. Subsequently he was left at Charleston as an exchange prisoner following Gen. Wise's retreat from the Kanawha Valley later in the month. This odd agreement left him in enemy hands and caused much confusion, detaining him from the 22nd Virginia until April of 1862.

During Patton's recuperation from his wound he was offered a teaching job at the Virginia Military Institute if unable to return to duty; he declined and was made colonel of the 22nd Virginia upon his return in April [records indicate Patton was appointed colonel March 3, 1863 to rank from November 23, 1861 and accepted January 15, 1863]. On May 10, 1862, he was slightly wounded at the battle of Giles Court House (Pearisburg), Virginia, and had to return home, only to discover his exchange had never been formally completed. After much legal argument he was finally exchanged and returned to command of the 22nd Virginia Infantry shortly after June 9. Patton remained with the regiment through the 1862 Kanawha Valley campaign and the Jones-Imboden Raid of 1863. Due to the absence of Gen. John Echols, Patton commanded Echols' Brigade at the battle of White Sulphur Springs August 26-27, 1863; where his own 22nd Virginia suffered at least 16 killed, 56 wounded, and eight captured, in a brilliant, hard-fought victory over Gen. William W. Averell. In fact, Patton often led Echols' Brigade, so much so that it was more often referred to as Patton's Brigade, and he would once again be placed in such command at Droop Mountain, as Gen. Echols would assume field command of all the Confederate forces.

Patton, nicknamed "Frenchy" due to his pointed beard (although most photos show him with a full beard), was known to be arrogant, a smart dresser, displayed classic chivalry toward the ladies, never missed a camp prayer, and upon formation of his Kanawha Riflemen; was considered one of the most promising of the early militia commanders, and a stern disciplinarian.[13]

Since Col. Patton was in command of Echols' Brigade at Droop, and Lt. Col. Andrew Russell Barbee of the 22nd Virginia was on medical furlough recovering from serious wounds sustained at White Sulphur Springs, the 22nd Virginia would be led at Droop Mountain by Major Robert Augustus ("Gus") Bailey, the son of Circuit Judge Edward P. Bailey of Lewisburg, (West) Virginia. He was born in 1839 and in 1860 was an attorney and resident of Fayette County, (West) Virginia. Before the war he served as a captain of the 142nd Regiment Virginia Militia, 27th Brigade, 5th Division. On June 6, 1861, Bailey enlisted at Charleston, (West) Virginia as captain of the Fayetteville Riflemen (Company K, 22nd Virginia Infantry). He was promoted to major of the 22nd Virginia November 23, 1861, and often led segments of the regiment while Col. Patton was in charge of Echols' Brigade. He was particularly noted for his bravery and leadership at Fayetteville, (West) Virginia September 10, 1862. Records indicate he may have also served as lieutenant colonel and colonel of the 14th Virginia Cavalry during a portion of 1862. He was Major in command of the Department at Lewisburg March 1, 1863, and received praise for his gallantry in leading a portion of the 22nd Virginia in the fight at White Sulphur Springs. Despite such displays of courage, at least one member of the Fayetteville Riflemen implied Bailey, who never married, was not particularly popular with the men. In the battle at Droop Mountain he would fall mortally wounded while attempting to rally the retreating Confederate army.[14]

The 23rd Battalion Virginia Infantry, also known as the 1st Battalion and as Derrick's Battalion, was first organized January 15, 1862, with five companies (A to E), under the command of Major David Stuart Hounshell, but the command never fully existed as a cohesive unit until reorganization on May 21, 1862, and was led by West Point graduate Lt. Col. Clarence Derrick. "Changes were made in company letters in accordance with the dates of commissions of their respective captains. Other companies were added until April 1, 1863, when the battalion consisted of eight companies A to H." Most of the men were from the southwest Virginia counties of Smyth, Tazewell, and Giles, while one company was from Mercer County, (West) Virginia, and another from Stokes County, North Carolina. The battalion experienced it's first combat in the 1862 Kanawha Valley campaign in (West) Virginia, then contributed significantly to the defeat of Gen. William W. Averell at White Sulphur Springs, August 26-27, 1863, where the battalion lost, at least, seven killed, 15 wounded, and one deserted.[15] According to Col. Patton, the 23rd Battalion Virginia Infantry would number "about 350 strong" at Droop Mountain,[16] although Micajah Woods of Jackson's Battery would give the smaller figure of 200.[17] In the battle the battalion would be led by Major William Blessing, apparently "filling in" for Lt. Col. Derrick.

Major William Blessing was born about 1825 and was a merchant in Smyth

Major William Blessing (center) commanded the 23rd Battalion Virginia Infantry at Droop Mountain. Armed to the teeth, these three Smyth County, Virginia rebels of the 23rd Battalion are (l to r): Capt. Robert Henry Hubble, Co. A; Blessing and Capt. (later Chaplain) John T. Kincannon, Co. A. SMYTH COUNTY HISTORICAL SOCIETY MUSEUM

County, Virginia when the war opened. He began to organize a company of infantry, which was mustered in at the Holston Woolen Mill in Smyth County September 14, 1861, and Blessing was elected captain. The company was originally assigned as Company A, 3rd Regiment of Gen. John B. Floyd's Brigade, but as organization of the 3rd Regiment was never completed, the company was then assigned as Company A (Smyth County Sharpshooters), 23rd Battalion Virginia Infantry and Blessing was promoted to major of the battalion March 5, 1863. He soon afterward distinguished himself in the clash at White Sulphur Springs.[18]

The 26th Battalion Virginia Infantry originated in the Spring of 1862 when Capt. George Mathews Edgar received authority to raise a battalion out of five companies from the mountain counties of Monroe, Greenbrier, and Mercer in (West) Virginia. The earliest known organization of these companies, known as Edgar's Battalion, was on April 29, 1862, and consisted of a number of men who had seen previous service in the 59th Virginia Infantry, the militia, conscripts and raw recruits. Edgar's Battalion was officially organized and mustered into Confederate service May 20, 1862, and was comprised of "those members of the 59th Regt. who escaped capture at Roanoke Island, North Carolina February 8, 1862, together with some recruits. It appears to have been raised for the purpose of forming part of that regiment when the captured members had been exchanged, but the companies were detached from that regiment to be organized into a separate battalion . . . dated October 18, 1862. This order appears to have been in confirmation of what had already been done in the field, and to prevent General Henry A. Wise from transferring the companies to his brigade from General Echols' Army, where the battalion was reported to have been serving."[19]

On May 23, 1862, only three days after official organization, the new battalion was engaged at the battle of Lewisburg, (West) Virginia, where a large segment of the men fled the field in panic, resulting in heavy casualties being inflicted upon the other Confederate units. Particularly hard hit was the 22nd Virginia Infantry, which thereafter refused to march with the battalion. Later in the year Edgar's Battalion added two new companies, one from Kanawha County, (West) Virginia and another from the Monroe and Greenbrier counties region, and officially became the 26th Battalion Virginia Infantry. The men began to redeem themselves in the 1862 Kanawha Valley campaign and with an intelligent victory over superior numbers at Tuckwiller Hill near Lewisburg in early 1863. Later in the year, at the battle of White Sulphur Springs, the 26th Battalion won back the respect of their comrades with a brilliant defense against the Union forces of Gen. William W. Averell, and lost five killed, 26 wounded, and one missing. The 22nd Virginia Infantry, which had suffered so heavily at Lewisburg due to Edgar's panic-stricken men, once again agreed to march alongside Edgar's Battalion. Company I of the battalion, although formed in 1863, remained on guard duty at the Confederate supply depot of Dublin, Virginia and did not join the battalion in the field until early 1864, thereby missing the Droop Mountain campaign. According to Micajah Woods of Jackson's Battery, the 26th Battalion Virginia Infantry numbered about 400 men at the time of the Droop Mountain battle, although the entire battalion would not be engaged in the affair.[20]

Lieutenant Colonel George Mathews Edgar, born March 1, 1837 at Union, Monroe County, (West) Virginia, received his early education at the town academy and entered the Virginia Military Institute at Lexington, Virginia January 14, 1853, graduating sixth in a class of thirty-three on July 4, 1856. He spent a year teaching at Union, followed by a one-and-a-half year stint as Assistant Professor at the Virginia Military Institute, where he served under the noted professor of Chemistry and Geology, Major William Gilham. It was during this time that Edgar and Gilham gave the first laboratory instruction in Chemistry ever presented in a southern college. In 1859 Edgar became Professor of Natural Philosophy and Astronomy at the North Carolina Military Institute at Charlotte. During the fall of 1860 Edgar resigned and took the Chair of Natural Sciences for the Florida State Seminary at Tallahassee, and was so engaged in February of 1861 when Florida seceded from the Union. Edgar promptly enlisted on April 2 as a private in the Leon Rifles and was immediately appointed sergeant, although he desired a higher rank.

Lt. Col. George Mathews Edgar of the 26th Battalion Virginia Infantry. General Echols detached his battalion to protect a road on the extreme right Confederate flank. As a result, Edgar's men were never engaged in the battle and were nearly cut off in the southern retreat. AUTHOR'S COLLECTION

The Leon Rifles were sent to Fort Barrancas, below Pensacola, and served under Gen. Braxton Bragg. Edgar's company was incorporated into the 1st Florida Regiment and he was appointed Sgt. Major. But with strong loyalties to Virginia, George obtained a discharge and returned to Monroe County to assist in raising a volunteer company, of which he was soon after made captain. The company served with the 59th Virginia Infantry, under Gen. Henry A. Wise, in western Virginia in 1861. Although most of the command was later captured at Roanoke Island, North Carolina in early 1862, Edgar avoided the debacle as he had contracted typhoid fever while in (West) Virginia and had remained behind. Following recovery from his illness he organized Edgar's Battalion on March 29, 1862, which evolved into the 26th Battalion Virginia Infantry, and he was elected Major. During the battalion's baptism of fire at Lewisburg, May 23, 1862, Edgar was seriously wounded (shot through the chest) and captured. As a result, his battalion was led by a succession of temporary commanders until his return during the 1862 Kanawha Valley campaign, and in November he was promoted to lieutenant colonel. Edgar went on to gallantly lead his men in the fights at Tuckwiller Hill and White Sulphur Springs in 1863.[21] Gen. John C. Breckinridge would later refer to Edgar as "a fine little officer."[22]

Rounding out Echols' Brigade was Capt. George Beirne Chapman's Company Virginia Light Artillery, from Monroe County, (West) Virginia, which was organized April 25, 1862, at Lewisburg, (West) Virginia. At that time the company numbered about 150 men, the majority from Monroe County, about 33 from Allegheny County, Virginia, about 15 from Roanoke County, Virginia, and a small number from Greenbrier County, (West) Virginia. Following muster into the Confederacy, the battery moved to Jackson's River Depot, Virginia and received armament of one 24-pounder howitzer, two brass 12-pounder howitzers, and two 6-pounder rifled pieces. Chapman's Battery entered it's first fray at the battle of Giles Court House (Pearisburg), Virginia May 10, 1862, and afterward fought at Lewisburg and the 1862 Kanawha Valley campaign. In August of 1863 it was the only Confederate battery on the field at White Sulphur Springs, losing one killed and 5 wounded. During the battle the Federal artillery damaged one of Chapman's pieces when their shot "twice struck the axle-body of the gun carriage." Although later repaired, the gun would also prove troublesome at Droop Mountain, forcing Chapman to abandon the piece along the Confederate line of retreat. During the Droop Mountain campaign Chapman would have four guns: two 3-inch rifled pieces, one brass 12-pounder howitzer, and one 24-pounder howitzer. One of the 3-inch rifles would be detached along with the 26th Battalion Virginia Infantry to guard another approach, leaving Chapman with three guns in the fight, including the one injured at White Sulphur Springs. Chapman's numerical strength in the battle is unknown but in the next major engagement in which the battery participated, at New Market, Virginia seven months later, Chapman could count about 123 men.[23]

Captain George Beirne Chapman, the son of Gen. Augustus A. Chapman, was born June 23, 1841, at Union, Monroe County, (West) Virginia. It was said that "as a boy Beirne Chapman was of blithe and joyous disposition, but as he grew older he became more reserved, apparently communing deeply with his thoughts. From a child he was always popular, for few could resist the movings of his kindly, generous nature." The same writer noted he was "endowed with a bright mind . . . an apt and diligent student . . ." at Reverend S. R. Houston's Academy at Union. He attended Washington College (later Washington and Lee) at Lexington, Virginia, where he studied law and was talented at debate and oratory. Chapman was described as "a handsome young man, not tall but well proportioned, with broad shoulders and easy, graceful carriage. His countenance showed strength and resolution, candor and kindness," and he was "... of impulsive temper," which was usually kept under control. He was additionally described as "high-spirited, ambitious and sensitive to dispraise and withal endowed with an abiding sense of justice and fair play." In 1861, while yet a student at Washington College, he enlisted as a lieutenant in Capt. William M. Lowry's Company Virginia Light Artillery and served in the 1861 Kanawha Valley campaign of Gen. Henry A. Wise. Chapman quickly became known as "an efficient and excellent disciplinarian." Following nearly a year's service with Lowry he resigned, and on April 25, 1862, organized Capt. George B. Chapman's Company Virginia Light Artillery. He became known for his great compassion for his men and attempted to instill in them many of his own excellent qualities. Additionally, he showed deep concern for the horses of his battery.[24]

As noted, Echols' Brigade numbered about 1,558 only seven days prior to the battle, and he claimed he had no more than 1,700 men directly under his command actually in the fight.

Jackson's Brigade: The Huntersville Line

Colonel William Lowther ("Mudwall") Jackson, also known as ("Bill") Jackson, was born at Clarksburg, (West) Virginia February 3, 1825. He was orphaned at the age of 10, became self-educated, studied law, and was admitted to the Bar in 1847. Shortly afterward he was elected to the office of commonwealth attorney for Harrison County, (West) Virginia. Jackson became a distinguished jurist and public official, was twice elected to the Virginia House of Delegates (1851 and 1853), served twice as state auditor, and superintendent of the Virginia library fund, sat as lieutenant-governor of Virginia for one term, resigning in order to serve as judge of the 19th judicial circuit of Virginia in 1860. While so engaged as the latter at Parkersburg in Wood County, (West) Virginia during April and May of 1861, he became involved in a number of confrontations at the courthouse between loyal Unionists and Confederate sympathizers, and in one fisticuffs encounter was described as an "excellent pugilist." Jackson attempted to rally Wood County men to the southern cause, but as he had no such authority to recruit for the militia or otherwise, he met with little success. During a controversial case in which he refused to jail Confederate sympathizers, he was literally run out of Parkersburg by loyal Unionists and joined the Confederate army as a private.

In May of 1861, Jackson, who stood six feet tall, weighed 200 pounds, and had dark red hair and blue eyes, was recommended to Gen. Robert E. Lee to serve as Confederate military commander at Parkersburg, and was described as "a gentleman of great personal popularity, not only with his own party, but with those opposed to him politically, and devoted to the interests of Virginia, to the last extremity." On May 10, 1861, he was appointed a lieutenant colonel of Virginia volunteers, and on June 15, 1861, formed the 31st Virginia Infantry at Huttonsville, (West) Virginia, the first regiment raised in western Virginia. He subsequently was given command of the 31st Virginia, which moved to guard the Laurel Hill pass near Belington, (West) Virginia. Jackson declined re-election and on December 14, 1861, at the close of the (West) Virginia operations, went to the Shenandoah Valley of Virginia to serve as a volunteer aide-de-camp on the staff of his second cousin, Gen. Thomas Jonathan ("Stonewall") Jackson, from June 4 to September 9, 1862. He was with "Stonewall" throughout the battles of Port Republic, the campaign before Richmond, the 2nd Manassas campaign, and the Maryland campaign, including the battles at Harpers Ferry and Antietam.

On February 17, 1863, "Mudwall" Jackson [reportedly so nicknamed by the northern troops in August of 1863 because they felt he would not stand and fight, as well as to distinguish him from his more famous cousin] received authority to raise a regiment for the provisional army from within the enemy lines in (West) Virginia. By April 11, 1863, he had organized the 19th Virginia Cavalry and was elected colonel of the regiment June 5, 1863, to rank from April 11. His command was placed with the cavalry brigade of Gen. Albert Gallatin Jenkins and participated in the Jones-Imboden raid in April of 1863. Later his men clashed with Federal soldiers at Beverly, Huttonsville, and, on October 13, 1863, lost a sharp contest at Bulltown in Braxton County, West Virginia. By this time Jackson's command had increased to the size of a small cavalry brigade and he was placed

Col. William Lowther Jackson of the 19th Virginia Cavalry commanded a cavalry brigade at Mill Point and Droop Mountain. *WEST VIRGINIA REVIEW*

in charge of a vast area known as the Huntersville Line, with his headquarters based at the small hamlet of Mill Point in Pocahontas County, West Virginia.[25]

Indeed, Jackson kept a watchful eye for any Federal movement, as indicated in a letter he wrote from Mill Point to Capt. Jacob W. Marshall on October 22, 1863. The dispatch read, in part, "Have you any news from Beverly? . . . I have sent Lt. [George W.] Siple to Dunmore . . . I expect the movement of Genl. Lee has compelled the abandonment of the [first] Averell raid."[26] Within Jackson's cavalry brigade could be found his own 19th Virginia Cavalry, led by Lt. Col. William P. Thompson; the 20th Virginia Cavalry of Col. William Wiley Arnett; Capt. Warren S. Lurty's Company Virginia Horse Artillery, and a number of independent companies under Major Joseph R. Kessler, of the 19th Virginia Cavalry, which would become the 46th Battalion Virginia Cavalry in early 1864.

The 19th Virginia Cavalry "was organized April 11, 1863, with ten companies (A to K), which were composed principally of former members of the 3rd Regiment Virginia State Line, which had been disbanded about March 31, 1863." The regiment also contained a number of men from the 31st Virginia Infantry, as well as former guerrilla companies recruited in the Little Kanawha region by Col. William L. Jackson. Some sixteen West Virginia counties contributed members to the 19th Virginia Cavalry, including Calhoun, Ritchie, Wetzel, Braxton, Wirt, Jackson, Roane, Marion, Harrison, Gilmer, Pocahontas, Randolph, Kanawha, Greenbrier, Clay, and Nicholas. As Col. Jackson was commanding the brigade, leadership of the 19th Virginia Cavalry fell to Lt. Col. William P. Thompson.[27]

Lieutenant Colonel William P. Thompson, son of the highly distinguished Judge George W. Thompson, was born January 7, 1837, in Wheeling, (West) Virginia and received his early education at the Linsly Institute, also at Wheeling. He subsequently attended Jefferson College in Pennsylvania, studied law, and was admitted to the Bar in 1857. Shortly thereafter, Thompson began law practice in Marion County, (West) Virginia, in which capacity he continued until the outbreak of the war. "He was opposed to the war and made some very earnest speeches in many parts of the district in favor of some reasonable adjustment of the differences that then existed between the North and the South." When war became inevitable he enlisted his company, the "Marion Grays" or "Marion

Lt. Col. William P. Thompson of the 19th Virginia Cavalry commanded the Confederate left flank at Droop Mountain. AUTHOR'S COLLECTION

Guard," at Fairmont, (West) Virginia on May 17, 1861, in the then "Army of Virginia," which eventually was incorporated into the Confederate army.[28]

Thompson's "Marion Guard" was designated Company A, 31st Virginia Infantry, and he was appointed captain of the company July 1, 1861. At the battle of [Top of] Allegheny in (West) Virginia on December 13, 1861, Thompson ". . . conducted himself most gallantly at the head of his company losing from his side a younger brother, the gallant Lieutenant Lewis S. Thompson, and having his own clothes pierced with bullets."[29] He resigned his commission on February 8, 1862, because a private in the regiment, rather than Thompson, was promoted to Major. He subsequently served as a Major with the 25th Virginia Infantry until May 1, 1862. On March 31, 1863 Col. William L. Jackson recommended Thompson, who had been serving as Jackson's adjutant, to raise a regiment in northwestern Virginia; on April 9, 1863 the Confederate Secretary of War granted Thompson permission "to raise a battalion for the Provisional Army within the enemy's lines in Northwestern Virginia. If practicable . . . [he could] . . . on the nucleus of the battalion, enlarge it to a Regiment." When completed, the organization was to be mustered into the service of the Confederacy "and to form a part of the command which Colonel W. L. Jackson . . . [had] . . . authority to raise."[30]

Thompson never formed the battalion, writing from Camp Harold, Headquarters, Huntersville Line, on April 16, 1863, that, ". . . having been elected [on June 15, 1863 to rank from April 11, 1863] Lieutenant Colonel of the 1st Regiment [19th Virginia Cavalry] organized for Colonel Jackson's command . . . declines to raise battalion or regiment to other worthy men."[31] Thompson, who stood 6 foot tall, was regarded as "a man of nerve and courage," and he was "genial, frank and courteous." He "inherited from his parentage a broad, comprehensive, penetrating mind. These endowments, developed to their fullness by his rigid legal and military training, together with his remarkable personal magnetism . . . made him conspicuous as a leader of men and thought."[32] Col. Jackson described him as an "able and gallant officer" who "raised the first company in North Western Virginia and was the first to occupy Grafton & Fetterman."[33] He was also known as a "gentleman of high character" and of "gallantry and intelligence."[34]

Colonel William Wiley Arnett's 20th Virginia Cavalry "was organized August 14, 1863 with ten companies, (A to K), which had previously been raised by Col. William L. Jackson as new companies composed of 'North Western Virginians,' or by Major John Buford Lady, afterwards of this regiment, to form a part of 'Col. William L. Jackson's' second, or new Regiment, under authority of the Secretary of War." The first current rolls of these companies were received August 1, 1863, "and were forwarded by Col. Jackson, evidently for the purpose of obtaining official recognition of the regimental organization, and for the purpose of raising a brigade of which he expected to obtain the command." The regiment consisted of many former members of the 31st Virginia Infantry and the 19th Virginia Cavalry, as well as some men who had been enlisted by Major David Boston Stewart for a battalion he attempted to raise in late 1862. Most of the men came from fifteen different West Virginia counties, including Monongalia, Marion, Wetzel, Doddridge, Randolph, Preston, Barbour, Upshur, Harrison, Lewis, Pleasants, Ritchie, Calhoun, Wirt, and Gilmer.[35]

Colonel William Wiley Arnett of the 20th Virginia Cavalry and his men defended the Confederate center and right at Droop Mountain. In this crude postwar sketch, purportedly taken from a photograph, he is apparently whittling on a piece of wood and chomping on a cigar. *WHEELING DAILY INTELLIGENCER*

Colonel William Wiley Arnett, born October 23, 1843, in Marion County, (West) Virginia, prepared at Fairmont Academy for Allegheny College in Meadville, Pennsylvania, from which he graduated in 1860. Arnett studied law both before and after his college term and was admitted to practice law at Fairmont in 1860, but closed his office in order to enlist as a private in Company A ("Marion Guard"), 31st Virginia Infantry. Directly after this, Virginia's governor appointed him a lieutenant colonel of a battalion, which was afterwards merged into the 25th Virginia Infantry. He resigned his commission, returned to the ranks of his old regiment and company, and was selected captain of Company A, 31st Virginia Infantry May 2, 1862. Arnett, who like Col. Jackson and Lt. Col. Thompson, stood six-foot tall, remained in such command until he was elected colonel of the newly formed 20th Virginia Cavalry October 7, 1863, to rank from August 14, 1863. Twice during the war he would be elected by the "refugees and camp voters" to represent Marion County in the Virginia Legislature. His obituary would state, "His courteous, kindly ways and even temper have given him a peculiar hold upon our affections . . ."[36]

Major Joseph R. Kessler of the 19th Virginia Cavalry, who had seen prior service in the 3rd Virginia State Line [and was wounded in December of 1862 at Prestonsburg, Kentucky] and as captain of Company C, 19th Virginia Cavalry, was appointed major of the 19th Cavalry June 5, 1863, to rank from April 11, 1863. In mid and late 1863 Kessler led a few independent cavalry companies [four organized prior to the Droop Mountain battle, and two after] which, in early 1864, would comprise the 46th Battalion Virginia Cavalry. This detachment participated, as did the 19th Virginia Cavalry, the nucleus of the 20th Cavalry, and Lurty's Battery, in the October 13, 1863, battle of Bulltown in Braxton County, West Virginia. On February 26, 1864, the six independent companies, all organized in 1863, would become the 46th Battalion Virginia Cavalry, with Major Kessler serving as lieutenant colonel. It is known that Major Kessler led a detachment of men in the Droop Mountain battle, although they suffered no recorded casualties.[37]

Completing Col. William L. Jackson's brigade was Capt. Warren S. Lurty's Company Virginia Horse Artillery, comprised of two 12-pounder howitzers "drawn by horses." This light artillery battery, "which maneuvered with the cavalry, with the cannoneers mounted on horseback," completed it's organization "on October 8, 1863 and was composed partly of transfers from other regiments." Reportedly, the battery had served as an independent company in Lunceford L. Lomax's Horse Artillery Battalion and afterward in Major James Walton Thomson's Horse Artillery, "both field organizations composed of independent batteries." Lurty's Battery was comprised of men from the West Virginia counties of Harrison, Webster, Ritchie, Upshur, Lewis, Calhoun, Marion, Greenbrier, and Gilmer, as well as the Virginia counties of Bath and Roanoke. They saw early action in the October 13, 1863 battle at Bulltown, West Virginia.[38]

Captain Warren Seymour Lurty, born May 18, 1839, in Clarksburg, Harrison County, (West) Virginia, was a cousin of Gen. Thomas Jonathan ("Stonewall") Jackson. In 1859 he was an attorney and "judging by trust deeds," his law practice apparently only lasted several months, from the summer of 1860 until early 1861. In 1861 he enlisted in Harrison County and saw his first service as a dispatch rider for "Stonewall" Jackson at Harpers Ferry. Lurty enlisted in the Staunton Artillery on August 16, 1861; and transferred to Company D, 6th Virginia Cavalry January 27, 1863. He was commissioned a 1st lieutenant and adjutant of the 37th Battalion Virginia Cavalry and the 19th Virginia Cavalry, and then promoted to captain of Lurty's Battery October 8, 1863. Lurty was recognized as "a man of commanding appearance and a speaker of unusual fluency."[39]

Jackson's Brigade [Huntersville Line] numbered an aggregate present of 938, with an aggregate present and absent of 1,611, with a total of 777 effectives, in November of 1863. At Droop Mountain his brigade would number 750.

Ferguson's (Jenkins') Brigade

Resting in the vicinity of Lewisburg could be found a cavalry detachment of Gen. Albert Gallatin Jenkins' Brigade, temporarily under the command of Col. Milton J. Ferguson of the 16th Virginia Cavalry. Gen. Jenkins had been wounded at Gettysburg; placing Ferguson in charge, who reported to his superior, Gen. Echols at Lewisburg. Ferguson had under him the 14th Virginia Cavalry; his own 16th Virginia Cavalry, and Capt. Thomas E. Jackson's Battery.

Colonel Milton Jameson Ferguson, referred to as "Jameson," was born about 1833 in Wayne County, (West) Virginia. In 1855 he was sworn in as attorney at law for Cabell County, (West) Virginia and soon afterward "established a lucrative practice in the settling of estates and managing trusts." In 1861 he served as Cabell County prosecutor and by mid-year was colonel of the 167th Virginia Militia from Wayne County. In early July of 1861 he led the militia in the battle of Mud River (Barboursville), Cabell County, (West) Virginia, and during the same

Colonel Milton Jameson Ferguson of the 16th Virginia Cavalry performed brilliantly with the Confederate rear guard during the retreat from Droop Mountain. This carte de viste of Ferguson was taken by Gihon of Philadelphia in 1864 when Ferguson was a prisoner of war at Fort Delaware. MAHLON P. NICHOLS, BEDFORD, VIRGINIA

month was captured by the 5th (West) Virginia Infantry, and remained a prisoner of war until January or February of 1862. During 1862, he served as commander of Capt. Milton J. Ferguson's Battalion Virginia Cavalry, or Guyandotte Battalion; and on February 10, 1863, was elected colonel of the 16th Virginia Cavalry, to take effect as of January 15, 1863. Attached to Gen. Albert Gallatin Jenkins' Cavalry Brigade, Ferguson assumed command of the brigade after Jenkins was wounded during the second day of fighting at Gettysburg.[40] One source described him as "... Jenkins' senior regimental commander ... the battle-scarred Colonel Ferguson,"[41] while another said he "was evidently a stunning figure when leading his regiment into battle. He had a long flowing beard that parted in front and flew over his shoulders as he rode at the head of the regiment."[42] Micajah Woods of Jackson's Battery said Ferguson's "soldier like qualities cannot be too highly commended."[43] At the battle of Droop Mountain Ferguson would be instrumental in serving as rear guard to the retreating Confederate army.

The 14th Virginia Cavalry, led by Col. James A. Cochran, was organized September 5, 1862, "with nine companies, some of which had previously served in a field organization known as Jackson's Squadron Virginia Cavalry; the tenth company was formed of surplus men of the other companies. Two of these companies failed to join the regiment and others were assigned in their places." The majority of the men were from the West Virginia counties of Greenbrier, Pocahontas, and Calhoun, as well as the Virginia counties of Charlotte, Augusta, and Rockbridge. Most of the companies had participated in the 1861 campaign in western Virginia, and later rode with Gen. Albert Gallatin Jenkins on his raid through Ohio in the Spring of 1862. Late in 1862 the 14th Virginia Cavalry served in the Kanawha Valley campaign and spent the winter of 1862-63 "on scattered outpost duty" in Greenbrier County. During 1863 the regiment fought dismounted at the Rummel barn at Gettysburg, protected Gen. Robert E. Lee's long wagon train during the retreat from Pennsylvania, served outpost duty along the Rappahanock River, fought at Brandy Station; then returned to the mountains of West Virginia. At Droop Mountain, portions of the regiment would fight dismounted on the front line and the command would also help cover the Confederate retreat. Regimental numbers at Droop are unknown, but according to a late 1863 report, the 14th Virginia cavalry, attached to Echols, had 16 officers and 172 enlisted men. Adding an approximate 25 men lost at Droop, the regiment may have had about 213 men in the battle, although many were detached, and Micajah Woods of Jackson's Battery said the 14th Virginia Cavalry had 600 at Droop Mountain.[44]

Colonel James Addison ("Jim") Cochran, known as "a gallant soldier," was born in 1830 in Augusta County, Virginia and became a farmer near Churchville, Virginia. On April 19, 1861, he enlisted as a 2nd lieutenant in Company I (Churchville Cavalry), 14th Virginia Cavalry and served as such until elected captain of the company May 15, 1862. On February 12, 1863, Cochran was promoted to colonel of the 14th Virginia Cavalry, and he is credited with commanding the 14th Virginia Cavalry at Droop Mountain although some believe the regiment may have been led in the fight by Major Benjamin Franklin Eakle.[45]

Major Benjamin Franklin Eakle of the 14th Virginia Cavalry was born August 7, 1826, in Augusta County, Virginia and in 1847 moved to White Sulphur Springs,

Major Benjamin Franklin Eakle of the 14th Virginia Cavalry commanded at least a portion of the regiment at Droop Mountain. BILL TURNER, CLINTON, MARYLAND

Flag of the 14th Virginia Cavalry, C.S.A. was captured April 9, 1865 by Sgt. John Donaldson, Co. L, 4th Pennsylvania Cavalry at Appomattox Court House. This flag may have been used at Droop Mountain. BOB DRIVER, BROWNSBURG, VIRGINIA

(West) Virginia. He worked as a merchant at Lewisburg, (West) Virginia and enlisted as a 2nd lieutenant in Company A (1st), 14th Virginia Cavalry May 23, 1861. Eakle remained in such position until November 22, 1861, when the company disbanded. He was elected captain of Company A (2nd), (Greenbrier Cavalry), 14th Virginia Cavalry, until elected Major of the regiment February 13, 1863. He was in command of the 14th Virginia Cavalry at Gettysburg, where he was wounded and had his horse shot out from under him in the third day's fight. His service record states he was in command of the regiment during November 1863, therefore, he may well have led the 14th Virginia Cavalry at Droop Mountain, although various battle reports mention Col. Cochran's presence.[46]

Colonel Milton J. Ferguson's 16th Virginia Cavalry "was formed January 15, 1863, by the consolidation of six companies of Captain Milton J. Ferguson's Battalion Virginia Cavalry, with four companies of Major Otis Caldwell's Battalion Virginia Cavalry." Both battalions had been formed in 1862 and Ferguson's consisted mostly of men from Wayne, Cabell, Putnam, and Kanawha counties of (West) Virginia, whereas Caldwell's men hailed primarily from Tazewell and Russell counties of Virginia. The 16th Cavalry was placed with [Gen. Albert G.] Jenkins' Cavalry Brigade, which "was the vanguard of forces that preceded the Confederate infantry on the march to Gettysburg." The regiment also fought Gen. George A. Custer's Michigan cavalry on the third day at Gettysburg. During the Confederate movement to Droop Mountain the 16th Virginia Cavalry was left behind at Lewisburg, under the command of Major James H. Nounnan, in order to cover the western approaches to town. The regiment later served as rear guard to the retreating Confederate army after rejoining Echols during his flight through Lewisburg.[47]

Major James H. Nounnan, born in Virginia about 1834; enlisted May 8, 1861, as a private in the Kanawha Riflemen (Company H, 22nd Virginia Infantry) and remained with that company until August of 1861. He transferred to Capt. James Corns' Cavalry Company, which eventually evolved into a company of the 8th Virginia Cavalry. In October of 1862 he departed in order to form his own cavalry company in Ferguson's Battalion. On February 10, 1863, Nounnan was promoted to Major of the 16th Virginia Cavalry, to take effect from the date of the regiment's organization.[48]

The third organization in Ferguson's command was Capt. Thomas E. Jackson's Battery Virginia Horse Artillery [2nd organization], which came together May 2, 1863. "Composed primarily of Jackson's 'Old' Battery [Capt. John Peter Hale's Kanawha Artillery], which was captured at Fort Donelson; recruits from the Virginia State Line; and transfers from other organizations" (particularly the 8th Virginia Cavalry). The men were principally from the (West) Virginia county of Kanawha, as well as some other surrounding counties. The original organization of the battery had fought in (West) Virginia in 1861 at Scary Creek, Cross Lanes, Carnifex Ferry, and Sewell Mountain, and was captured in 1862 at Fort Donelson, Tennessee. The second organization was placed with the cavalry brigade of Gen. Albert Gallatin Jenkins and participated in the great battle of Gettysburg. At the Droop Mountain fight, the battery would consist of two guns: one 10-pounder Parrott, commanded by 1st Lt. Randolph Harrison Blain; and one 3-inch rifle, in the charge of Lt. Micajah Woods. According to Woods, these were the only "long range" guns Echols had at Droop, and were, therefore, the only two to have much effect upon the enemy.[49]

Based upon an August 8, 1863, supply report, Lt. Blain procured four guns at Staunton for the battery, including the two used at Droop. The whereabouts of the other two 3-inch rifles is unclear. The report also listed some 151 rounds of ammunition, varying from "18 ten pound Parrott shells" to "18 three inch rifle canister" to "40 ten pound Parrott case-shot" and "75 ten pound Parrott percussion shell & charges." As the battery is not believed to have been engaged between this time and the Droop fight, it is assumed most of this ammunition was expended at Droop. Other supplies that could be found on the list included 255 fuses, 50 igniters, 200 friction primers, 36 Whitworth charges, and 30 other charges.[50] On October 18, 1863, Lt. Woods reported: "Our Battery is in good order, having only yesterday [October 15] received a full complement of the best ammunition." The numerical strength of the battery at Droop is unknown but at the battle of New Market some seven months later there were 93 men in Jackson's Artillery.[51]

Captain Thomas Edwin Jackson was born July 30, 1834, in Fredericksburg, Virginia, the son of William A. Jackson, who served as lieutenant colonel of the 22nd Virginia Infantry from December 21, 1861, until the reorganization in May of 1862. Thomas is believed to have been a resident of the Kanawha Valley when the war broke out and had been touring the world for some years previous. Lt. Woods of the battery would write of Jackson, "He has made the circuit of the world, spent some time in California, Chile, Patagonia, China, Australia, and especially in France and England. His voyage consumed six years and he returned shortly before the war." At the outbreak of the conflict he apparently formed Capt. Thomas E. Jackson's Horse Artillery, which contributed greatly to the Confederate victories at Scary Creek, Cross Lanes, and the draw at Carnifex Ferry. The battery apparently merged with the Kanawha Artillery and was captured, as was Capt. Jackson, at Fort Donelson. Upon his release he was commissioned "Major" commanding a Battalion of Light Artillery, Virginia State Line, until disbanded about March of 1863.[52] Lt. Woods, who also served in the State Line, wrote: "Jackson is a great favorite with Genl. [Albert Gallatin] Jenkins and [Gen. John B.] Floyd and indeed with everyone who knows his bravery and gallantry are proverbial, and . . . [is] . . . about 24 or 25 years of age."[53] On May 2, 1863, the date of the battery's organization, Jackson was commissioned captain.

Jackson's Battery was led by men of great competence, as evidenced by lieutenants Randolph Harrison Blain and Micajah Woods. Blain, who was a distant cousin of Gen. John C. Breckinridge, was born August 16, 1842, at Williamsburg, Virginia. He was privately tutored and was a member of the Washington College [Washington and Lee] class of 1861. Blain also attended a course of military instruction at V.M.I. He enlisted June 24, 1861, at Yorktown as a private in the 3rd Company, Richmond Howitzers and discharged September 9, 1862. He then served as Sr. 1st Lt. in Jackson's Battery, having been commissioned May 2, 1863.[54]

Micajah Woods was born May 17, 1844, near Ivy Depot in Albermarle County, Virginia. He attended Lewisburg Academy, the military school at Charlottesville, and the Bloomfield Academy. Although only 17 years of age, he joined the Confederate Army in August of 1861 and served with the 51st Virginia Infantry as an

Lieutenant Micajah Woods of Capt. Thomas E. Jackson's Battery Virginia Horse Artillery was in command of a three-inch rifled gun at Droop and according to him, personally fired and sighted every shot, "throwing 120 shell." MICAJAH WOODS PAPERS, SPECIAL COLLECTIONS DEPT., MANUSCRIPTS DIVISION, UNIVERSITY OF VIRGINIA LIBRARY

aide-de-camp to Gen. John B. Floyd. As he was under military age he spent the winter of 1861-62 at the University of Virginia at Charlottesville. In May of 1862, Woods enlisted in the 2nd Virginia Cavalry and fought with "Stonewall" Jackson at Port Republic; and with Gen. J.E.B. Stuart at Second Manassas, Crampton's Gap, and Antietam. In October of 1862 he was appointed 1st lieutenant of cavalry in Gen. John B. Floyd's newly-organized Virginia State Line, and spent the following winter campaigning in (West) Virginia and eastern Kentucky. In April of 1863 he became a 1st lieutenant in Capt. Thomas E. Jackson's Battery Virginia Horse Artillery, and remained in that capacity until the close of the war.[55]

On November 6, 1863, the date of the Droop Mountain battle, Gen. Sam Jones reported that, "Col. Ferguson has two regiments of Jenkins' brigade in front of Lewisburg, with 72 officers and 810 enlisted men present for duty."[56]

Artillery Battalion

In command of all Confederate artillery on the field at Droop Mountain (the batteries of Lurty, Chapman, and Jackson) was Major William L. McLaughlin, born January 6, 1828, near Bell's Valley in Rockbridge County, Virginia. McLaughlin graduated from Washington College (now Washington and Lee) in 1850 and LL.D. [Doctor of Laws] in 1851. He became a lawyer and surveyor at Lexington, Virginia and a member of the Virginia Convention. On April 29, 1861, he enlisted as a 2nd lieutenant in the 1st Rockbridge Artillery at Lexington. He was promoted to captain of the battery August 14, 1861, and served in such capacity until promoted to Major and Judge Advocate April 10, 1862. Reportedly, he briefly led the 26th Battalion Virginia Infantry following their defeat at Lewisburg, (West) Virginia May 23, 1862, and was made Major and Quartermaster on the staff of Gen. John Echols July 28, 1862. He subsequently was promoted to Major of Artillery and Chief of Artillery, commanding an artillery battalion with Gen. Echols October 9, 1862. He held this command at Droop Mountain.

Major William L. McLaughlin commanded all Confederate artillery on the field at Droop Mountain. ROBERT DRIVER, BROWNSBURG, VIRGINIA

Brigadier General Robert S. Garnett, under whom McLaughlin served early in the war, said McLaughlin was "active, zealous and intelligent in the execution of . . . [his] . . . duties. [His] Battery did gallant and efficient service at the battle of Kernstown . . . especially at the close of the action—when it doubtless assisted largely in inflicting on the enemy the heavy loss he sustained that day, and in checking his advance at this critical period." Gen. John Echols said of him, "I do not know of one single individual who contributed more to the glorious results of these two days (Bull Run and Kernstown) than did Capt. McLaughlin. His courage, his efficiency, his skill as an artillerist was conspicuous and entitle him to honor and gratitude from all." When McLaughlin declined to run for re-election as Capt. of the 1st Rockbridge Artillery in 1862, a member of the battery said: "Captain McLaughlin was a good officer of extraordinary information on all subjects, somewhat indolent physically, but cool and discreet in action. He had not the appearance & dress of a soldier according to the common stand out. He was tall but crooked and careless in his appearance & dress and was an awkward rider, and whilst he had many of the most intelligent men as his supporters, it was generally thought that he was not as rigid in enforcing discipline in camp as was desirable considering the fact that there were some pretty large proportion of 'hard cases' in the company." One Confederate general later described him as "a vigorous and gallant officer."[57]

CHAPTER 4

Diary of a Campaign

November 1-The Departure

As Brigadier General Averell prepared to march on the clear and warm morning of Sunday, November 1, 1863, the quaint little town of Beverly in Randolph County, West Virginia became a hub of military activity. By 8:00 A.M. the regimental and battery commanders had filed their field returns containing "the exact number of men and horses in their column" and, afterwards, called in their respective pickets.[1] Averell's "4th Separate Brigade" finally began to depart town at 10:00 A.M., the 28th Ohio Infantry in the vanguard. Following Col. Moor's Germans came the 10th West Virginia Infantry, with Company B serving as Provost Guard throughout the entire campaign. Company I of the 10th West Virginia, as earlier noted, was on detached duty at Petersburg in Grant County, West Virginia and, therefore, did not accompany the regiment on the expedition. Bringing up the rear of the infantry detail could be found batteries B (Keepers') and G (Ewing's) of the 1st West Virginia Light Artillery.[2]

The mounted regiments of the brigade left town at 12:00 noon, the 14th Pennsylvania Cavalry in the advance. Behind Col. Schoonmaker's troopers were, in order, the 8th West Virginia Mounted Infantry, the 3rd West Virginia Mounted Infantry, the aggregate 395 members of the 2nd West Virginia Mounted Infantry, with a "portion" of Gibson's Independent Cavalry Battalion bringing up the rear. Also with the column was the signal corps detachment of the 68th New York Infantry; although their services would not be required until reaching Huntersville.[3] Due to a toothache suffered throughout October, Lt. Col. John J. Polsley of the 8th West Virginia Mounted Infantry did not accompany the column, remaining at Beverly in charge of the post. The 32-year-old Polsley had been promoted to lieutenant colonel March 1, 1863 and was known as a "very conscientious officer...[who] raged at lack of food and equipment, poor organization, and ineffective campaigning."[4] He also had a long-standing grudge against Col. Oley and would later unsuccessfully charge him with cowardice in the Droop Mountain battle.

The mounted regiments had been instructed to carry no more than one day's rations of forage on their horses, not exceeding 12 pounds in weight. The wagons were to haul their own rations, not over three day's worth, and were to "move with the commands" to which they were attached. Afterwards, they were to move with the brigade train.[5] Averell's Chief Quartermaster, Capt. W. H. Brown, had fallen ill, so 1st Lt. Alexander J. Pentecost, Quartermaster of the 2nd West Virginia, replaced him on the expedition.

The scene was exemplified by Capt. Edgar B. Blundon, Company F, 8th West Virginia Mounted Infantry. Blundon, who had been wounded in the fight at White Sulphur Springs, hastily wrote a letter at noon, which read in part: "We are now ready to start upon the long anticipated expedition. I do not expect to have an

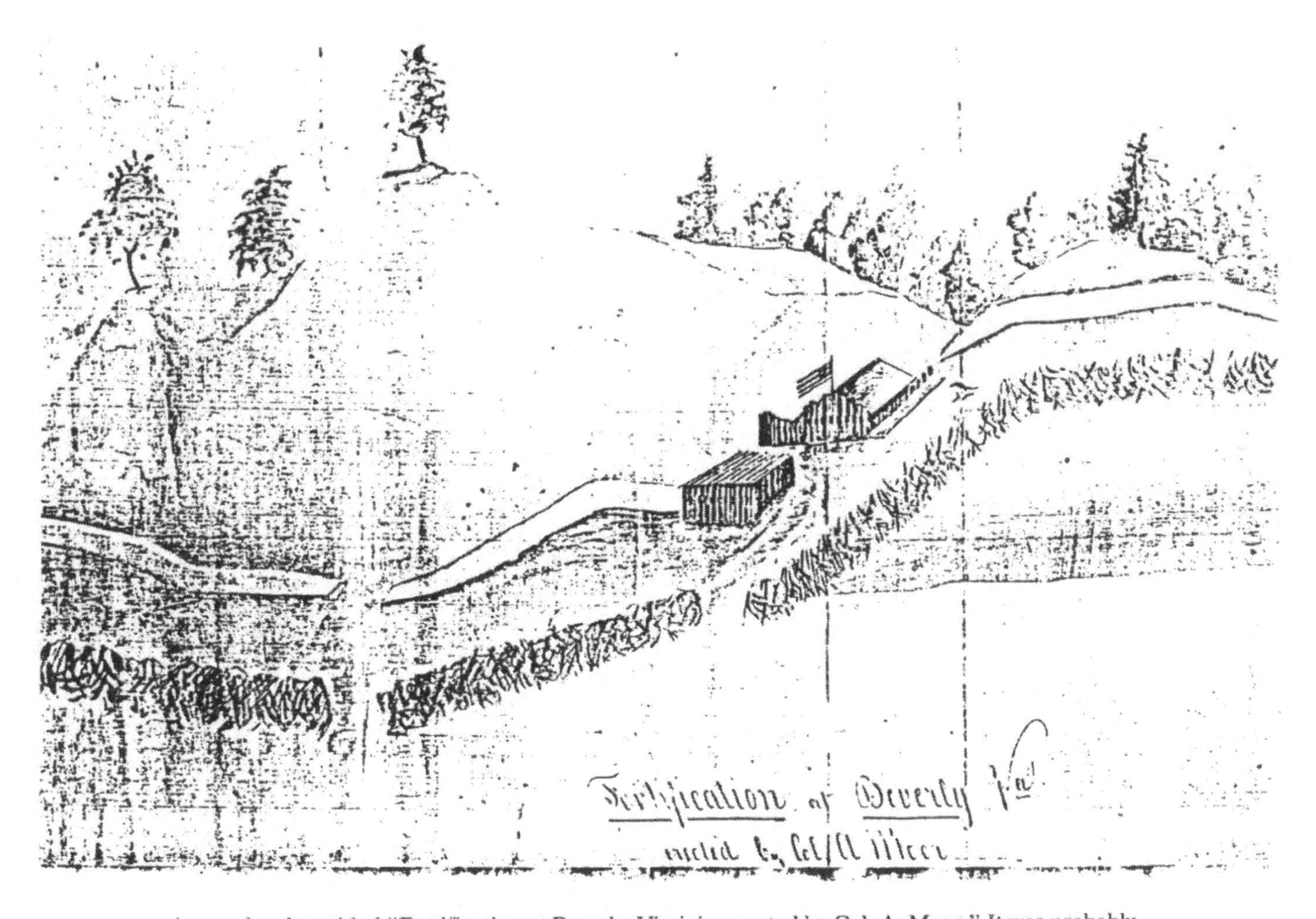

A rare sketch entitled "Fortification at Beverly, Virginia, erected by Col. A. Moor." It was probably made while Col. Augustus Moor of the 28th Ohio Volunteer Infantry commanded the post between July and November of 1863, and is signed "Heer" [left corner], probably Capt. Arnold Heer, 28th O.V.I. AUGUSTUS MOOR PAPERS, RATTERMAN COLLECTION, UNIVERSITY OF URBANA-CHAMPAIGN, URBANA, ILLINOIS

opportunity of writing again before we return. When that will be I can not tell or have no idea...the bugle sounds "to horse." Goodbye for the present, dear Sallie. May the Lord preserve us both and bring us together in peace."[6] A correspondent, probably with the 28th Ohio, added, "...an early order reached us in camp saying, provide three day's rations in your haversacks...so quiet had our destination been kept, that none, save our commanding officers, knew to what point we were being sent."[7]

Twelve miles were covered during the first day's march, moving due south by way of the Staunton and Parkersburg Turnpikes, going into camp at about 4:00 P.M. at and near Huttonsville. One participant described Huttonsville as "...made up of a frame barn, and a place for a house; the proprietors being absent. Colonel Moor made himself at home and provided for the boys as a western pioneer would have done, on first squatting on what was to become his own territory, and I am sure none enjoyed the joke better than the Colonel himself. We slept upon the ground and were kindly covered by nature's own covering..."[8]

November 2-A Scenic March

The morning of Monday, November 2, broke as a "fine Indian summer day." At 7:00 A.M. Averell's Brigade forded the Tygart River and marched southeasterly, while one squadron of the 8th West Virginia Mounted Infantry was detached and sent in advance of the infantry. The remainder of the 8th West Virginia continued with the mounted segment of the column, while the 14th Pennsylvania Cavalry rode in "the rear of the command."

After covering two miles, the column arrived at the foot of Cheat Mountain, where the men viewed the rebel fortifications which had been constructed there in 1861. They also saw evidence of the battle which had taken place there during that period. Ascending the western slope of the mountain on the Staunton and Parkersburg Turnpike, "the brigade presented an animated and picturesque appearance" as it covered the winding, six-mile road to the summit.[9] While engaged in climbing the mountain, the column encountered a refugee family from White Sulphur Springs, who claimed they were escaping from "rebeldom."

At noon Averell's men reached the summit of Cheat Mountain and "halted to rest and close up the column before...[the] descent."[10] Here the men showed the first signs of fatigue, but after a search of their haversacks, "all was merry."[11] While at the summit, the soldiers could view the beautiful Cheat valley and mountain, the Alleghenies to the east "covered with dark fir forest," and the valley of the Greenbrier River stretching to the southeast. Also visible were the 1861 Federal earthworks on Cheat Mountain, as well as Gen. Robert E. Lee's old works at the Top of Allegheny mountain, each position staring defiantly at the other.[12] One correspondent wrote, "The scenery on Cheat Mountain is most delightful; the large pines, decked with evergreen, lifting themselves; high above the mountaintops, seemed to wave us onward; the beautiful mountain rivulets gave forth such a gentle murmur as led us to believe they were not complaining when we found it necessary to cool our lips, or slake our thirst, at their crystal founts; while the lofty heights and deep valleys all conspired to make it a grand and interesting scene; while here and there a little cottage, rudely built upon the mountain sides, or upon a hill-top, and a few sheep or cattle grazing nearby..."[13]

Having covered the one mile descent to the Cheat River at the eastern foot of the mountain, the column came to the Gum farm; a noted bushwhacker hang-out where "a large party of guerrillas [had] recently blockaded the road behind a ten man scouting party of the 8th West Virginia," and attempted to capture them. The corporal of the 8th West Virginia and his nine men had cut through, with one man wounded and one horse killed.[14]

Continuing eastward, the command crossed over the Cheat River valley and another mountain, "which is the divide between" the Cheat and Greenbrier rivers. Between 6:00 and 8:00 P.M. the column went into camp near Traveler's Repose, a former "hotel hidden away in the valley" at (Camp) Bartow. The day's march had covered approximately 21 miles, and found the men in a "valley of utter desolation" in which "human habitation and fences [were] all gone," presenting "a mournful solitude."[15] Sgt. Thomas Rufus Barnes, Company K, 10th West Virginia Infantry, wrote in his diary that his regiment "crossed the Cheat and camped in the Greenbrier valley in Pocahontas County."[16] Corp. George Washington Ordner, Company B, 2nd West Virginia Mounted Infantry, scribbled in his diary: "10 P.M.–Camped at Greenbrier River for the night. 22 m.[iles covered]."[17] One writer claimed that opposite Averell's camp was a small grove of evergreens from which bushwhackers fired upon the men, apparently with no effect.[18]

November 3–First Contact

Brigadier General Alfred N. Duffié's column, consisting of the 34th Ohio Mounted Infantry, the 2nd West Virginia Cavalry, and one section of Capt. Seth J. Simmonds' Kentucky Battery, totaling an aggregate of 970 officers and enlisted men, and 1,025 horses; departed headquarters at Charleston in the Kanawha Valley at 6:00 A.M., Tuesday, November 3. With Lewisburg over 100 miles distant, Duffié would march eastward on the James River and Kanawha Turnpike and cover some 29 to 30 miles before going into camp for the night. At Fayetteville, in Fayette County, West Virginia, the 12th and 91st Ohio volunteer infantry regiments, which were to join Duffié en route, received marching orders. Pvt. James Ireland, Company A, 12th Ohio, wrote in his diary: "Cloudy and rainy all day We are making preparations for a raid to Lewisburg in junction with Genl. Averell. We are to have four days rations in haversack and ten in wagons."[19]

Averell's men were up early on the morning of November 3 as well, with a 100 man squadron of the 2nd West Virginia Mounted Infantry, under 1st Lt. Arthur J. Weaver, leaving Bartow at 7:00 A.M. Close on the heels of Weaver's squadron came the 28th Ohio, the 10th West Virginia, and Keepers' Battery, all under the command of Col. Augustus Moor, leaving Lt. Col. Gottfried Becker in charge of the 28th Ohio. At 9:00 A.M. the remainder of the 2nd West Virginia Mounted Infantry moved out of Bartow, followed closely by the 3rd West Virginia Mounted Infantry, Ewing's Battery, Gibson's Independent Cavalry Battalion, the 14th Pennsylvania Cavalry, and then the ambulances and brigade train.[20] A portion of Averell's "4th Separate Brigade" marched eastward to Allegheny Mountain and

Gen. Lee's 1861 summer fortifications, known as either "Camp Allegheny" or "Top of Allegheny." According to a postwar historian, Averell had "learned that a number of cabins, built and abandoned at Top [of] Allegheny, were still there. For some reason he was determined to destroy them..."[21] These troops were to rejoin the main body at Arborvale, near Green Bank. On Top of Allegheny the Federal detachment met two additional refugee families from White Sulphur Springs, and a scouting party was detached and sent to Fort Baldwin at the summit, where numerous fires were built, all hay in the vicinity procured and accommodations acquired for some half-dozen impromptu "generals."[22]

Later in the day, near Green Bank, the squadron of the 8th West Virginia, which had been sent on a circuitous route the previous day; was spotted by Pocahontas County native Lt. George W. Siple, Company F (Capt. William L. McNeel's Company), 19th Virginia Cavalry. Siple's command, estimated by Federal authorities as some 300 to 400 men, was stationed at Glade Hill just northeast of Dunmore in upper Pocahontas County. According to a postwar historian, this encounter took place at Boyer, south of Bartow. Two of Siple's men moved too close to the Federals, resulting in the wounding of Pvt. John Adam McNeil, who suffered a broken leg when his horse was shot out from under him.

According to McNeil, as he attempted to elude the Yankees, he arrived at the home of Mrs. Adeline E. Brown in Green Bank at 11:00 A.M., the enemy firing their weapons at him. Mrs. Brown, who said the people of Green Bank were completely unaware of Averell's earlier presence at Bartow, remembered, "we heard a great hallooing and shooting, and when we ran to the door a young boy [McNeil] in Confederate uniform dashed by...I never saw so cruel and thrilling a sight in all my life as when I saw those men trying to kill that one boy."[23] McNeil, who was attempting to relay the information of Averell's arrival to Gen. John Echols, was "desperately hurt" when his horse fell on him and he was unable to walk for three months, and "only then with crutches until after the war." Three other rebels were wounded or captured in the confrontation, including Pvt. Wilson Pugh, who was caught. The Federal account of the skirmish differed somewhat, as the 8th West Virginia claimed to have met the enemy pickets, wounding one, capturing five, and destroying the rebel camp. In the afternoon, Averell's main column marched southward and rejoined with the squadron of the 8th West Virginia at Green Bank.

The small clash at Green Bank, about 20 miles north of Huntersville, was the first knowledge the Confederates had of Averell's presence, which prompted Lt. Siple to rush off a dispatch to Col. William L. Jackson and Col. William W. Arnett. Jackson was then headquartered at Mill Point, just north of Hillsboro in Pocahontas County, with the main body of the 19th Virginia Cavalry and Capt. Warren S. Lurty's Company Virginia Horse Artillery. At 6:00 P.M., Jackson received the message from Siple that the "enemy had appeared in force at Green Bank," and took immediate action. First by sending the information to Gen. John Echols at Lewisburg and also to Col. Arnett, who was with his 20th Virginia Cavalry at Marlin's Bottom [Marlinton], eight miles north of Mill Point, preparing log cabins for winter quarters. The same information was relayed to Capt. Jacob Williamson Marshall, Company I, 19th Virginia Cavalry, who was at Edray, about four miles north of Marlinton, watching the Marlin's Bottom and Huttonsville

Turnpike. Arnett had apparently already received the information from Siple and had earlier relayed it to Capt. Marshall. Jackson took an additional precaution by recalling to Mill Point a detachment of 120 mounted men from the 19th Virginia Cavalry, under Lt. Col. William P. Thompson, which had departed earlier on an expedition to Nicholas County, traveling as far as the foot of Cold Knob Mountain in Greenbrier County. Jackson also directed Lt. Siple to keep check on Averell's "intentions and strength" and, if pressed, to fall back via the Beaver Creek road through Huntersville to Mill Point, blockading the road en route. However, as a result of this order, Siple encountered Averell's detachment coming down Allegheny Mountain to Arborvale, near Green Bank, and became cut off in Pocahontas County, unable to make contact with Jackson until after reaching Union in Monroe County following the Droop Mountain fight. Siple [McNeel] would manage to escape Averell by moving up Galford's Creek, crossing Allegheny Mountain to the waters of Back Creek, and then continuing south to Callahan's.[24] Siple's inability to communicate with Col. Jackson during the campaign would cause "Mudwall" to be "blind" to Averell's movements, intentions, and strength; and also to continually underestimate, and overestimate the danger confronting him.

For added insurance, Col. Jackson sent 30 troopers to Huntersville via the Beaver Creek road, and ordered Col. Arnett at Marlinton to also send his scouts to Huntersville. While the Confederates maneuvered about, Averell's Brigade continued to march southward. Correspondent "Irwin" accompanied the troops and recalled the brigade marching down the valley by way of Green Bank, "a fine country not yet devastated by war...fine mansions...with appendages for Negro huts...[and] passed through a magnificent forest of white pine." He said the brigade, or at least the portion he traveled with, camped for the night at Matthews Mills, near Dunmore, where an abundance of hay and corn for the horses was located.[25] The 14th Pennsylvania Cavalry, having covered some 25 miles, camped at Cobb's Meadows, where plenty of forage was located for their mounts. The detachment of the 2nd West Virginia Mounted Infantry, under Lt. Weaver, which had moved in advance of the column, recorded the capture of two prisoners during the day's march [although the regimental historian says this took place November 4]. They then went into camp about 15 miles north of Huntersville. Sgt. Barnes of the 10th West Virginia Infantry claimed his regiment camped at "Warwick's Mill on Sittington's Creek in Pocahontas County."[26] Corp. George Washington Ordner, 2nd West Virginia Mounted Infantry, claimed his regiment camped at Mors [Morris] Mills and covered 15 miles during the day's march.[27] Adjutant John Lang of the 28th Ohio Infantry filed a field report from "camp near Dan Morris" stating his regiment then consisted of 23 commissioned officers and 582 enlisted men, for a total aggregate of 605 men. He also listed 25 horses present with the 28th Ohio.[28] Averell's Brigade spent a "cold, frosty night" but kept warm with huge fires.

November 4-Holding the Line

There was a flurry of activity in the various camps on the morning of Wednesday, November 4. Averell and Jackson were now acutely aware of each other's presence, making it imperative that Jackson at least delay Averell until he could be reinforced by Gen. Echols from Lewisburg, as well as to bring together his various scattered detachments to Mill Point.

Near the foot of Cold Knob Mountain, Lt. Col. William P. Thompson received Jackson's dispatch at 3:00 A.M. and immediately ordered his detachment of 120 troopers from the 19th Virginia Cavalry to fall back to their old camp on the John H. Kellison farm, known as Camp Miller [formerly Camp Northwest near Huntersville]. However, after arrival at the old camp ground and removing saddles, Thompson received another dispatch ordering his detachment to hurry forward to Mill Point. Thereupon, Thompson moved his cavalry out and ordered Capt. James W. Ball, Company E, 19th Virginia Cavalry, to take the dismounted men at the old camp and move by a regular march to Mill Point. In the course of the day, Company I of the 19th Virginia Cavalry received an issue of clothing at Mill Point.

Other Confederate officers also received Col. Jackson's communication during the early morning hours of November 4. At Lewisburg, Gen. Echols was handed a message from Jackson claiming 1,000 Federals had appeared in front of Green Bank. What transpired next is a bit confusing as the various Confederate reports list conflicting time frames and information. According to Col. Arnett at Marlinton, he received Lt. Siple's message of Averell's presence on *November 4* [possibly a mistake in the original printing of Arnett's correspondence] and immediately relayed it to Jackson at Mill Point; although Siple had apparently warned Jackson the previous day. Arnett said he also notified Capt. Jacob W. Marshall at Edray, to whom Jackson had apparently already sent this notification, as he said this took place November 3. A short time thereafter, Arnett relayed Jackson's message to Siple, along with a message of his own to keep him informed of Averell's movements. After waiting for a sufficient time and getting no response, Arnett correctly assumed Siple had been cut off by Averell's rapid movement, and Arnett had to send his own scouts out to report on the Yankee whereabouts.

Averell had begun marching at 7:00 A.M. with the 3rd West Virginia Mounted Infantry in the lead, followed by Gibson's Independent Cavalry Battalion, the 14th Pennsylvania Cavalry, Ewing's Battery, the 8th West Virginia Mounted Infantry, and the 2nd West Virginia Mounted Infantry. At 7:30 A.M. the remainder of the column moved out, with one of the infantry regiments in the advance [records do not indicate which regiment], then Keepers' Battery, the other infantry regiment, and the ambulances and trains. Col. Moor of the 28th Ohio had also been ordered to furnish a Capt. Brown "with a small party of pioneers, about ten men, to march at the head of his train" and "also 50 men to march in the rear of the train." All wagons were to move with the train and the entire column was told "hereafter there will be no unnecessary yelling or shouting in camp or on the march."[29] During the morning's movement the column burned a rebel camp, then proceeded to ignite another enemy campsite near Huntersville.

General Duffié's men were on the road early, too, marching to Gauley Bridge, where they crossed the Gauley River in small ferry boats and marched eastward throughout the day on the James River and Kanawha Turnpike. Upon departing Gauley Bridge a number of delays were created by numerous unexpected blockades the enemy had placed in the road for the first eight miles. Under the supervision of Capt. Alexander H. Ricter, of the 2nd West Virginia Cavalry, acting inspector general of Duffié's Brigade, the obstacles "were removed, or bridges built, or roads dug around them." Pvt. James Ireland, Company A, 12th Ohio, marching from Fayetteville to join with Duffié, wrote in his diary: "Clears up this morning. Warm and pleasant for the season. Reveille early. Take breakfast. Prepared to start on the raid by 7 o'clock A.M. Arrive at Boyers [Bowyers] Ferry by 12 A.M. (8 miles from camp on New River). Cross over on poor temporary pontoon raft and camp on the banks of the river. Witness the drowning of two horses through the recklessness of their masters."[30]

As Gen. Averell approached Huntersville, news of his advance prompted the break-up of a court in town which had just levied a $1,000 tax on the county to support the destitute families of rebel soldiers, and another $1,000 for the benefit of Confederate soldiers wounded in the October 13, 1863, battle at Bulltown, West Virginia. Correspondent "T.M." wrote, "This [fine] is about equivalent to $200 a bushel, as corn sells there from five to seven dollars a bushel, and everything in proportion. So you see that treason, not only against the United States, but also against the state of West Virginia, is being openly practiced within striking distance of us."[31] The head of the Federal column reached Huntersville, "deserted, save by two families," sometime between 11:00 A.M. and noon [correspondent "Irwin" says 11:00 A.M. while Averell says noon]. The 2nd West Virginia Mounted Infantry and 14th Pennsylvania Cavalry rode into town at 1:00 P.M.; and as the advance of the brigade halted, Huntersville was perceived as a "forsaken, desolate place...the saddest picture of the punishment that has [had] overtaken the poor, deluded people."[32] Correspondent "T.M." noted, "loyal men of Pocahontas and Greenbrier counties" had long since been driven from their homes, but he felt they would return once the Federal army made it safe for them. He felt most of the people that remained were "bitterly disloyal...few men [remained] except the very aged and decrepit...women were plenty, but their beauty was marred by the impress on their countenance...of the treason that lodged within and their manners were as badly spoiled as their beauty."[33] Finding no opposition at Huntersville, Averell ascertained that Lt. Col. William P. Thompson of the 19th Virginia Cavalry was at Marlinton [actually it was Col. Arnett] with about 600 men, and developed a plan to cut him off.

Averell's strategy was to get in advance of the Confederates and blockade the road at Stephen's Hole Run near Marvin Chapel, where the Marlinton [Lewisburg] and Cackleytown [Beaver Creek/Mill Point] roads unite. Less than an hour after his arrival in Huntersville, Averell ordered Col. Schoonmaker to take his 14th Pennsylvania Cavalry and the 3rd West Virginia Mounted Infantry, advance on the Cackleytown [Mill Point] road, and intercept the enemy if it appeared feasible. The designated route, the Beaver Creek road, was a narrow trail "used for years as a shortcut from the Little Levels district [Hillsboro and vicinity] to the county seat of Huntersville." This was "a well-used county road that went to the

The Beaver Creek-Huntersville Road (right center) as it connects with the Marlinton-Mill Point Road (modern (Rt. 219). This view is looking north and Marvin Chapel is just out of the picture to the right on Rt. 219. BRIAN ABBOTT

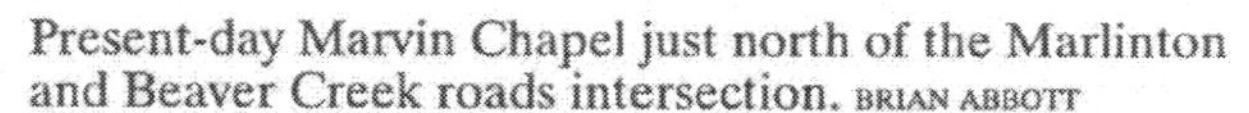

Present-day Marvin Chapel just north of the Marlinton and Beaver Creek roads intersection. BRIAN ABBOTT

Little Levels area;" a road in continual use until about the time automobiles arrived in the area.[34] As Col. Schoonmaker galloped down the road 16-year-old resident Agnes Sharp said, "the road was blue all day" with Federals, and citizen Bill Buckley's mother recalled, "Union soldiers hurrying down the road."[35] Josiah Davis of the 3rd West Virginia Mounted Infantry scribbled in his diary, "the 3rd West Virginia and 14th Pennsylvania Cavalry advanced by Beaver Lick [Creek] road," which was the direct route to Mill Point.[36]

At about 4:00 P.M., Col. John H. Oley's 8th West Virginia Mounted Infantry reached Huntersville, where he received orders from Averell to take his own 8th West Virginia, the 2nd West Virginia Mounted Infantry (which had managed a few hours rest), and a section of Ewing's artillery, and to move in the direction of Marlinton as part of his overall plan to entrap the retreating Confederates before they could reach Mill Point. Timing was crucial, as around noon Col. William L. Jackson had gained knowledge from his scouts at Huntersville of the Federal approach on the Beaver Creek road, and immediately ordered Col. Arnett at Marlinton to "draw in Captain Marshall" from Edray, slightly north of Marlinton, and fall back to Mill Point, blockading the road, "at all practicable points," with felled trees.

Arnett was already in motion, his scouts having returned with the news of the enemy at Huntersville, only six miles east of his camp. He had ordered his wagons loaded and moved out on the road to Mill Point, about eight miles south. Arnett also dashed off a dispatch to Col. Jackson informing him of Averell's location, and sent orders to Capt. Marshall to move "by a mountain path intersecting the [Lewisburg] road," on which Arnett would also retreat, to Mrs. Kee's, at the top of Price Hill, one mile distant from Arnett's camp at Marlinton. A postwar historian claimed Arnett moved from Elk Mountain by way of Onoto and Green Hill road to the top of Price Hill where a Mrs. Waugh resided in 1958.[37] While the Federals were yet occupied at Huntersville, Arnett followed orders and dug away at the sliding hillside and blockaded portions of the road over Price Hill, just west of Marlinton, as effectively as possible, until joined at Mrs. Kee's by Capt. Marshall's detachment.

Preparing to make a stand at Mill Point, Col. Jackson suspected the main Federal thrust would be on the Beaver Creek road; therefore, he verbally ordered Lt. Col. Thompson, who had arrived at Mill Point with his portion of the 19th Virginia Cavalry earlier, to go up the Beaver Creek road, blockade the intersection of the Cackleytown [Mill Point] road and Lewisburg pike, one mile north of Mill Point, and stall the enemy until Col. Arnett could pass through the crossing. Thompson complied and took his men to the designated point, where his dismounted soldiers hid in ambush behind blockades. At 3:00 P.M., Col. Schoonmaker's force arrived at the disputed intersection and immediately skirmished with Thompson's men; while pioneers from the 14th Pennsylvania Cavalry and the 3rd West Virginia Mounted Infantry attempted to cut through the blockades while constantly under severe enemy fire. At 5:00 P.M., as the bullets flew in the stand-off between Schoonmaker and Thompson, the 10th West Virginia Infantry arrived at Huntersville, according to Sgt. Thomas R. Barnes of the regiment.[38] The 28th Ohio supposedly reached town at about 6:00 P.M.

As the sun began to go down, Col. Arnett and Capt. Marshall departed Mrs.

A rare 1890s view of Marlinton. Col. John H. Oley's men camped here on the night of November 4, 1863. Photo by Dr. N.R. Price. POCAHONTAS COUNTY HISTORICAL SOCIETY

A rare view of the bridge at Marlinton taken in the mid-1890s. POCAHONTAS COUNTY HISTORICAL SOCIETY

Kee's for Mill Point, leaving a small squadron under Major John Buford Lady, to keep watch on the Federals. Arnett had already received several dispatches from Col. Jackson warning him of the enemy force moving down the Beaver Creek road and realized it was imperative he move rapidly in order to avoid being cut off. Col. John H. Oley's 8th West Virginia, along with the the 2nd West Virginia and Ewing's Battery, as ordered, reached Marlinton at about dusk, where Oley found only one to two enemy pickets and learned Col. Arnett's 20th Virginia Cavalry and "[Capt. Jacob W.] Marshall's [Company I, 19th Virginia Cavalry] and [Capt. Elihu] Hutton's [Company C, 20th Virginia Cavalry] companies" had fallen back toward Hillsboro three or four hours previously, blockading the road en route. In compliance with orders, the Federals began to cut the blockades out and went into camp near the *bridge*.

Also at dusk, at about the same time Oley's men entered Marlinton; Arnett's command passed the disputed Cackleytown and Lewisburg pike intersection, where Lt. Col. Thompson's men were yet blazing away at the Yankees, "disputing every inch of ground," as they fell back, the Federals "constantly pressing on with great pertinacity." The skirmishing lasted until several hours after nightfall and Col. Arnett had safely passed through the intersection "with the residue of the infantry." Lt. Col. Thompson had lost one man killed in the affair and said the "enemy suffered considerably," although records do not indicate any damage to the Federals. Realizing they were facing overwhelming odds, Thompson and Arnett slowly withdrew to nearby Mill Point, where Thompson took up a defensive position on the "hill south of Mill Point...with a view to ascertain whether they [the Federals] had artillery." Col. Jackson, undoubtedly relieved all his detachments had arrived safely at Mill Point, ordered Col. Arnett to "the hill southeast of Mill Point" while Capt. Warren S. Lurty's Battery was sent to a hill south of Mill Point, "on an eminence, covering the road."

Sometime during the day, Col. Milton J. Ferguson, 16th Virginia Cavalry, commanding Jenkins' Cavalry Brigade that was resting in Greenbrier County, received a note from Gen. Echols informing him of Averell's movement toward Mill Point. Ferguson responded by ordering a portion [some sources say six companies] of the 14th Virginia Cavalry, under Col. James A. Cochran [or possibly Major Benjamin F. Eakle], to advance to Mill Point and cooperate fully with Col. Jackson. Since Ferguson had not received any orders from Gen. Sam Jones, he reported directly to Gen. Echols, his immediate superior, and the two concluded it would be hazardous to move Ferguson's entire force from the Lewisburg area. Accordingly, Ferguson directed his 16th Virginia Cavalry, under Major James H. Nounnan, to move to Bunger's Mill, about 5 miles west of Lewisburg; while a squadron of the 14th Virginia Cavalry, led by Capt. James Alexander Strain [Company H], was to remain on picket and outpost duty on the road to Nicholas County. These units where to guard against any enemy approach from the direction of Kanawha and Nicholas counties. Although no enemy was discovered, the command did notice "increased activity in that direction [which] caused apprehension such a movement would be made in conjunction with Averell."

As the Federal troops began to settle into their various camps for the evening, 2nd Lt. Abraham Clarkson Merritt, 68th New York Infantry, signal corps detachment, arrived and was sent to "the Knob" near Huntersville, where he was to

Bungers Mill also known as Hutsonpillar's Mill on Milligan's Creek. This area served as a military campground throughout the was and was picketed by the 16th Virginia Cavalry during the Droop Mountain affair. Located just west of Lewisburg, the white house in the background is the only remaining structure and is now a private residence. GREENBRIER COUNTY HISTORICAL SOCIETY

Capt. James Alexander Strain of Co. H, 14th Virginia Cavalry. He led a squadron of his regiment on picket and outpost duty on the road to Nicholas County, West Virginia. CAL SEEBERT, SPARTANBURG, SC

observe and report by rocket signals any information from Lt. Denicke, who had accompanied Col. Schoonmaker's detachment to Marvin Chapel near Mill Point. A signal was supposed to have been sent by Denicke at 8:00 P.M., but none came due to the fact that Col. Schoonmaker didn't actually arrive at Marvin Chapel until 11:00 P.M., delayed by the skirmishing with Thompson. Josiah Davis, 3rd West Virginia Mounted Infantry, verified the problem, writing in his diary, "reached junction of the road with Marlin's Bottom and Lewisburg road that night about 9:00 P.M., after encountering numerous blockades...was fired into by the enemy and instantly took a defensive position and waited for daylight."[39]

Around 9:00 P.M., Gen. Averell received a report from Col. Oley, 8th West Virginia, that he had reached Marlinton and Lt. Col. Thompson's [Arnett's] rebels had retired toward Mill Point, blockading the road.

At 10:00 P.M., Lt. Merritt had yet to receive a rocket signal from Lt. Denicke and, with Averell's consent, called in the signal station. Col. Schoonmaker, as noted, had been delayed for three hours by the skirmishing and blockades in the road; but as soon as his pioneers were able to cut their way through the obstructions, the command came within full view of Col. William L. Jackson's "Huntersville Line" at Mill Point. At 11:00 P.M. Schoonmaker and Jackson both realized they were about to have a hard fight on their hands. Schoonmaker prepared accordingly by dismounting his men and "placing them in the most available position." In the opposing camp Col. Jackson placed Col. Arnett in charge of all the infantry, which was posted by detachments in the best possible locations along the southern side of Stamping Creek, a small stream which flowed in a ravine between the opponents.[40] The rebels would remain in such defensive position throughout the night.

After settling into camp on the north side of Stamping Creek, Col. Schoonmaker had Lt. Denicke send a rocket signal to Averell to inform him of the situation, but got no response as Averell had already called in his signal station. Despite this, some of Col. Jackson's rebels later claimed they were able to view the rockets and they *were* answered "from a point near Huntersville," as well as on the "Beverly road beyond Marlin's Bottom [Marlinton] Bridge." Gen. Averell did receive a dispatch from Schoonmaker at about midnight, informing him that Col. Jackson and Lt. Col. Thompson had managed to elude the trap and link up at Mill Point, and were then in position in his front threatening an attack.

Colonel Jackson actually had little intention of assaulting Schoonmaker, but he had deployed his men to repel any attack on his line; as he had reportedly discovered the Federal force confronting him had artillery, although this was false, as Schoonmaker had no artillery. Jackson had his men build campfires in the rear and dispatched another message to Gen. Echols advising him of the situation.

As the evening closed, Jackson and Schoonmaker, while "aligning troops far into the night," went into their respective camps, which, according to Lt. Col. Thompson, were only 300 yards apart; or as Col. Jackson said, the Yankee camp was in "plain view." The two opposing colonels knew a battle at Mill Point would be imminent at first light.

General Averell spent the night at Huntersville with his two infantry regiments and Keepers' Battery, while Col. Oley, having covered approximately 22 miles during the days march, camped at Marlinton with his 8th West Virginia, the

2nd West Virginia, and Ewing's Battery. Frank Reader, 2nd West Virginia, recalled that at Marlinton "the hills were filled with bushwhackers, who made things lively..."[41] 2nd Lt. Andrew P. Russell, Company H, 2nd West Virginia, who was "on picket duty during the night, destroyed a considerable quantity of small arms and also burned the [rebel] quarters consisting of very comfortable log cabins."[42] Many years later an area writer would remember, "the firewall of one of those cabins is in my front yard. My father would not allow it to be destroyed in his lifetime and it will be preserved. A few years ago we found a big machete [probably a Bowie knife] buried there"[43]

General Duffié's command, which had marched eight miles past Gauley Bridge, quartered for the night at Hamilton's near Hawk's Nest on Gauley Mountain, in Fayette County. The two Ohio infantry regiments coming to join Duffié, camped on the north bank of the New River at Bowyers Ferry, where James Ireland, 12th Ohio, passed "a sleepless night."[44] Gen. Echols and Col. Ferguson were yet at Lewisburg, while Gen. John Daniel Imboden was in the Shenandoah Valley of Virginia covering Col. Jackson's extended right.

General John Daniel Imboden was born February 16, 1823, near Staunton, Virginia and became known as a "man of varied interests, and five wives" [Not at the same time]. He attended Washington College (now Washington and Lee) for two years and later, taught school and practiced law in Staunton. He was involved in politics and "served 2 terms in the state legislature but failed to be elected a delegate to the Virginia secession convention, perhaps because of his strong secessionist views." Another writer said Imboden, "had been a candidate for a seat in the convention, but was defeated by the candidate of the Union party." Reportedly, Imboden advocated "independent secession, and the maintenance of an independent State that could mediate between North and South and lead in the formation of a new Union, with local rights more clearly defined." He organized the Staunton Artillery and led them in the capture of Harpers Ferry at the outbreak of hostilities. He fought gallantly with his battery at 1st Bull Run. It was said, "he and other young and enthusiastic leaders were the forerunners of the spirit which was to dominate Virginia for four years, but at that moment they were coldly received by the majority of the people, not yet aroused."

In 1862, Imboden was commissioned a colonel and participated in Gen. Thomas J. ("Stonewall") Jackson's Shenandoah Valley Campaign, taking part in the battle of Port Republic. At Staunton he organized the 1st Virginia Partisan Rangers [which in early 1863 became the 18th Virginia Cavalry] and on January 28, 1863, was promoted to brigadier general.

General Imboden was effective in the "Jones-Imboden Raid" through (West) Virginia in early 1863 and was noted for his actions guarding Gen. Lee's left flank in the advance during the Gettysburg Campaign and subsequent retreat. On July 21, 1863, he was assigned command of the [Shenandoah] Valley district and on October 18, 1863, captured the garrison at Charlestown, West Virginia during Gen. Lee's Bristoe Station campaign.[45]

General Imboden received a note in the evening of November 4, 1863, from 1st Lt. John Thomas Byrd, Company G, 18th Virginia Cavalry, who was commanding a detachment at Hightown in Highland County, Virginia. The message informed Imboden of Averell's *5,000* troops in the vicinity of Bartow. A few hours

Brigadier General John Daniel Imboden commanded the (Shenandoah) Valley District, C.S.A., during the Droop Mountain campaign and made an unsuccessful pursuit of Averell. U.S. ARMY MILITARY HISTORY INSTITUTE, MOLLUS COLLECTION

after the receipt of this knowledge, Imboden was given another dispatch stating Averell had moved in the direction of Huntersville; although by this time, Averell was threatening Col. Jackson at Mill Point.

November 5-Prelude to Disaster: The Fight at Mill Point

As early as 3:00 A.M., Thursday, November 5, after having set fire to the Confederate winterquarters, Gen. Averell moved the 10th West Virginia Infantry, the 28th Ohio Infantry, and Keepers' Battery out of Huntersville en route to Mill Point, which was Col. Jackson's storage and supply depot, and so named for the series of "over-shot" mills established there dating back to 1778 when Valentine Cackley located in the vicinity.[46] Records are sketchy as to whether Averell took Oley's or Schoonmaker's route. According to Sgt. Thomas R. Barnes, 10th West Virginia, his regiment was "ordered out and marched to Shackleytown [Cackleytown] 15 miles in 4 hours."[47] The distance given would seem to indicate he moved by way of Marlinton. John D. Sutton, Company F, 10th West Virginia, in a postwar tour in which he retraced his steps during the campaign, said his regiment "crossed the Greenbrier River at Marlinton [bridge location] and waded the cold waters."[48] He added that Col. Thomas M. Harris, being an educated physician, ordered the men to remove their shoes and stockings and then put their shoes back on, cross the river, then place their stockings back on, in an effort to prevent chills and sickness.[49] Despite these two reports of veterans, postwar historian Calvin Price claimed Averell moved down the Beaver Creek road and that Sutton's regiment crossed the Greenbrier River eight miles below Marlinton at the mouth of Beaver Creek.[50] Whichever route Averell took, the 10th West Virginia Infantry filed a field return for the day indicating 569 men present for duty, one straggler lost, and 23 horses (12 private and 11 public). This figure would remain the same the following day at Droop Mountain.[51]

Averell had also sent orders ahead to Col. Oley to cut out all blockades in the road and advance upon Mill Point. Having received the message at about 6:00 A.M., Oley had the 2nd and 8th West Virginia regiments, as well as Ewing's Battery, in the saddle by daylight and moving to Col. Schoonmaker's relief.

Daylight also brought about renewed activity on the Mill Point front. Col. Schoonmaker "...found the enemy had taken a very strong position and was waiting on attack," and realized Col. Jackson's force outnumbered his own "three to one." Additionally, Jackson had the two 12-pounder howitzers of Capt. Warren S. Lurty set up in a strong position on a hill just south of Mill Point. Indeed, Jackson had been up all night posting his men in strategic defensive positions along Stamping Creek, which prompted Schoonmaker to place his own command in "as strong a defensive position" as possible, while he sent another communication to Averell to hurry forward.

At daylight, as expected, the fight at Mill Point between colonels Jackson and Schoonmaker commenced. With dawn breaking Schoonmaker's skirmishers and sharpshooters advanced against Jackson, obviously testing the strength of the rebel

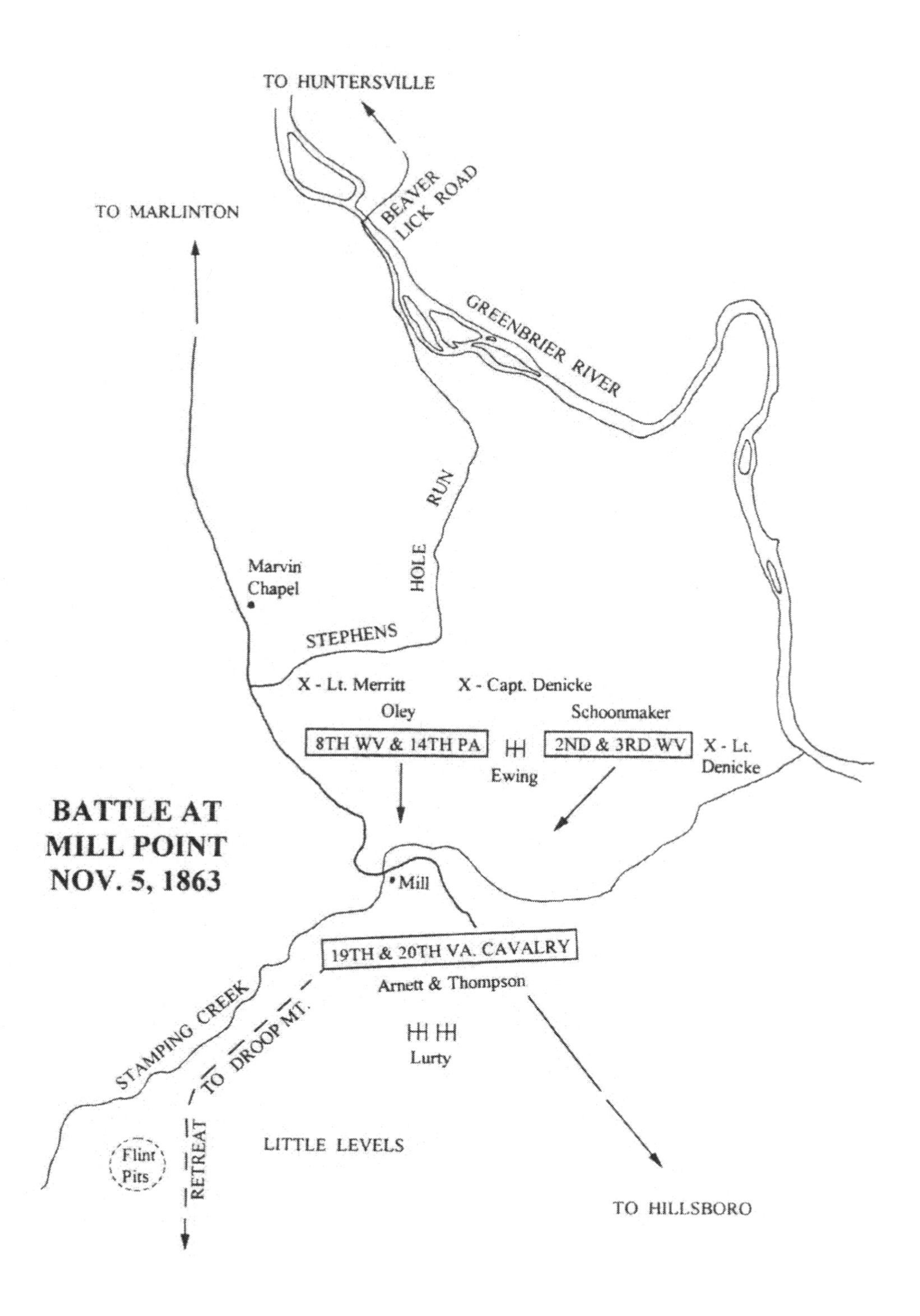
TO HUNTERSVILLE
TO MARLINTON
BEAVER LICK ROAD
GREENBRIER RIVER
HOLE RUN
Marvin Chapel
STEPHENS
X - Lt. Merritt
X - Capt. Denicke
Oley
Schoonmaker
8TH WV & 14TH PA
Ewing
2ND & 3RD WV
X - Lt. Denicke
BATTLE AT MILL POINT NOV. 5, 1863
Mill
19TH & 20TH VA. CAVALRY
Arnett & Thompson
STAMPING CREEK
TO DROOP MT.
Lurty
RETREAT
Flint Pits
LITTLE LEVELS
TO HILLSBORO

Mill Point mill located at the junction of U.S. 219 and State Route 39, was built immediately after the war to replace a series of mills dating back to 1778, and remained in operation until 1941. BRIAN ABBOTT

View looking north from field on top of hill just north of Stamping Creek at Mill Point. Federal forces occupied this location during the battle. Shrapnel and bullets have been excavated from the field. BRIAN ABBOTT

View looking south across Stamping Creek at Mill Point. Lurty's Battery is believed to have been posted at the top of the hill near the tree at center of photo. BRIAN ABBOTT

Field excavations indicate Lurty's Battery was posted just to the left of the dead tree in the center of photo. This position is on the crest of a hill just south of Stamping Creek at Mill Point. BRIAN ABBOTT

Spent friction primers used to fire cannon, found at the position of Lurty's Battery at Mill Point. RICHARD ANDRE PHOTO, MIKE SMITH COLLECTION

line. Eight-year-old C. L. Stulting lived on the farm of Hermannus Stulting and his wife, Johannah, from Utrech, Holland about a mile south of Mill Point [Stulting says Hillsboro]. He recalled the battle, writing, "about six o'clock we were all sitting around the breakfast table unaware of there being any soldiers in our neighborhood [young Stulting possibly was kept uninformed but his family undoubtedly knew of the danger] when we heard the firing of army guns just outside of our house."[52] Col. Jackson's men opened fire on Schoonmaker's skirmishers and maintained such "for some half an hour." Receiving no return fire from the Federal skirmishers, Jackson ordered Capt. Lurty to fire his artillery at the main enemy force "on top of the hill."

Colonel Jackson, taking note of the sound of the Federal sharpshooters and movement in the Federal camp, suspected the Federals were up to something, and selected a detachment of 30 men, under Lt. Laban Radcliff Exline, to make a reconnaissance and drive back the enemy. Exline, of Fairmont, West Virginia, had previously served as a captain in the 31st Virginia Infantry, and "with his arm shattered, [had] led two charges at [the battle of] South Mountain," Maryland, in 1862.[53]

Lacking any artillery to respond to Lurty's, Schoonmaker's skirmishers were forced to fall back and seek shelter in a masked position which protected the main Federal force, although Lurty's pieces succeeded in causing "some confusion" in the Yankee camp. Schoonmaker claimed his retrograde movement duped Col. Jackson into suspecting he was retiring from the field, and Jackson retaliated by advancing his rebel infantry, although there is nothing in Jackson's reports to indicate he ever took the offensive. Col. Schoonmaker knew this was the critical point and admitted Jackson could have driven him off if not for the timely arrival of Col. Oley's command.

Just outside Marlinton, Col. Oley had heard the artillery at Mill Point, realized Schoonmaker had no cannon, and hurried his men forward at a trot in order to cover the nine to 10 mile distance. Upon arrival Col. Oley found the 3rd West Virginia Mounted Infantry and the 14th Pennsylvania Cavalry in line of battle and being pressed hard by the enemy. Frank Reader of the 2nd West Virginia said, "Jackson had drawn up his men in line of battle and was firing at our advance."[54] Oley reported to Schoonmaker, who immediately ordered him to dismount the 2nd West Virginia Mounted Infantry, the 8th West Virginia Mounted Infantry, and deploy them accordingly. Schoonmaker would lead the Federal left, with the 2nd and 3rd West Virginia regiments; Oley would command the right, with the 8th West Virginia Mounted Infantry and the 14th Pennsylvania Cavalry. The one section of Capt. Chatham T. Ewing's Battery was positioned on a hill in the center. Lt. Merritt of the 68th New York Infantry, signal corps detachment, was ordered to the extreme right and Lt. Denicke to the left. The two were to communicate with the center station during the battle.

Colonel Jackson was well aware of the Federal reinforcement as Capt. Exline's party, having successfully driven back the remaining Federal sharpshooters, had returned with information on the arrival of the Yankee artillery, which Jackson felt had been successfully delayed by the blockading of roads. He knew the howitzers of Lurty's horse artillery couldn't contend with the long-range guns of Ewing and opted to fall back to the crest of nearby Droop Mountain, just south of Hillsboro and 24 miles north of Lewisburg. Jackson issued orders to Lt. Col. Thompson and Col. Arnett to retire from the field at the very moment Ewing's artillery was positioned, with instructions to Thompson to cover the rear with his cavalry, while Arnett was to take "the safest route to Droop" with the infantry.

1st Lieutenant David Poe, Company A, 20th Virginia Cavalry recalled the Federals were "evidently preparing to surround our command. [Col.] Jackson prepared [to] fall back, leaving Company 'A' 20th regiment as a skirmish line along the bank of Mill Creek [Stamping Creek]. The Company that day numbered from sixty to eighty men, all the officers were present. We were directed to remain until we had orders to fall back. That was done to give time for Jackson's men to move their trains back to Droop Mountain across the Litle [sic] Levels of Pocahontas, a distance of three or four miles. We had to keep up a brisk firing all along our line owing to our weakness and the distance we were from the main body of our command."[55]

Averell claims he reached the field at 8:00 A.M. and found Oley had just arrived and the Confederates were retiring. Josiah Davis of the 3rd West Virginia Mounted Infantry, giving a time frame that conflicted with Averell's, said his regiment routed the Confederates at Mill Point at 9:00 A.M. and then "waited for the balance of the U.S. troops to come up."[56] Even Col. Schoonmaker, in his official report, implied Averell arrived later in the day, shortly after 11:00 A.M., with his two infantry regiments. Frank Reader of the 2nd West Virginia Mounted Infantry did not give a time, but wrote, "...General Averell pushed his men rapidly to the front."[57] Despite the time Averell reached Mill Point, it was imperative Col. Jackson get his rebels to a better defensive location quickly, or he would be facing the entire Federal army.

Little did Col. Jackson realize, help would soon be on the way. During the morning Gen. Echols, at Lewisburg, had received another dispatch from Jackson informing him of the critical situation at Mill Point on November 4. Jackson had estimated Averell's force, then about three miles from Mill Point, at about 2000 strong, and said he planned to halt any further enemy advance, and again requested reinforcements from Echols. If unable to hold Averell in check, Jackson said he would fall back toward Lewisburg. Echols found himself faced with a dilemma. If he refused to support Jackson, and Jackson was defeated, he would be soundly criticized by the southern press. On the other hand, if he moved to Jackson's assistance, he might be cut off from the rear by another Federal column approaching from the west. Indeed, Gen. Duffié was already so engaged in such a maneuver.

Echols decided to race to Jackson's assistance and quickly sent a dispatch to him to inform him of his decision. Echols then began to organize his brigade for the march north. Lt. Randolph Blain, Capt. Thomas E. Jackson's Company, Virginia Horse Artillery, Jenkins' [Ferguson's] Brigade, recalled he was awakened just before dawn by Major William L. McLaughlin, "who gave orders to be ready to move by eight [A.M.] o'clock. At that hour we started on the Frankford road making 15 miles by dark."[58] McLaughlin said his own command consisted of Chapman's Battery (four pieces) and Capt. Jackson's Battery (two pieces), and departed Lewisburg on the morning of the 5th along with the infantry. Pvt. Joseph R. Perkins, Company E, 26th Battalion Virginia Infantry, Echols' Brigade, said his command "received orders at Lewisburg to meet Jackson at Academy [Hillsboro]."[59] During the morning, Lt. William Frederick Bahlmann, Company K (Fayette Rifles), 22nd Virginia Infantry, Echols' Brigade, who had been serving on a court-martial at Lewisburg, departed his boarding room for the courthouse, but "felt something in the air that made" him suspect something "was stirring." Upon arrival at the courthouse, Bahlmann found the 22nd Virginia had received marching orders, and most of the soldiers in the room joined with their commands as they marched through town. En route to his regiment Bahlmann stopped at a Mrs. Graves and asked the young ladies for some lunch. They responded and gave him biscuits with some type of red preserves on them. When Lt. John P. Donaldson, Company H, 22nd Virginia, came by, Bahlmann shared his food with him as they moved out.[60]

By 9:00 A.M., Echols had his entire command on the march and would cover some 14 to 15 miles the first day. During the morning Echols was joined en route by Col. Milton J. Ferguson. As anticipated at Mill Point, Capt. Chatham T. Ewing's artillery [which Col. Jackson claimed consisted of 5 pieces] opened fire at 11:00 A.M.; while Col. Oley was ordered to make a rapid advancement against Jackson, which his men did "with enthusiasm." Col. Arnett reported the Federal artillery barrage caused no confusion in his ranks as his men proceeded to retreat "under cover of the hills and through timber, not withstanding the shells of the enemy [which] burst in numbers over our heads and near our ranks." Lt. Col. Thompson, following Col. Jackson's directive, formed his part of the command under a "severe and accurate shelling of the enemy batteries" as he sent the bulk of his men to the rear. Reportedly, one Yankee artillery round "entered a large barn near Mill Point, and, according to an old gentleman, the shell 'explored' [exploded]."[61] Lincoln S. Cochran, a Pocahontas County resident and Confederate soldier who was not present in the Droop Mountain campaign, remembered "a cannonball passed through the top of James E. Moore's barn," where Lee

Moore would reside in 1935.[62] 1st Lt. David Poe, Company A, 20th Virginia Cavalry, wrote, "I thought time very long for the message to come bidding us to retire. When it did come it came to the end of the line that I had charge of. Adjutant [Benjamin M.] Smith, who brought the message for us to fall back, stood on a hill quite a distance from our line, and called to me loud enough to be heard by the enemy. To save further delay I arose from my cover and gave command, 'attention, company on right and left take intervals in retreat, double quick march.' Before I got through the command company 'A' was getting out of that valley just as fast as their feet could carry them. About three regiments of bluecoats were coming after us yelling and shouting. We had to travel at least three miles before we were safe. We then took a position on Droop Mountain where we remained until the next morning."[63]

In order to cover Col. Jackson's retreating forces, Lt. Col. Thompson kept Capt. John W. Young [Company E, 20th Virginia Cavalry], 1st Lt. John W. Coffman [Company I, 19th Virginia Cavalry], and about 30 men with him. For about forty minutes this detachment held a "hill under a terrific shower of shell and grape." One historian would later write, "...the whole of the Levels raked with shot and shell by the Northern batteries."[64] As soon as Lt. Col. Thompson felt the infantry was in a "comparatively safe" position, his men began to retire from the field. The retrograde movement of Jackson's rebels was recalled by a writer years later, stating, "When the guns began to thunder it occurred to Jackson that his battle line was just the right distance from the Federal batteries to be in range of grapeshot and he withdrew his army by having them slip silently up the stream until they were hid by the bend of the mountain and he took them out by the flint pits."[65]

Colonel Schoonmaker, having ordered a general advance, had his men firing briskly on Jackson's troops, pressing the enemy with his skirmishers and the main column. Col. Oley said Schoonmaker instructed him "to advance rapidly and drive the enemy." His "men started with enthusiasm, but the rebels made no stand and precipitately retreated." The Federals attempted a cavalry assault in column against the Confederate rear guard under Lt. Col. Thompson, but Capt. John W. Young, by a "skillful and determined movement" of his detachment, repulsed the Yankees, who "never got within 200 yards" of the Confederate rear between Mill Point and Droop Mountain. Young C. L. Stulting, huddling in the family house built in 1858, watched the armies pass, writing, "...we saw the advance [rear] guard of the Confederate Army on horseback firing back and slowly retreating across the field while the Federal army on foot were shooting from behind some trees near there."[66] He continued, "...we heard afterward that the advance [rear] guard of the Confederate Army was tolling the Federal Army to the breastworks they had built on the top of Droop Mountain. The advance [rear] guard of the Confeds formed a line of battle when they got across this field in front of our house. The Yankee troops formed a line of battle four men deep behind the line fence between the Grattan Miller land and the Davis Poague farm. One side of our yard fence formed a part of this line fence. Our house had a large stone chimney on the side next to where the Yankees were stationed. Our family and several Yankee officers stood behind this chimney so that bullets shot by the Confeds at the Yankees would not hit us, for there were some balls hitting the house nearby all the time...this skirmishing around the house and across the

This home, known as the Stulting residence during the Civil War, is located just north of Hillsboro. It was later the residence of noted author Pearl S. Buck. BRIAN ABBOTT

State historic highway marker located at the Stulting House. BRIAN ABBOTT

field lasted about one hour when the Federal army commenced firing a few cannon from a hill near our house and then the Confederate troops retreated to their breastworks on Droop Mountain."[67]

During this period, Gen. Averell reached the field [although, as noted previously, some sources give his arrival as early as 8:00 A.M.] with the two infantry regiments and Keepers' Battery of four guns, which Lt. Col. Alexander Scott's 2nd West Virginia was ordered to support. A participant, believed to be with the 28th Ohio, added, "We, having marched in almost a double-quick for thirteen miles, came only in time to see the rear guard of the rebels going over the hills, near Mill Point and our batteries sending a few anxious shells after them. We followed them six miles and then encamped..."[68] Major Thomas Gibson's Independent Cavalry Battalion, marching with Averell's train via Marlinton, also arrived at Cackleytown [Mill Point]. Col. Schoonmaker turned over command to Gen. Averell, who realized he was only 34 miles north of Lewisburg and if he pressed Jackson too hard, the enemy would retreat too quickly through Lewisburg and deprive Gen. Duffié of the opportunity and time necessary to get in the Confederate rear. Although Averell was not scheduled to arrive at Lewisburg until 2:00 P.M., Saturday, November 7, he felt he could yet trap Jackson's entire command and ordered Col. Schoonmaker to take a portion of the 14th Pennsylvania Cavalry and the 3rd West Virginia Mounted Infantry and pursue the Confederate rear [Averell says in his report he sent *three* mounted regiments after Jackson].

While Col. Schoonmaker chased Lt. Col. Thompson and the Confederate rear guard past the Stulting house, through Hillsboro, and toward Droop Mountain, as far as Little Coal Creek;[69] the remainder of Averell's Brigade came up and went into camp between Mill Point and Hillsboro at 2:00 P.M., the very hour Col. Jackson gained his first knowledge that Gen. Echols was moving to his support. Col. Oley said his 8th West Virginia followed the rebels, "unsuccessfully for 4 or 5 miles," before receiving orders to end the pursuit and go into camp.

Colonel Schoonmaker, although he said he followed the enemy with a "portion of the 14th Pennsylvania Cavalry and the 3rd West Virginia as far as Hillsboro," actually covered the seven miles of "beautiful cleared valley" south of Hillsboro, and caught up with Lt. Col. Thompson's men at the northern base of Droop Mountain at 3:00 P.M. Thompson's detachment would have been cut off if not for Capt. Warren S. Lurty, whose battery was already posted on Droop Mountain with Jackson's infantry. Taking note of the situation, Lurty fired a "few well-directed shells" that drove off the Federal troopers in a "total route." This brought about an end to the affair at Mill Point. Capt. Denicke of the 68th New York Infantry, signal corps detachment, stated, "the enemy having withdrawn to the summit of Droop Mountain, I ordered Lt. Merritt to open communications with Gen. Averell's headquarters from our advanced pickets." Averell put it bluntly, recording, "My advance was withdrawn from the fire of his artillery, and the attack postponed until the ensuing day."

Although much gunpowder was expended by both sides during the Mill Point battle, casualties were extremely light. Averell admitted a "trifling loss on either side," while Lt. Col. Thompson reported one Confederate wounded and none killed or captured. Col. Schoonmaker disputed Thompson's report and claimed he captured "a few prisoners" during the pursuit of Thompson's rear guard. Frank

Tree under which Gen. Robert E. Lee slept September 15, 1861, at Mill Point. Although not pertinent to the Droop Mountain battle this 1928 view ran in the Droop Mountain commission booklet and was often mistaken by many as the better known "Lee Tree" on Sewell Mountain. According to an old newspaper clipping, General Lee "came near to being a guest of the Beard family. Only he slept under a tree in the front pasture and would not eat any food other than that cooked by his own darkey." The tree was cut down in the 1940s by the electric company against the objections of the landowner. DROOP MOUNTAIN BATTLEFIELD COMMISSION BOOKLET

Modern view of the location of the Lee Tree at Mill Point.
MIKE SMITH, DROOP MOUNTAIN STATE PARK

State historical marker at Hillsboro describing the Droop Mountain battle and the Salem Raid in 1863. BRIAN ABBOTT

Reader, 2nd West Virginia, noted, "three men of our command were wounded in the little fight."[70] Corp. George Washington Ordner, also of the 2nd West Virginia, barely made note of the action in his diary, writing: "Had a skirmish this A.M."[71] Schoonmaker summed up the entire event by praising Col. Oley and Lt. Col. Alexander Scott, who drove the rebels off "scarcely fifteen minutes from their arrival" on the field.

Colonel Jackson, having reached the summit of Droop Mountain with his command, had immediately deployed his men in a defensive line, which included sending Col. Arnett, with his 20th Virginia Cavalry, to "a high point" Jackson had selected "adjacent to the road," and where they would remain until the following morning. From the 3,100 foot crest of Droop Mountain Col. Jackson could clearly view the Federals' camps sprouting up around Hillsboro. This fact was confirmed by John D. Sutton, Company F, 10th West Virginia Infantry, who said, "...our camp before the [Droop Mountain] battle was in fair view of the Confederates, and a rebel later told him they could look down and see what we were cooking for supper."[72]

While the Mill Point fight had been in progress Gen. Duffié's Federals had continued marching 22 miles eastward to Tyree's Tavern on Sewell Mountain, "building a bridge and cutting away a number of obstacles" en route. At 4:00 P.M. Duffié was joined at Tyree's by Col. Carr B. White with his 12th and 91st Ohio Volunteer Infantry regiments. Col. White and his foot soldiers had marched from Fayetteville in Fayette County, "delayed by the blockades," as well as a "number of their wagons containing subsistence stores," which had broken down. Pvt. James

Rare postwar photo of Hillsboro looking to the west, showing the town much as it looked during the war. Date and photographer unknown. BILL MCNEEL AND THE POCAHONTAS COUNTY HISTORICAL SOCIETY

Rare postwar winter photo of Hillsboro, taken from approximately the same location as the photo above. The two story house in right foreground can be seen in both photos. Col. Moor's flanking party moved on this road through upper left of photo, then cut south (left) after reaching the mountain. BILL MCNEEL AND THE POCAHONTAS COUNTY HISTORICAL SOCIETY

Ireland, Company A, 12th Ohio, wrote of the day's march in his diary: "warm and cloudy, threatens rain all day. Rise early. Get breakfast having no cooking utensils but our tin cups. Our Regt. left in the rear. Ascend an uncommon and rough and difficult hill. At the top of the hill my Co. left as rear guard. Find a very bad road which hinders us very much. Reach the turnpike from Gauley to Lewisburg. Move up it a couple of miles and camp at the foot of big Sewell Mt. on the farm of Terie [Tyree]-an old secesh bushwhacker. The boys go for his poultry and apples. At this point we are joined by Genl. Duffey's [Duffié's] forces consisting of two mounted Regts. and one section of artillery."[73]

General John D. Imboden's Confederates had been on the move as well, having broke camp four miles from Bridgewater, in the Shenandoah Valley region, at daybreak, and marched to Buffalo Gap, where Imboden ordered six day's rations of hard bread and bacon be sent by railroad cars to Goshen Depot. Echols, with his brigade, had received another dispatch from Jackson during the day informing him of engaging the enemy and again requesting his support. Echols managed to cover approximately half the distance to Droop Mountain from Lewisburg and went into camp in the vicinity of Spring Creek, near the Greenbrier River. Lt. Micajah Woods, Jackson's Battery, recalled, "...Thursday we camped three miles beyond Frankford."[74]

Operating from the Federal camps around Hillsboro, Col. Schoonmaker "placed pickets on the main roads and by-roads" and returned to town at 7:00 P.M. 1st Sgt. George H. Mowrer, Company A, 14th Pennsylvania Cavalry, said, "...We went into camp about three miles from the *bridge*."[75] Apparently the bulk of Averell's mounted men spent the night at Hillsboro; while the infantry camped to the rear at Mill Point although George Washington Ordner of 2nd West Virginia Mounted Infantry, said his regiment camped at Mill Point.[76] John D. Sutton, 10th West Virginia Infantry, wrote, "...the army went into camp on the levels between Mill Point and Hillsboro."[77] Sgt. Thomas R. Barnes, also of the 10th West Virginia, recorded in his diary his regiment "camp[ed] for the night [at Cackleytown]–the rebs are 5 miles in front."[78] Correspondent "M" wrote in the *Wheeling Intelligencer*, "we accordingly bivouacked for the night on what is known as the 'Little Levels,' and is about three miles distance from the enemy position,"[79] while another source said the Federal army "bivouacked in the Edmonston field on the top of the hill as you start up the mountain."[80] Postwar historian Calvin Price claimed the Yankee army camped "along the road between Mill Point and Hillsboro, in the fields...[later] owned by M. J. McNeel [2nd Lt. Matthew John McNeel, Company F, 19th Virginia Cavalry] and the Capt. Edgar estate. In plain view of his camp was the large, brick home of Colonel Paul McNeel, the member for Pocahontas County in the convention at Richmond that declared secession."[81] Also from the *Wheeling Intelligencer*, correspondent "Irwin" claimed Averell's men, at least the portion to which he was attached [probably the mounted West Virginia regiments], "went into camp on the *morning of the 6th* on the McNeil [McNeel] farm, owned by Col. Paul McNeel, whose son was a Confederate captain in McNeill's Rangers."[82] He added the command located plenty of corn, oats, and hay for the horses, the men rested well, and some of the men purchased butter for 15 cents in postal currency [equivalent to $5 in Confederate money]. Reportedly, the old boys around Hillsboro, in postwar years, would say that on

A crude sketch of the Academy at Hillsboro. According to an historical marker, "about April 1841, Rev. Joseph Brown opened the first school of higher education. In 1842 the Academy was charted [chartered?] by the Presbyterian General Assembly for the University of Virginia. The brick building was constructed near this site." Reportedly, Federal artillery parked on the grounds the night prior to the battle of Droop Mountain. BRIAN ABBOTT

Modern view of the Academy site at Hillsboro. The structure is believed to have been approximately where the white building on the left is located. BRIAN ABBOTT

the evening prior to the battle the Federal "artillery parked on the academy lot," which was "adjacent to the Rev. M. D. Dunlap's place,"[83] although M. A. Dunlap later said "evidently Keepers' battery was not parked on the old acadmey lot the night before."[84] Dunlap's ancestor charged Averell's "staff and officers camped on the old academy lot and in the brick [academy] building itself."[85] From these various accounts there can be no doubt but that Averell's Brigade camped in assorted locations between Mill Point and Hillsboro the night before the battle of Droop Mountain, with only pickets in advanced position beyond Hillsboro, although John Blagg, 10th West Virginia Infantry, wrote [probably incorrectly] "...we camped in front of the Joe Beard house,"[86] which is on the southern outskirts of Hillsboro.

Averell's personal location for the night varies according to source. M. A.

Rare postwar photo of "The Manse," home of Rev. Mitchell D. Dunlap at Hillsboro. Reportedly, Gen. Averell spent time here the night before the battle of Droop Mountain. LANTY MCNEEL COLLECTION

Dunlap later claimed Averell spent the night at "The Manse," the home of his [Dunlap's] uncle, Rev. Mitchell D. Dunlap. Averell reportedly told the reverend he was going to give battle the next day. Of Averell, Rev. Dunlap said "no more perfect gentleman...ever lived."[87] On the other hand, postwar historian Calvin Price said Averell made his headquarters camp "along the hill on the western edge of the Levels and about where Confederate veteran M. J. McNeel" would reside in 1935. Price said Averell was a guest of Col. Paul McNeel that night, and although all the area townsfolk were southern sympathizers, all who met Averell "were charmed by him."[88] Reportedly, prior to his visit, Averell sent three young, gentlemanly officers to the McNeel home to advise the family that he had heard it was the home of an elder in the Presbyterian church, and as Averell was also an elder in said church, he wished for the family to know that they would not be under any apprehension or harm. One officer, acting as spokesman, also told the family that the officers were to remain and guard the house; which they did until the following morning. This kindly act caused the McNeel family to feel gratitude toward Averell.[89] Historian Price added, the only individuals present at the McNeel house at this time were the women, children, and slaves, as the area men had hidden to avoid capture.[90]

Among those avoiding being caught was Cornelius Stulting, 20-year-old brother of young C. L. Stulting. Although one source says Cornelius "hid in a log cabin on Droop Mountain to avoid Confederate service,"[91] C. L. Stulting remembered, "My brother [Cornelius] was about 18 years old [when the war broke out] and had just returned from Marlinton where he had enlisted in the Southern army [the only Stulting on Confederate records is Nicholas Stulting, Company F, 19th

Virginia Cavalry, who enlisted April 1863 at Hillsboro at the age of 44] and when he saw the Yankees all around the house he did not know what to do to keep them from taking him with them. He asked my father [Hermannus] for a chew of tobacco to make him sick and went upstairs to bed. Several Yankees went upstairs and looked at him and found him a very sick boy. When the balls were hitting the house [during the pursuit of the rebels from Mill Point] he came downstairs with a blanket around him and stood with the rest of us before the chimney. The Yankees did not take him therefore the chew of tobacco had the desired effect."[92]

Another amusing incident during Averell's encampment was told by postwar Federal Judge George W. McClintic, who was a young boy living in Hillsboro at the time of the battle. Many years later, in a dedication speech for the Droop Mountain Battlefield Park, he would relate how the Union army requisitioned the family cow on his farm near Hillsboro to feed the soldiers, and how his mother made a personal appeal to Averell to return the cow, which was "desperately needed for the milk for the family of young children." Gen. Averell granted the request. The cow lived for many years, and from that time on was known as "Old Averell."[93]

On the top of Droop Mountain, Col. Jackson spent most of the night arranging his line of defense, noting "no apparent enemy movement," although he claimed he spotted "large campfires in the neighborhood of Huntersville." Based upon his own conjecture, as well as information brought to him from scouts, Jackson estimated the force confronting him at 3,500 strong. Later that night, or the next morning, he convinced himself Averell had received reinforcements, bringing the enemy total to 7,500. Averell did not receive any reinforcements, but his army did vastly outnumber that of Jackson's, which then only came to about 750, with the remainder located "on the Locust Creek road or cut off in Pocahontas County." Jackson would get a little help in the early morning hours when Col. James A. Cochran arrived with the 14th Virginia Cavalry and "kindly released...[him]...from all picket duty." It was help well needed as "part of the [Federal] army...[would] take up position before daylight." The lines had now been drawn and Col. Jackson intended to make a hell of a stand on the crest of Droop Mountain the following day.

A 1928 view "looking from breastworks to Hillsboro, showing Yankee Flats and white house [barely discernible] to left, General Averell's headquarters on night before the battle." DROOP MOUNTAIN COMMISSION BOOKLET

Col. Paul McNeel house at Hillsboro. General Averell reportedly spent time here the night before the battle of Droop Mountain. MIKE SMITH, DROOP MOUNTAIN STATE PARK

Hill above Matthew John McNeel's spring house, believed to be where Averell's men camped the night before the battle at Droop. MIKE SMITH

CHAPTER 5

The Battle of Droop Mountain
November 6, 1863

Throughout the night of November 5 and into the early morning hours of Friday, November 6, 1863, dark, gloomy clouds threatened to bring rain to the mountains of Greenbrier and Pocahontas counties of West Virginia. Gen. Echols, at his camp near Spring Creek, 14 miles south of Droop Mountain, had continued to receive dispatches from Col. Jackson describing the Confederate defensive position at Droop Mountain, his intent to hold it, a request for aid, and an estimation of Averell's strength at 3,500. Echols needed no coaxing, as he was already well aware of Jackson's precarious situation, and resumed the march to Droop Mountain possibly as early as 2:00 A.M. [times vary according to source]. Pvt. Samuel D. Edmonds, Company B, 22nd Virginia Infantry, recalled, "...about 2 o'clock in [the] night the long roll beat and we had to get ready and march for Droop Mountain." 1st Lt. William Frederick Bahlmann, Company K, 22nd Virginia Infantry, said the men were rousted from their sleep at 2:30 A.M. Records of Company A, 23rd Battalion Virginia Infantry, state the group "marched from Camp Arbuckle in Greenbrier County to Droop Mountain."[1]

Since Col. George S. Patton had been placed in charge of Echols' Brigade, Maj. Robert Augustus ("Gus") Bailey took charge of the 22nd Virginia Infantry. According to James Henry Mays, Company F, 22nd Virginia: "how proud he

State highway historical marker alongside Route 219 at Droop Mountain Battlefield State Park. This park was actually established in 1928.
BRIAN ABBOTT

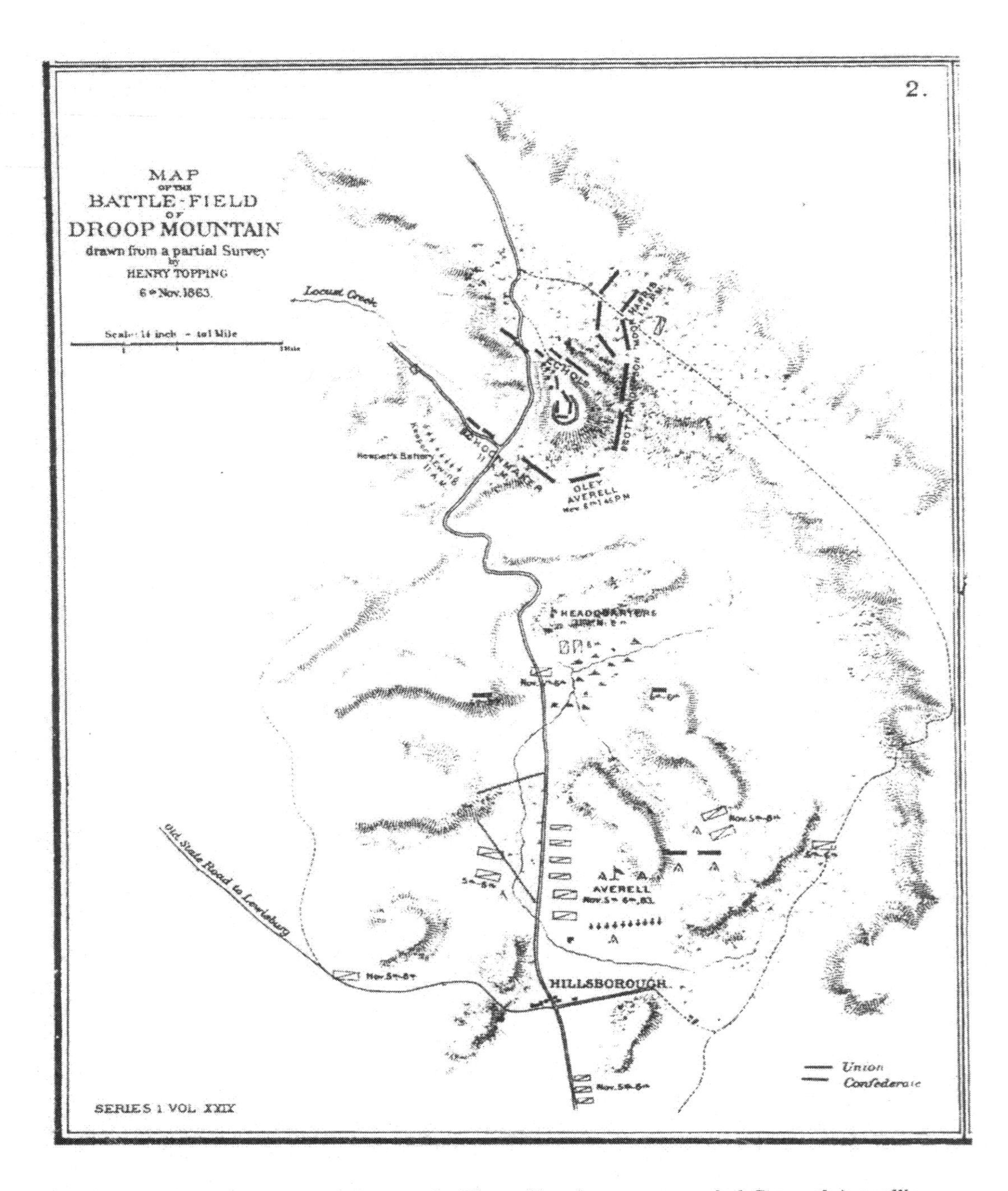

This map, drawn from a partial survey by Henry Topping, accompanied General Averell's report of the battle. The map is fairly accurate except that it shows Echols with only five artillery pieces, when, in fact, he had seven cannon on the field (and an eighth piece on a detached movement). This Federal map is the best and most well-known of the battle, and many variants of it have been used by historians and writers. WEST VIRGINIA STATE ARCHIVES

[Bailey] seemed to be of his little army as he rode along reviewing it as we were lined up ready for the march to Droop Mountain." Capt. John Clairborne McDonald, Company K, 22nd Virginia, was given command of Echols' brigade guard, which left 1st Lt. William F. Bahlmann in charge of Company K throughout the battle. Bahlmann noted that he and 2nd Lt. Martin L. Smiley were the only two officers of the company present, and that he received orders to detail two men from the company to cook. For this task Bahlmann selected Privates James Sarver, Jr. of Craig County, Virginia and Jackson Burwell Huddleston from Fayette County, West Virginia. This decision was not difficult to reach as both men were without shoes and therefore unable to march.[2]

Despite Major Bailey's optimism many of the soldiers in the ranks remained cautious, concerned they were possibly being led into a trap. Even Col. Patton, "who was never accused of being afraid or unwilling," reportedly expressed his displeasure with the maneuver, having stated at an earlier council of war his belief that the command should have remained at Lewisburg, where the open fields and rolling hills provided terrain more suitable to waging combat.

Pvt. Jackson Burwell Huddleston of Co. K (Fayette Rifles), 22nd Virginia Infantry. Being barefoot and therefore unable to march he was detailed to cook for his company at Droop Mountain.
JOHN ARBOGAST, WHITE SULPHUR SPRINGS, WEST VIRGINIA

After covering only two miles Echols realized the "old road from Hillsboro to Falling Springs [Renick]" was open, thereby exposing Jackson's distant right flank. In order to keep the route [which somewhat follows present-day Brownstown Road] guarded Echols detached the 26th (Edgar's) Battalion Virginia Infantry, as well as one rifled piece from Chapman's Battery under the command of Lt. John Campbell. According to a postwar source, Edgar's Battalion then numbered about 600 men [although Micajah Woods of Jackson's Battery gave a figure of 400], and Echols stationed them on the Locust Creek road. The exact position of Edgar's men is not clear, but Pvt. Joseph R. Perkins, Company E, 26th Battalion Virginia Infantry, would later write his command moved up the "old road" to where Winters Cochran would reside in 1935. Another account states Edgar's men "halted between the Davy Cochran farm and the Droop Church." One southern correspondent would later say "...as the enemy could easily get in our rear by taking a road on our right flank, it was necessary to detach the 26th Battalion to blockade it." Although this move was a strategic necessity, as it gave protection to Jackson's far right flank at Droop Mountain, it would also prove a veritable tragedy as most of Edgar's men were from Greenbrier and Monroe counties and possibly knew the terrain better than anyone else in Echols' command. As Averell never did threaten Jackson's distant right, the knowledge and firepower of Edgar's men, as well as the extra cannon, could have proven beneficial in the impending battle. But this is a moot point as Echols had no way of knowing Averell would not utilize the approach in

Old Droop Church on Brownstown Road. The 26th Battalion Virginia Infantry was stationed near this location during the battle. MIKE SMITH

General position on the Brownstown Road held by the 26th Battalion Virginia Infantry during the battle. MIKE SMITH

question.[3]

Having detached Edgar and the one artillery piece, Echols resumed the trek northward. Along the march Pvt. Mason ("Mase") V. Helms, Company K, 22nd Virginia Infantry, reported to 1st Lt. William F. Bahlmann and said, "I am sick." Bahlmann asked of Helms, "Why don't you go to Dr. [Gabriel C.] McDonald," the brigade surgeon, as "...he is the only man that can excuse you?" Helms, "a most efficient soldier," for some unknown reason, refused to report to the medical officer and would fall in action later that day at Droop Mountain. The anticipation of battle also caused many in the ranks to become nervous. One such example could be found in Privates Lewis M. Perry and Alexander Hall, both of Company B, 22nd Virginia. Neither had ever been in combat and were closely watched over by veterans Samuel D. Edmonds and Pvt. John Bennett of the company. Edmonds said Hall kept shaking profusely. When asked if he was cold, Hall replied in the negative; Edmonds then knew Hall had a case of the "Yankee chills." Reportedly, during the ensuing engagement, Perry ran to a nearby log house and nearly got shot. Edmonds had his own problems as well during the march, as he slipped and fell into a mudhole; since his rifle had been strapped over his shoulder, the muzzle of the weapon became clogged with mud.[4]

On the crest of Droop Mountain Col. Jackson had been occupied throughout the night and early morning hours aligning his troops, "...very advantageously beyond the crest and on the side of the mountain." Col. William W. Arnett had previously dismounted his 20th Virginia Cavalry and placed them in an advanced spot at about the center of the rebel line, on a projecting, elevated spur, which commanded the Lewisburg pike to their right. His men had spent the night erecting temporary log and rock breastworks, creating a horse-shoe shaped defensive barrier around the crest of the knob. Averell would later refer to the enemy's "rude breastworks of logs, stone, and earth."

Possibly as early as 6:00 A.M. the 14th Virginia Cavalry, or at least a portion of the regiment, had arrived at Col. "Jackson's camp at the eastern base of Droop" and a squadron was detached to guard the old, or Locust Creek road; another group relieved Jackson's pickets, and the remaining six companies would fight dismounted on the Confederate extreme left (four companies under Lt. Col. John Alexander Gibson) and center (two companies under Captains William Thomas Smith and Alpheus Paris McClung). Pvt. James Steel McClung, Company K, 14th Virginia Cavalry, recalled his regiment arrived at Droop on the evening of November 5, and were sent out on picket duty; then "next morning...were placed on the skirmish line, on the face of Droop Mountain."[5]

Private George Henry Clay Alderson, Company A, 14th Virginia Cavalry, claimed his regiment was "on the point of the mountain where the enemy [would evidently be] on three sides" of his company. Alderson also noted, "General Averil's [sic] campfires could be seen brilliantly burning in the valley below [near Hillsboro]." To protect the Confederate left and rear Jackson sent two companies of the 14th Virginia Cavalry to the extreme left flank. With this group was Capt. Edward Edmunds Bouldin, Company B, 14th Cavalry, who said his men were situated on the left, next to the mountains, where he sent two men to the woods on the mountainside in front of the squadron to "picket and reconnoiter the enemies lines." When the two men returned they reported finding a little

Capt. Edward Edmunds Bouldin of Co. B, 14th Virginia Cavalry. He claimed his men discovered the obscure road on the Confederate left and reported that information to General Echols. AUTHOR'S COLLECTION

known approach [Lobelia-Caesar's Mountain road] by which Averell could flank Jackson's left. Bouldin reportedly later relayed this information to Gen. Echols, whom he claimed "failed to act upon it."[6]

Also covering Jackson's left and rear could be found the dismounted members of the 19th Virginia Cavalry, under Lt. Col. William P. Thompson. Thompson would be assigned the task of preventing Averell from flanking the command and gaining the Confederate rear. Rounding out Jackson's 750-man defense was Lurty's Battery, placed on the far right of the line, to the rear of the 20th Virginia Cavalry, in a position Jackson described as "self protecting."

Previous fears of rain subsided as daylight broke sunny and clear, "with a high wind blowing." Apparently undaunted by Capt. Bouldin's report of the obscure road on the left, Col. Jackson ordered Lt. Col. Thompson to relieve the squadron of the 14th Virginia Cavalry which had been posted to the left. Thompson accomplished this by sending 10 men, under 1st Lt. William W. Boggs, Company I, 20th Virginia Cavalry; to fill their place on the line, and shortly afterward bolstered the position with 15 men serving with Lt. Mark V. Jarrett, Company G, 19th Virginia Cavalry. Jackson, however, realized the left was yet undermanned as it "was an important point of attack from the enemy," and ordered Lt. Col. Thompson to send to the left 100 cavalrymen, led by Capt. Jacob Williamson Marshall, Company I, 19th Virginia Cavalry, who was to dismount the men and send their horses to the rear. Marshall was instructed to hold the left "at all hazards." About 20 minutes later, Lt. Col. Thompson received another message from Col. Jackson ordering him to personally report to the left "with the residue of the cavalry" and hold it while awaiting further orders. Despite Col. Jackson's best efforts

Capt. Jacob Williamson Marshall of Co. I, 19th Virginia Cavalry. A pre-war resident of Randolph County, Marshall was appointed Captain on March 19, 1863, and temporarily kept the Federal attack on the Confederate left at Droop Mountain in check. His commanding officer said that during the affair he "distinguished himself for his coolness and calm disregard of danger." He survived the war, became a merchant, hotel operator, farmer and Census Deputy Collector for the Internal Revenue Service. He passed away in 1899. POCAHONTAS COUNTY HISTORICAL SOCIETY

to form an effective defense, M.A. Dunlap would later remark: "He did not have enough men to form a picket line across his front. If he had his men evenly distributed across his front they would have stood more than a rod apart."[7]

Averell's Movement

General Averell was also up early in the morning plotting his strategy. During the night he had sent orders to Major Thomas Gibson to detail at 6:30 A.M. one squadron of his battalion as an advance guard for Col. Augustus Moor. At the same hour, Moor received orders from Gen. Averell to move up from Mill Point to Hillsboro with the 28th Ohio Infantry, 10th West Virginia Infantry, and Keepers' Battery. Confirming this movement was John A. Blagg of the 10th West Virginia who remembered the infantry brigade "broke camp at dawn." After covering the distance to Hillsboro, Col. Moor's command halted for about an hour-and-a-half to rest.

At 7:30 A.M., in the following order; Gibson's Battalion, the 14th Pennsylvania Cavalry, the 8th West Virginia Mounted Infantry, Ewing's Battery, the 2nd West Virginia Mounted Infantry, the 3rd West Virginia Mounted Infantry, the ambulances and trains moved forward. The regiment in the rear, the 3rd West Virginia, was to provide a squadron to march in the rear of the train. Among this contingent was Francis Reader of the 2nd West Virginia, who recalled that "shortly after sunrise on the 6th Gen. Averell marched to Hillsboro at Droop Mountain..." Correspondent "M" also took note of this, writing, "...a little after sunrise, the brigade was moved forward to the little village of Hillsboro, where a few skirmishers were thrown forward and diverse maneuvers executed for the purpose of mystifying the enemy in regard to our numbers and intentions. The enemy in the meantime wasting a great deal of ammunition firing at our troops, whenever they uncovered themselves, but doing no damage whatever."[8]

It was also shortly after daylight that a group of Averell's officers left headquarters and walked to a small hill south of Hillsboro to reconnoiter the enemy. While so engaged some rebel snipers, hidden in the timber on the opposite side of the hill, opened fire upon the officers. Federal skirmishers were called forth and quickly drove the rebels back from the wooded area, which covered the entire south side of the Davis Poague farm, about where Henry Beard would reside in 1935.[9]

At about 8:00 A.M., Col. Moor received an order from Gen. Averell to "feel [the] enemy along [the] Lewisburg road." For this assignment Moor detached three companies of the 28th Ohio as skirmishers. They immediately advanced southward, skirmishing and driving back the enemy pickets. Reportedly, Capt. Charles Drach, Company A, 28th Ohio, "was the first, on that eventful day, to come into contact with rebel bullets" during this skirmishing movement. Col. William W. Arnett of the 20th Virginia Cavalry remembered this, claiming that as the Federals sent "small squads forward from time to time, slight skirmishing ensued." Intermittent firing continued across the Levels for about an hour, until the three companies of the 28th Ohio managed to clear the area to the foot of Droop Mountain, and the rebels fell back to their naturally fortified position on the summit. Correspondent "Union" said this skirmishing lasted three hours, and then Capt. Drach returned with his company of the 28th Ohio, having located the rebel position. Averell would relate, the Confederate position was "defined by a skirmishing attack of three companies of the 28th Ohio Infantry." As the Federal skirmishers waited at the base of the mountain for further orders, Gen. Averell and his staff rode quietly across the Levels to the base of the mountain, carefully surveying "every knoll and ravine and strip of woods." Occasionally halting to take note of the Confederate positions from the various standpoints, Averell carefully formulated his strategy. During Averell's reconnaissance he never spoke to his staff, except for the issuing of necessary orders, which he did "in that cool, deliberate manner peculiar to him, and so well calculated to inspire confidence." He took one final glance at all the enemy positions and approaches, then galloped away to formulate his plan of attack.[10]

Partial view of Averell's objective looking south-southwest from hill behind Confederate veteran M.J. McNeel's house at Hillsboro. The modern lookout tower designates the approximate right of the Confederate line, the finger ridge the center, and the park headquarters the left. It's believed Averell's men camped in these fields the night before the battle. MIKE SMITH, DROOP MOUNTAIN STATE PARK

Having ascertained the Confederate strength and positions based upon the skirmishing action and his personal examination of the terrain, Averell decided upon his troop dispositions "with wonderful promptness, there being no hesitation and no consulting with subordinates." Averell knew a direct frontal assault would be suicidal and formulated a plan of "textbook simplicity" involving three phases. First, Col. Augustus Moor would make a flanking movement with his infantry brigade, totaling 1,175 men and comprised of some 605 men (with 16 horses and nine mules) of his own 28th Ohio Infantry, under Lt. Col. Gottfried Becker; the 569 men (and 23 horses) of the 10th West Virginia Infantry of Col. Thomas M. Harris; Capt. Julius Jaehne's Company C, 16th Illinois Cavalry; and

A modern view from near the Beard house at Hillsboro, which illustrates the type of terrain the Federal troops had to cross in order to assault the mountain. The Confederate right, the area of the present lookout tower, is the cleared knoll on the top of the mountain near the center of this picture. BRIAN ABBOTT

Hill south of Hillsboro reportedly used by General Averell as the Federal command post to direct troop movements during the battle. The center of the Confederate position and left flank are barely visible on the left horizon. MIKE SMITH, DROOP MOUNTAIN STATE PARK

Lt. Abraham C. Merritt of the 68th New York Infantry, signal corps detachment. Merritt was to communicate with the center and left stations, as evidenced by Capt. Denicke, who reported, "I ordered Lieutenant Merritt to report to Colonel Moor...Lieutenant Merritt was instructed to communicate both with [the] center [Capt. Denicke] and [the] station on our left [Lt. Denicke]. I pointed out to the lieutenant the direction that the force was to take, and the nature of the ground was plainly visible." Pvt. John D. Sutton of Company F, 10th West Virginia Infantry, later claimed Moor took two pieces from Ewing's Battery but there is no valid, corroborating documentation to support this, although Sutton was indeed present. Another account states Moor took two of Ewings pieces and one Company of the 14th Pennsylvania Cavalry, although the cavalry notation is probably an indirect reference to Capt. Jaehne's Illinois cavalry, which was a part of Gibson's Battalion. Excavations have uncovered some shrapnel on the Confederate left flank but this may be nothing more than pieces of shells which burst nearby.

Moor was to fall back about a mile and move the column over an obscure six to nine mile circuitous route in which they would "ascend a range of hills which ran westward from Droop Mountain" and strike the unprotected enemy left flank, the true weak point in the Confederate line. While Moor's infantry brigade was en route, Col. James M. Schoonmaker would take his 14th Pennsylvania Cavalry and the Federal artillery to the left in order to make a demonstration against the Confederate right and center, which was intended to distract attention from Moor. In conjunction with this Lt. Denicke was to communicate with Lt. Merritt and the center signal station.

The third and final part of the plan was for the 2nd, 3rd and 8th West Virginia mounted infantry regiments, fighting dismounted [and probably under the direct command of Col. Oley], to make a direct frontal assault upon the Confederate center and right once Col. Moor had made contact with the rebel left. Capt. Denicke, of the signal corps, in the center; was to occupy various positions throughout the day, always accompanying Averell. With the exception of Gibson's Battalion, and a portion of the 8th West Virginia, which were held in reserve, almost all the Federals at Droop would fight on foot.

Colonel John H. Oley, soon after taking his place in the column, received orders to advance his 8th West Virginia Mounted Infantry past Hillsboro and relieve the three skirmishing companies of the 28th Ohio so they could join Moor in the flanking movement. Oley complied by sending forth three companies of the 8th West Virginia under the direct supervision of Major Hedgeman Slack, and gave support to Slack's detachment with five other companies. Within a short matter of time, Oley had cleared the hills of Confederate skirmishers and pickets up to the foot of Droop Mountain [although reports indicate the three companies of the 28th Ohio had already accomplished this].

Averell knew his proposed assault of Droop Mountain would be difficult. Describing the formidable enemy position on the mountain; Averell said, "the main road to Lewisburg runs over the northern slope [of Droop] which is partially cultivated nearly to the summit a distance of two-and-a-half miles from the foot. The highway is partially hidden in the views from the summit and base in strips of woodland. It is necessary to pass over low rolling hills and across bewildering ravines to reach the mountains in any direction." Indeed, Averell's mounted infantry, fighting dismounted, would have to cross an open area known as the Levels [later referred to as the Yankee Flats], an expansive farmland between Hillsboro and the mountain's base. For protection from enemy fire, the soldiers would have to utilize every rock, ravine, hollow, tree, stump, and piece of woods for cover. Averell further described Droop as "[our] position [was] similar to that upon South Mountain in Maryland, but stronger from natural difficulties and breastworks."

Averell did not have to worry too much about the civilian population. Historian Calvin Price recalled, "Nearly all the Levels' homes were occupied that day by the women and children. Nearly all the non-combatant men were hiding in the woods." William T. Collins, a baby at the time, whose father was away fighting in the war, was removed from the area by his mother, Nancy Lambert Collins. She purportedly "took the children in a canoe to get away and go to her family."[11]

Colonel Moor received his flanking order at 9:00 A.M. and gathered together his aggregate infantry force of 1,175 men. Meanwhile, Major Thomas Gibson, as previously ordered, had detached Capt. Julius Jaehne's Company C, 16th Illinois Cavalry, to assist Moor and moved his Independent Cavalry Battalion to Hillsboro, where he sent out two pickets of 20 men each one mile to the right and left of town. The remainder of his battalion served as a reserve.

In the meantime, Col. Schoonmaker had received his orders to move against the Confederate extreme right with his 14th Pennsylvania Cavalry and Keepers' Battery, with instructions to maintain a steady fire on the enemy to distract attention from Col. Moor. In order to hold communications with Averell, who would remain with the center signal corps station at various times throughout the day; Lt. Denicke, 68th New York Infantry, signal corps detachment, was to accompany Schoonmaker and remain in contact with the center station throughout the affair.

Arrival of Echols

At 9:00 A.M., the very same hour Moor and Averell's other field commanders were beginning to implement their orders, Gen. Echols and Col. Milton J. Ferguson arrived at Col. Jackson's position on the "western extremity of Droop." Pvt. Milton Butcher, Company B, 20th Virginia Cavalry [later Company B, 46th Battalion Virginia Cavalry], claimed Col. James Addison Cochran and his 14th Virginia Cavalry also made their appearance at about the same time as Echols, although most companies of the 14th had apparently arrived during the early morning hours. Jackson's men were obviously greatly relieved to see the reinforcement, which was met with loud rebel cheers, a display of battleflags, and the music of a band, all clearly audible and visible to the Federals in the fields below. Lt. Micajah Woods of Jackson's Battery said Echols found "...Jackson (W.L.) was holding the enemy in check with 600 men on a very powerful position on Droop Mtn." Averell, however, was apparently unperturbed by the increased enemy numbers and did not make any changes in his battle plans.[12]

Echols immediately assumed field command, briefly studied and approved Jackson's troop dispositions and placed Col. Patton in charge of the 1st [Echols'] Brigade, which arrived on the field a few minutes after Echols.

Patton, following Echols' orders, instructed Major William Blessing's 23rd Battalion Virginia Infantry to take position "to the right of the turnpike road at the summit of the mountain," which placed them on the right of the rebel line. Major Blessing described the position as being on "the right of the main road leading from Mill Point to Lewisburg." Shortly after arriving at the designated post, Blessing received orders to deploy his two flanking companies, Company A (Smyth County Sharpshooters) and Company F (under Capt. William Killinger), and advance them as skirmishers about 400 yards in advance of the battalion. About a half hour later, at about 9:30 A.M., the skirmishers were recalled. Blessing was also ordered to detach Company C, from Tazewell County, Virginia (under Capt. Francis M. Peery), for picket duty, and for them to report to Col. Patton for instructions.

Echols placed all of the artillery on the field under Major William L. McLaughlin, who was told to move Chapman's and Jackson's guns "to the front, just beyond the summit of the mountain, near a point where Col. Jackson had placed" Capt. Lurty's two pieces. After making a quick reconnaissance of the position, McLaughlin said he found Lurty's two howitzers "on a projecting spur of the mountain commanding the approaches from the front." McLaughlin decided to put Capt. Jackson's two guns with Lurty and ordered Capt. Chapman to take his three pieces "to the hill in rear and in easy supporting distance, one piece" so placed as to cover all approaches to the right. A postwar historian said the rebel artillery was "posted upon both sides of the road at the crest of the mountain." Lt. Randolph Blain of Jackson's Battery said, "...at once our pieces, one Parrott and one three-inch rifle, the first commanded by myself and the second by Lieutenant [Micajah] Woods...were ordered on the field...our position was upon the brow of the mountain & I think a most elegant one." Micajah Woods agreed with Blain and wrote, "The position for artillery was splendid..."[13]

According to William F. Bahlmann, and others, the 22nd Virginia Infantry entered the affair 550 strong and under the immediate command of Major Robert Augustus ("Gus") Bailey. As earlier noted, a portion of Company D was not present. The regiment was placed to the rear of the artillery as a support. Pvt. Samuel D. Edmonds, Company B, 22nd Virginia Infantry, remembered being stationed "behind a knob," where, while awaiting further orders, he cleaned the mud acquired earlier in the day from his gun barrel. This completed Echols' battle line and according to a Federal source, after Col. Patton viewed the situation he boasted that the Confederate position was so strong his 22nd Virginia Infantry alone could hold it against Averell's entire command, although it is quite doubtful Patton ever made such a claim. Adding to this doubt is a story that Patton felt the estimate of enemy strength at 3000 by Capt. McNeel's scouts was too low as there were "too many campfires." Throughout most of the day Echols would command the Confederate right, Col. Jackson the center, and Lt. Col. Thompson on the left.[14]

While the 22nd Virginia stood deployed in the woods, Adjutant Noyes Rand came to Lt. William F. Bahlmann of Company K and ordered him to assume command of two volunteers from each company and "cover a road." Bahlmann flatly refused, stating he was the only officer present to command his company. Delegating the assignment to others, Bahlmann turned to the two tallest men

This photo, taken in Richmond, Virginia, in January 1862, denotes two officers of Co. H (Kanawha Riflemen), 22nd Virginia Infantry. To the left is 2nd Lt. John P. Donaldson, who was wounded in the side and shoulder at Droop Mountain. On the right is Adjutant Noyes Rand, who was praised by Colonel Patton for his gallantry in the fight. LOUISE TESTERMAN COLLECTION, UNIVERSITY ARCHIVES COLLECTION, VIRGINIA POLYTECHNIC INSTITUTE AND STATE UNIVERSITY LIBRARIES

standing near him, Pvt. Hiram Hill Painter, who was 5'10", and Pvt. Edward T. Bones, height unknown. Bahlmann told them, "Boys, you would better go." The two asked no questions, threw their guns to right shoulder arms, and left with Adjutant Rand. Bahlmann later received information that this detachment of approximately 20 enlisted men and two lieutenants "scarcely got under fire."[15]

According to Echols, his men combined with Jackson's brought the total engaged Confederate strength to about 1,700 men and seven artillery pieces (previously eight, with one having been detached with Edgar's Battalion), and he believed he was facing 7,000 Federals. Unfortunately for Echols, much of his manpower was comprised of dismounted cavalry, good soldiers but not generally reliable in a stand-up infantry-style fight. Yet Echols remained confident of his position, where he explained he had an extensive partial view of the Little Levels, Hillsboro, and the enemy lines at a distance of about two or two-and-a-half miles. Quartermaster Thomas Algernon Roberts of the 22nd Virginia Infantry recalled, "During the battle I was with General Echols on an eminence on the right of the road overlooking the battlefield."[16]

Although Droop Mountain was a virtually impregnable, natural fortress, Echols seemed oblivious to the obscure country road which ran across the west side of the mountain, despite various reports circulated about its existence. In addition to the earlier report by Capt. Bouldin of the 14th Virginia Cavalry, information on the route came from Capt. James Monroe McNeil, Company D (Nicholas Blues), 22nd Virginia Infantry, who had grown up on Droop Mountain hunting squirrels and probably knew the terrain better than any one person in the Con-

The battlefield as seen looking north from the park lookout tower. Confederates posted here could see the Federal artillery, which was in the general vicinity of the white-roofed barn in the distant right. Route 219 [Lewisburg Pike/Midland Trail] can be seen zigzagging down the mountain at left and then cutting to the right through the center of this picture, with Hillsboro in the far distance. GARY BAYS

The lookout tower located on the Confederate right flank. From this location the rebel army could clearly view the Federals at Hillsboro and the Federal artillery.
GARY BAYS

federate army. He was also an accomplished guitarist, song writer and poet whose compositions were often sung around the campfires of the regiment. McNeil reportedly came to his commanding officer and informed him of the little known road on the left by which the Federals could flank the rebels. The commanding officer purportedly told McNeil "that if the Virginia command wanted any advice from a Captain they would let him know." Another version states McNeil was advised, "when we want your advice, we'll promote you to the rank of General." It is not known if such stories are hearsay or truth, but if they did indeed occur then Echols was certainly guilty of neglecting his left flank.[17]

Sometime following Echols' arrival on the field Col. Milton J. Ferguson, commanding a cavalry brigade detachment, led a daring reconnaissance of the Yankee pickets near Hillsboro. Utilizing ten volunteers, including Sgt. James Zechariah McChesney of Company H, 14th Virginia Cavalry, the detachment somehow managed to get by the Federal pickets at the base of the mountain without attracting attention, and approached within 250 yards of the Beard house at Hillsboro. A young lady exited the house to relay information to them as McChesney and two troopers rode toward the house to ascertain her message. To avoid being spotted by Yankee pickets, the young lady hid behind the corner of the spring house; but, one Yankee rode around, took notice of her and fired his weapon. The "ball struck the fence by her side," which compelled her to flee to the safety of the house, having already given McChesney "the points." Ferguson, McChesney, and the other troopers then returned safely to the Confederate lines.[18]

Sgt. James Zechariah McChesney of Co. H, 14th Virginia Cavalry. He made a successful scouting expedition to the vicinity of the Beard house on the morning of the battle. BILL TURNER, CLINTON, MARYLAND

Yankee Advance

While Echols had been occupied aligning his and Jackson's men, Col. James M. Schoonmaker had been advancing his 14th Pennsylvania Cavalry; Keepers' Battery; and Lt. Denicke's 68th New York Infantry, signal corps, toward the Confederate right, driving enemy pickets and pushing back the rebel skirmish line. Schoonmaker moved Keepers' six guns into the "best possible position," which Major McLaughlin of the rebel artillery said was "within about five-eighths of a mile" of his guns. A postwar historian said the Federal artillery was posted, "a short distance up [the] mountain and in a field on left of road where Mr. Bruffey now [1958] lives." Col. Schoonmaker remarked the opposing batteries were "scarcely 2000 yards apart," but McLaughlin's guns were on an elevation at least 500 feet higher than Keepers', which would make it nearly impossible for the Federal pieces to do effective work. A correspondent for the *Ironton Register* known only as "C," but probably a member of the 2nd West Virginia Mounted Infantry, wrote that: "The nature of the ground was such that our artillery could be of but little service. Capt. Keepers' however,...[would keep]...the enemy's artillery pretty busy." Confederate gunner Micajah Woods, Jackson's Battery wrote, "...ours were the only long range [Confederate] guns. So [Lt. Randolph] Blain & myself were masters of the situation on our side. The enemy...[would keep]...on us with a six gun Parrott Battery 2250 yards from us from the only position they could get, which was much inferior to ours, we having 12 degrees natural elevation." Woods contradicted his own estimation of the enemy artillery firepower in a later letter, writing the Federals had "four 10 Pdr. Parrotts 2300 yards distant." Also, in later correspondence, he wrote, "The guns of Lurty's and Chapman's Batteries were of not efficient range to reach them with effect so it fell to the lot of our Parrott & 3-inch Yankee guns to return the rude messengers sent us."[19]

A 1928 view of the position of Captain Ewing's Battery G, 1st West Virginia Light Artillery, "320 poles east of Lurty's Battery, by grove of trees." DROOP MOUNTAIN COMMISSION BOOKLET

View from the park lookout tower facing northeasterly. Keepers' Federal artillery is believed to have been posted somewhere in the vicinity of the white-roofed barn near the center of this photo. GARY BAYS

Locust Creek as viewed from the Confederate breastworks at the center of the line, circa 1928. DROOP MOUNTAIN COMMISSION BOOKLET

Traces of the original Locust Creek road can still be seen. MIKE SMITH, DROOP MOUNTAIN STATE PARK

One postwar historian says the 14th Pennsylvania Cavalry, fighting dismounted, was "near the Locust Creek bridge" and Keepers was on the "high ground above Beard's mill." Another historian says they were "on the left of the road around [the] head of Locust Creek...because of difficult ground, they failed to get into the battle." 1st Sgt. George H. Mowrer, Company A, 14th Pennsylvania Cavalry, said his company "was engaged in supporting the battery," while Corp. George W. Arrison, Company D, 14th Pennsylvania Cavalry, later claimed the group sent under Schoonmaker also included the 2nd West Virginia Mounted Infantry. However, there is no evidence to indicate the 2nd West Virginia went anywhere other than with the mounted infantry regiments, fighting dismounted, which were to attack the rebel center. Lt. Denicke, accompanying Schoonmaker, "took his station on a knob occupied by one of our batteries. From this point he [would keep]...constant communication with [the] center station..."[20]

Although no official correspondence could be located to verify the exact location of Ewing's Battery during this time, M.A. Dunlap, who was personally acquainted with veterans of the battle, said Ewing was placed "on the ridge on the Lockridge place on the ridge just east of the first cut in the road." Another postwar historian said Ewing was "on the left of the pike between Hillsboro and the foot of Droop Mountain." Apparently Ewing was at the same, or nearly the same, location as Keepers. According to M.A. Dunlap, Capt. Ewing ordered [Dr.] Pvt. William O. Hartshorne, Ewing's nephew and aide, to keep moving to avoid being shot by the enemy. Ewing purportedly threatened to slap Hartshorne with the flat of his sword if he continued to pause while they searched for a suitable location for the guns.[21]

Soon after Col. Schoonmaker got his men in position Sgt. John Evans and Pvt. William A. R. Davis, both of Company G, 14th Pennsylvania Cavalry, caught the aroma of fresh cooked roast beef and potatoes emanating from the rebel camp. Putting on captured [probably at Mill Point] rebel uniforms and sidearms, the two managed to nonchalantly walk into the rebel camp, steal some hot food, and run back to their regiment. Capt. William W. Murphy, Company G, 14th Pennsylvania Cavalry then distributed the meal among his company. After the battle it would be discovered the stolen meal had been prepared by Capt. William Lamb McNeel , Company F, 19th Virginia Cavalry, for his commanding officer. Davis said it "was a little mean to capture so good a feast provided by so distinguished a man as Captain McNeel" for his commander.[22]

Roar of the Big Guns

As earlier noted, at about 9:30 Major Blessing's skirmishers of the 23rd Battalion Virginia Infantry were recalled and remained in position for an hour, until about 10:30, when they were ordered to march to the support of the battery, about 400 yards to Blessing's left. According to Pvt. Milton Butcher, Company B, 20th Virginia Cavalry [later Company B, 46th Battalion Virginia Cavalry], at about 10:00 A.M., "lively skirmishing began, first on the eastern face of the mountain, continuing southward near the Locust Creek Mill, and north to the Black Mountain." By 11:00 A.M., Capt. John V. Keepers' Federal guns were in place, and,

An authentic reproduction of a Parrott gun used by the southern artillerists at Droop Mountain. Located on the Confederate left near the residence of the park superintendent. GARY BAYS

according to southern artillerist Major McLaughlin, opened upon the Confederate cavalry horses and batteries. Apparently, with this volley, the battle of Droop Mountain officially began, although a premature report in the *Pittsburgh Gazette* claimed, "Schoonmaker opened the fight with his regiment." Maj. Gen. Sam Jones had been in Tennessee when Echols had informed him of the Federal advance through Pocahontas County. As a result of this information, he hurried back towards West Virginia and arrived at Dublin "about the hour the firing commenced at Droop Mountain." Nine-year old Nickson McCoy, son of Pvt. George Washington McCoy, Company F, 19th Virginia Cavalry, had climbed a tree near his home at Hillsboro to watch the battle and fell from the branches when the first cannon was fired. Purportedly, some Confederate soldiers on Droop Mountain were just getting their campfires going good, cooking a pot of chicken soup when one of the first U.S. artillery shells either directly struck the cooking pot, or fell nearby, startling the rebs. Echols, on the other hand, claimed the rebel pieces fired the initial artillery rounds at the Federals "in the valley below," to which Keepers replied "vigorously and rapidly." Pvt. James Steele McClung, Company K, 14th Virginia Cavalry, was on Echols' skirmish line and reported, "Along about 10:00 o'clock the Yankees ran a piece of artillery out in the field and began to fire. The sergeant said, 'Scatter out boys, or we will all be killed.' We went up the mountain and moved fast too, and where we topped the mountain, there was Capt. Eagle [Eakle] who said, 'Boys, go back and get something to eat.'" The guns of Lurty, Chapman, and Jackson then engaged in a rousing artillery duel with Keepers [and apparently with Ewing as well] for "about a half hour," until around 11:30 A.M. Reportedly, 2nd Corp. Timothy McCune, Lurty's Battery, asked Capt. Lurty for permission to fire his gun at a group of Yankees and later learned the shell landed in or on a big pot of chicken soup they were cooking, apparently copying the incident the Federals had earlier fulfilled. Micajah Woods of Jackson's Battery stated, "we were being thoroughly protected by the brow of our position, upon which we would run our guns by hand, after each rebound which put it in between us & the enemy's shots & served as a complete cover for guns, man, and horses." He also later wrote, "Our position was several degrees higher than the enemy's and was furthermore advantageous in affording very complete cover behind the brow for horses, carriages, and men. Each rebound of our guns, which we would run by hand to the front, would bring us out of danger in loading." Lt. Blain, also of Jackson's Battery, said, "...in a few moments [after placing the guns] the enemy opened upon us with four pieces." When the band of the 14th Virginia Cavalry struck up "Nelly Gray" the Federal artillery ceased firing, but as soon as the tune ended, the Federal guns re-engaged "and sent the deadly shots into our midst," reported Sgt. James Z. McChesney of the 14th Virginia Cavalry.[23]

While Capt. Ewing was placing the guns of Battery G, 1st West Virginia Light Artillery, as well as afterwards, a small rebel cannon "greatly annoyed" his battery, and eventually killed most or all of Ewing's horses, as well as the mount of [Dr.] Pvt. Hartshorne. After the battle a wounded Yankee prisoner would tell Sgt. McChesney of the 14th Virginia Cavalry the rebel artillery "did the best firing they ever saw, that the first shot dismounted one of their pieces."[24]

Artillery would roar throughout the area during the day, and could be heard miles away. Pvt. Joseph R. Perkins, Company E, 26th Battalion Virginia Infantry, said his command was stationed in a field below where Winters Cochran resided in 1935 and, "with their field glass they could see the Yankee artillery and hear the rattle of small arms." Sixteen-year-old Agnes Sharp, who lived in a log house near the mouth of Coal Pit Hollow on Beaver Creek road, said "the roar of the cannons could be plainly heard." The Buckley brothers from the same area as Agnes Sharp, "climbed Buckley Mountain [which is about eight miles north of the Confederate artillery position] to watch the cannon fire on Droop Mountain." Confederate Capt. James Monroe McNeil's brother, Clairborne McNeil, of Buckeye, West Virginia, having served two years in Confederate service, was home on indefinite leave of absence and climbed a height near his home at Bridgers Notch, north of Stamping Creek, to watch the battle. In a letter John R. Woods wrote to Micajah Woods he stated, "Walter said he heard distinctly the cannon on Friday as he was on the road & after reach[ing] Millborough." Several families living in the low place formed by Locust Creek stayed there all day under the severe artillery fire. Author Pearl Buck, a descendant of Cornelius Stulting, wrote Cornelius had hidden in a log cabin on Droop to avoid Confederate service, and, "...all day and all night the cannon roared back and forth over the mountain and the [Stulting] family sat in fear, scarcely able to pray even, lest Cornelius be caught in his hiding place."[25]

Despite valiant efforts, the Federal gunners failed to reach the enemy batteries, while McLaughlin's artillerists were sending shells "right on target." Capt. Denicke noted that the Confederate artillery kept up a heavy fire at Keepers' Battery, doubtless attracted by the signal corps flag. Schoonmaker continued to

Various artillery shell fuses found on the Droop Mountain battlefield. (left to right): underplug for Bormann fuse; Read fuse and Bormann time fuse. GARY BAYS

At right: Confederate copper-brass time fuse adapter for the Read elongated shell of case shot. A number of these have been found in the general vicinity of Keepers' Federal battery, indicating the Confederate artillery was fairly accurate in the battle. BRIAN ABBOTT

Below: Remains of spent artillery friction primers, which were used to fire the cannon. These were found in the location of the Confederate artillery at Droop Mountain. GARY BAYS

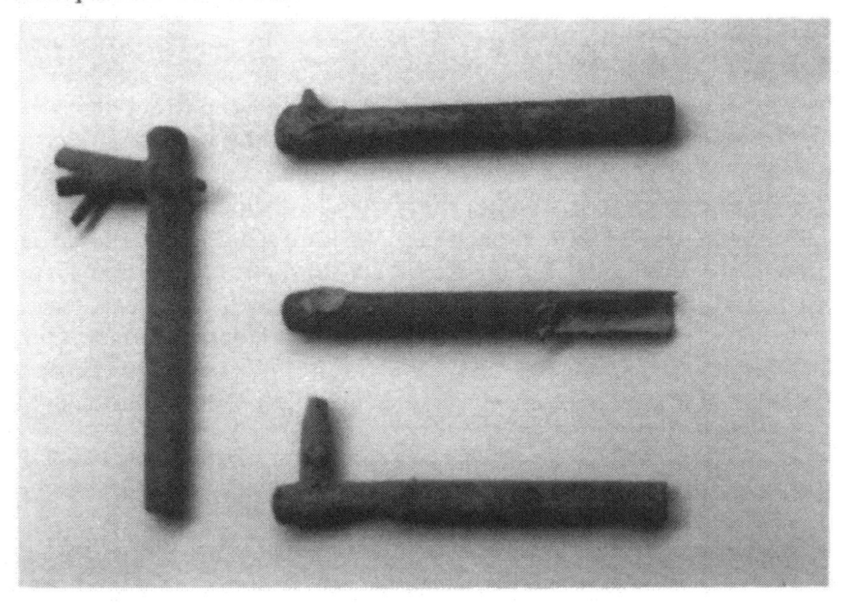

utilize Keepers' Battery, but concluded it was a waste of ammunition, withdrew two sections of the battery, and placed the remaining section behind protective cover, although the rebel gunners continued to pound his original position. Throughout this maneuver he maintained a "brisk fire, which occupied the attention of the enemy's battery entirely." Major McLaughlin apparently failed to realize Schoonmaker's intent and felt he had "silenced and driven rapidly from the field" Keepers' guns. Micajah Woods remembered, "The firing was very heavy, endeavoring to draw the attention of our forces while they were massing their infantry on our left under cover of the woods. In about an hour, around noon, the whole force, infantry and artillery, was engaged and we succeeded in driving the enemy's battery from the field entirely by our shell, which were thrown beautifully." One Confederate report said "Desultory fire [was] kept upon [the] enemy infantry and cavalry as they came within range." Capt. Chapman then moved his battery up and joined with Lurty and Jackson's artillery. After the war M.A. Dunlap said, "Almost everyone I have heard speak of the battle speak of the good shooting the Confederate artillery did."[26]

M.A. Dunlap claimed the Federal artillery shot "too high" [which would contradict official reports] and some shells passed over to the west side of the mountain and landed near the Sinks and the George Hill place. One shell in particular is known to have landed at George Hill's at the western foot of Droop and was discovered by some family members after the war. An Irishman there by the name of Mike took the shell inside the house while some young boys took out the fuse, placed it on the porch, and fired it up, an activity which attracted the family to the porch. Inside the house Mike poked at the shell with a piece of iron and it exploded, raising the upper floor of the house from the joist. Someone yelled inside and Mike answered, "Ol'm kilt! Ol'm kilt." When the smoke cleared Mike was found alive, his hair burned off and his big toe pinned to the floor by a piece of shrapnel.[27]

Purportedly, well over 50 years ago, a chestnut snag tree could be viewed along the right side of Rt. 219 [Lewisburg Pike] just as the road descended from the top of Droop towards Hillsboro. The tree had a hole in it caused by an artillery shell from the battle.

While the "brisk artillery duel" was in progress, Averell's skirmishers kept up a "frequent and heavy" fire against Echols' line and Col. Moor continued to move his infantry column toward the Confederate left. Echols recalled, "Soon after the opening of the artillery, skirmishing commenced along the line...."

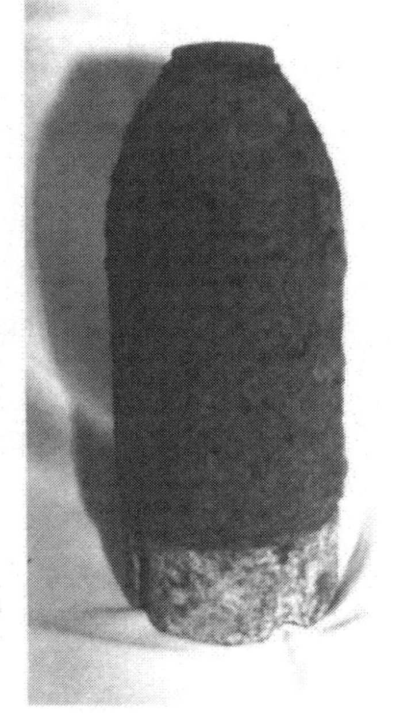

An unexploded Confederate Burton artillery shell recovered at the northern base of Droop Mountain by Mrs. Wayne Varner. At present this is the only known specimen of a Burton shell found at Droop Mountain. COURTESY WAYNE VARNER, BEVERLY, WEST VIRGINIA, BRIAN ABBOTT PHOTO

Three dirt-encrusted artillery rounds found at the foot of Droop Mountain. Left to right: 12-lb. Bormann cannonball; 12-inch Read shell and another 12-lb. Bormann cannonball.The fuse on both cannonballs remains unpunched, indicating they may have been used as solid shot or the cannoneer, in the heat of battle, may have forgotten to punch them. TIM MCKINNEY

At about 12:00 noon Lt. Col. Alexander Scott, having moved to the front, was ordered to dismount his 2nd West Virginia Mounted Infantry, then numbering about 395 men. He was also told to "detach one company and post them in an elevated position as a guard" for the horses. Postwar historian Calvin Price claimed the 2nd, 3rd, and 8th West Virginia mounted infantry regiments "were placed to the right of the [Lewisburg] pike about the Renick place, out of sight of the rebels...dismounted...every fourth man holding horses while the men lay on their arms awaiting orders." Following Scott's compliance with the order, and the detachment of some additional men, the 2nd West Virginia went into action with about 200 men, the "remnants" of the regiment. Scott soon afterward received orders to take position between the 3rd and 8th West Virginia regiments "and [to] act as a support for..." them.[28]

The battle began to heat up around 1:00 P.M. as Col. John H. Oley received orders to place his 8th West Virginia Mounted Infantry to the left of the 2nd and 3rd West Virginia. The three units, fighting dismounted, were to assault Echols' line at the center and center right, in conjunction with Moor's strike on the left and Schoonmaker's pounding of the right. Oley complained his location "was an exceedingly difficult one. The side of the mountain in our front was bare of trees, fences or any protection from fire. The ascent was very steep..."

Captain Francis Mathers, Company I, 8th West Virginia, recalled the movement of the three West Virginia regiments as they advanced from Hillsboro to Droop Mountain, writing, "the skirmishers moved off in splendid style, with the supporting line close behind them, and in a very short time the firing became brisk and animated...the regiments in front...moved slowly but it was a steady onward movement, over a hill, across a field, through the woods, and across ravines, the rebels retiring as if to husband their strength for the strong position."[29]

Lieutenant Colonel Scott's 2nd West Virginia managed to struggle to the foot of Droop Mountain and passed the 8th West Virginia, leaving Oley's men to their left. Company B of the 2nd West Virginia was deployed as skirmishers to relieve those of the 8th West Virginia on the extreme left of the 2nd West Virginia, "on the bluff in the woods." As the balance of the 2nd West Virginia advanced to the right of Company B, the company "filed right to join the 2nd West Virginia," then nearing the top of the mountain. In an attempt to locate the 3rd West Virginia, which had already partially advanced up the slope, Scott ordered his men forward "through briers, tree-tops, and obstacles of various kinds." Upon reaching an open piece of ground, Scott reformed the 2nd West Virginia and advanced farther up the hill, where he "formed in line at the left of the 3rd West Virginia." Simultaneously, at 1:00 P.M., Col. Schoonmaker correctly assumed Col. Moor would probably soon make contact with the Confederate left. Immediately returning the 14th Pennsylvania Cavalry, Schoonmaker moved his portion of the command at a double-quick, along with two artillery sections, and advanced toward the center of the Federal line, placing the artillery "in position so as to make several very effective shots on the crest of the hill before the point was carried..." Micajah Woods, Jackson's Battery, recalled this, writing, "They attacked in heavy force and determination bringing a Battery in position in our front to occupy the attention of the center...again the distance was too great for all the guns but those of Jackson's Battery. This time we literally knocked them from the field forcing them to leave two guns unattended by men or horses which we have good reason to believe were disabled—our shot and shell being thrown with great precision." Samuel D. Edmonds, 22nd Virginia Infantry, said he observed "one of our 24 pound shells appear to burst on their cannon" and made them hunt for another place.[30]

Lieutenant Colonel Scott, 2nd West Virginia, said Battery B, 1st West Virginia Light Artillery [Keepers' Battery] on the Federal left, opened up and an "interesting and lively artillery duel continued for some time." It was apparently sometime during this exchange that 2nd Lt. Joseph W. Daniels, of Battery B, was

Four base sections of artillery shells found on the Confederate right and the Federal artillery position at Droop Mountain. All are Read-Parrott shells except for a Hotchkiss in lower left. GARY BAYS

A Federal artillery Schenkl shell with a copper-brass percussion fuse. Remains of this type of shell have been recovered from the Confederate center and right on the Droop Mountain battlefield. GARY BAYS

killed instantly, decapitated by an artillery round while standing beside Col. Schoonmaker and "working his section manfully without fear of danger." Also standing near the lieutenant and witnessing this tragedy was his 19-year-old son, Pvt. [later Corp.] Winston M. Daniels of the battery. M.A. Dunlap claimed that while Capt. Ewing and [Dr.] Pvt. William O. Hartshorne were standing with their backs to the cut in the pike "a lieutenant [Daniels] was standing in the road in front of them talking" when a rebel artillery round took off his head. Sgt. John W. Worthington of the battery spoke very highly of the Christian virtues of Daniels. Sgt. James Z. McChesney, Company H, 14th Virginia Cavalry, said the Yankees later claimed "the rebel artillery shot at a single man on horseback and killed him," undoubtedly referring to Daniels. Micajah Woods related, "They [U.S.] admit that their artillery was completely whipped. Two of their Parrott guns were dismounted on the field. An aide-de-camp of Gen. Averell had his head carried with a shell—a lieutenant in one of their batteries was killed and 15 men besides killed & wounded," although Keepers' Battery would officially report only two killed and five wounded in the battle. Along with Daniels, the other member of the battery killed was Pvt. James Jackson, son of Joseph Jackson of Hanging Rock near Ironton, Ohio. Hiram Depew said these members of the battery "were killed and wounded, while discharging their duties upon the field of battle, the position we were placed in, was far inferior to that of the enemy's. And owing to the advantage which the enemy's artillery had over our Battery, it is, indeed, a mystery how we escaped as well as we did." Capt. John V. Keepers suffered a

Two specimens of iron canister balls (grape shot) fired from the artillery at Droop Mountain. The larger one is about the size of a golf ball. These were usually fired at short range targets. GARY BAYS

partial loss of hearing in both ears caused by the "concussion of the artillery" of his own battery.[31]

Major McLaughlin of the Confederate artillery would later claim the Federals advanced three guns to their "previously occupied position" and opened upon his batteries. Micajah Woods, Jackson's Battery, stated; "The enemy made desperate efforts to turn our right and reach the rear & capture the artillery. While the infantry were engaged hottest, again they brought a battery of 10 pdr. Parrotts and opened upon us furiously from the same position. Again we fairly knocked them from the field. [Lt.] Blain did good execution with his gun (or his gunner did) for he did not fire a single shot [himself]. My gun did most admirably, I fired it myself & sited every shot, throwing 120 shell. It drew the particular attention of all." Major McLaughlin confirmed Capt. Thomas E. Jackson responded with his two pieces, as did Capt. George B. Chapman "with his rifled piece," and the combination silenced the Federal guns.[32]

V.M.I. (Virginia Military Institute) cadet button found in 1986 by Michael Hively in the soil from a modern grave being dug in the private cemetery in the park. The Confederate artillery was posted in this location during the battle. Many former cadets in the Confederate army wore such buttons and the military school also issued material to many soldiers. RICHARD ANDRE

Standing near the Southern guns was citizen George Hill, who had been trapped within Confederate lines earlier in the day and walked to the Confederate artillery position. He said every time the cannon fired the vibration from the blast would knock him down on the ground. After experiencing this procedure repeatedly, Hill took note of the southern artillerists flexing their bodies to absorb the shock. Imitating this method Hill managed to retain his balance during the remaining artillery fire.[33]

While Schoonmaker and Keepers attempted to keep the rebel artillery busy; the 2nd, 3rd, and 8th West Virginia regiments slowly struggled to gain position in order to support Col. Moor when he made his appearance. Col. Oley remarked the Confederate guns "commanded every inch of the ground, and their sharpshooters were on the summit behind a breastwork of logs, consequently there was a slight hesitation of my men at the start and a disposition to get too far to the right, in the line of the 2nd West Virginia, where trees and brush offered some protection." Oley felt his best alternative was to keep his men to the left in order to occupy the attention of the rebels, as well as to keep the troops from massing too extensively.

Lieutenant Colonel John J. Polsley, 8th West Virginia, who was not present, would later interpret things quite differently, charging Oley with cowardice by saying, "the 2nd [West Virginia] was on the left [actually they were on the right] of the 8th [West Virginia] and became engaged first with the enemy. Col. Oley hesitated. [Major Hedgeman] Slack urged him to advance, but he still hesitated. The 2nd [West Virginia] at last became hotly pressed when Major Slack advanced the regiment forward. Oley halted it. The men were all eager to go on when Oley halted the regiment." On a later date Polsley would become more critical of Col. Oley, writing, "[Major] Slack ordered it [8th West Virginia] on again and took the greater part of it up just in the nick of time, and Oley followed with the balance. I have understood that the General was not at all satisfied with the Colonel's conduct on that day. I am satisfied that, as he has never done it, so he never will lead the Regiment in a fight when he can avoid it." Despite Polsley's charges, Capt. Francis Mathers, Company I, 8th West Virginia, said "the 2nd and 8th West Virginia moved forward until they got within point blank range of the enemy sharpshooters, the 8th exposed to a galling fire from the rebel breastworks, and right [directly] under the rebel battery that opened...with shot and shell." The regiment was partially protected by the woods, and Capt. Mathers said the men lay on the ground to permit the artillery projectiles to pass over top of them. Capt. Mathers also reported his company and the 8th West Virginia "had to charge up the open fall of the mountain, exposed to the grape and canister of the Rebel battery as well as the fire of the rebel sharpshooters."[34]

Having already partially advanced up the mountainside, Lt. Col. Scott's 2nd West Virginia halted for a brief rest. Scott consulted with Lt. Col. Frank Thompson of the 3rd West Virginia and the two agreed to "advance at once on the enemies works on the crest of the hill." Soon afterward, the entire line, including the 2nd, 3rd, and 8th West Virginia regiments "moved steadily up" as the rebel artillery fire and skirmishing remained constant. Pvt. Alfred McKeever, Company C, 3rd West Virginia, whose half-brother Capt. James McNeil led the Nicholas Blues (Company D, 22nd Virginia Infantry) in the opposing force, had been equipped with some type of new rifle. At the beginning of the battle McKeever, who had clerked at Hanley's store at Hillsboro as a young boy, had bragged that the rebs couldn't hit him at that distance. Immediately afterward, an enemy bullet cut his red suspender strap. After the war McKeever became a prominent minister.[35]

On the mountaintop, soon after the artillery had opened the battle, it had become apparent to Col. Patton that Averell was "concentrating on the Confederate center and left." To meet this threat Col. Jackson notified Patton to send a regiment to his "right and rear as a reserve...and to protect the right of the hill on which the artillery was posted." Patton honored this request by moving the 22nd Virginia Infantry forward to the left of McLaughlin's front, where they were able to give Col. Jackson's men good protection. Samuel D. Edmonds, Company B, 22nd Virginia, recalled being placed about 15 feet in advance of Chapman's 24 pounder and 10 pounders, companies B and H of the 22nd Virginia "just under the muzzle" of Chapman's 24 pounder gun. Edmonds said he felt like he was in front of "about ten cannon," as the rebel guns "commenced throwing 24 pound shell over...him...and the others, at the Yankee cannon in an open field about one-and-a-half miles distant." Each time the rebel guns fired, shells passed about four feet overtop of the heads of the 22nd Virginia.[36]

Major Blessing's 23rd Battalion Virginia Infantry, now consisting of six companies (Company C and another were on picket on roads covering the Confederate rear) situated about 400 yards to the right of the Confederate batteries, was ordered by Patton to fill the former position of the 22nd Virginia. When Blessing's men arrived at the designated post they were assigned a spot "under cover immediately in rear of the artillery," where they would remain for about 10 minutes. It was at about this time that Echols was informed of the Federal attack on his extreme left.

The 28th Ohio: First Contact on the Left

Colonel Augustus Moor had "skillfully and resolutely" marched his infantry column some four miles out of Hillsboro to the mountain's base, in a zigzag line in a northwesterly direction, "along ditches and behind fences," and side hills. M.A. Dunlap later claimed Col. Moor "added two miles to the distance from Hillsboro to the mountains." But the movement had proved to be more difficult than anticipated. As the column departed Hillsboro shortly after 9:00 A.M., Moor realized his biggest task was to move the men west, across the open valley without being spotted by the enemy. Taking every precaution to remain hidden, Moor ordered his riders, probably the officers and Company C, 16th Illinois Cavalry, to dismount prior to emerging from the woods around Hillsboro and to carry their arms at a trail. Reportedly, former area residents Austin Brown, Ike Brown, and Mose Stilley (Stillig) served as Moor's native guides, although Moor would later claim they misled him. John Barlow Kinnison, an area blacksmith, possibly too old to join the army, was also mentioned as one of Moor's guides, although descendants claim no knowledge of this. A postwar account claimed the Federals "forced four men who lived at Hillsboro" to guide them.[37]

Utilizing postwar landmarks, M.A. Dunlap said Moor's movement began at about 9:00 or 10:00 A.M. and moved "through Samuel Clark's yard." Dunlap stated that during the war Samuel Clark's place was in timber from "the Dunlap's almost to the Presbyterian Church lot" and the "Dunlap place had a tract of timber along the Clark line." Undoubtedly such wooded area provided Moor good coverage, and he would later say a wounded rebel officer told him Echols "had no idea of his approach." This was probably incorrect as Echols had an excellent view of the Federal lines at Hillsboro, and a number of Confederates recalled observing Moor's departure from town. One such witness was Lt. Randolph Blain of Capt. Thomas E. Jackson's Battery, who remarked, "During the [first artillery] fire I observed an immense force of cavalry file past Hillsboro to our left."[38]

From Samuel Clark's yard Col. Moor advanced "across to the Brown spring, thence through to Preston Clark's place, thence to the foot of the Little Mountain, thence around the foot of the Little Mountain to the McCorkle place, thence to the road in John B. Kinnison's sugar camp..." Kinnison, the same area blacksmith Col. Moor purportedly impressed as a guide, "shod horses for both armies" while in the vicinity "and lived at the base of Droop Mountain where the road coming west out of Hillsboro starts up towards Caesar's Mountain. He had a large sugar camp," the remnants of which are still visible. He reportedly "hid livestock in a low place near the house, which also is still there." From the sugar camp locale Moor moved up the mountainside until he arrrived at the old trail [present Caesar's Mountain/Lobelia/Viney Mountain road] on top. Former Confederate soldier, Lincoln S. Cochran, said the Federals "got to the road leading from Hillsboro to Hills Creek (now Lobelia). When they got to the top of Caesar's Mountain they struck the Bruffey's Creek road and came back and [eventually] attacked the Confederates in the rear near the west end of the glades..."[39]

Kinnison Sugar Grove and Camp located west of Hillsboro. This is where Colonel Moor's flanking movement ascended Droop Mountain. MIKE SMITH, DROOP MOUNTAIN STATE PARK

Moor claimed the 28th Ohio and the 10th West Virginia Infantry "marched up the same road until near the summit, the 28th Ohio in the lead." Some writers have claimed the 10th West Virginia joined with the 28th Ohio at the Dar place, so named for area pioneer Abraham Dar. The column turned southward and approached the Confederate extreme left, along present-day Lobelia road. Moor reported that from about 1:00 A.M. until 2:00 P.M. [his time frame reference was apparently slightly off] he "had been marching due south, forming nearly a semicircle of about nine miles from the starting point," driving Confederate skirmishers steadily. Pvt. John D. Sutton, Company F, 10th West Virginia, believed the entire movement "covered about 6½ miles on a back road." 2nd Lt. Henry Bender, also of Company F, 10th West Virginia Infantry, said that when Moor's column came near the top of the mountain the two infantry regiments "filed to the left until both regiments were the same distance on the mountain, when we formed and marched toward the enemy." Bender added that Company D of the 10th West Virginia "led the advance [of the regiment] at Droop Mountain, being deployed in front and flank, till the fight commenced, when it took the position assigned it."[40]

A 1928 view of the ridge over which the 28th Ohio and 10th West Virginia crossed and formed line of battle about 12:45 P.M. on November 6, 1863. DROOP MOUNTAIN COMMISSION BOOKLET

Colonel Moor complained he could have arrived on the enemy left sooner had he not been misled by what he called incompetent and unreliable native guides, although modern investigation of the route taken supports the course supplied by the guides. A postwar account claimed the "four" guides were in sympathy with the Confederates and intentionally took the wrong route, leading the Yankee's "several miles out of the direct line." When Col. Moor discovered the mistake he sent out his own scouts to locate the road. In the meantime, one of the guides escaped and reportedly found his way to the rebel lines and warned them of the impending attack on the left flank. But the information arrived too late for Echols to change his lines in time. A witness to the battle would later say, "If the Southern troops had been able to form [on the left flank], the battle would have been a victory for them. Despite this Moor said he traveled by the sound of cannon [although there was no artillery on the Confederate left] and the direction that the flying pickets took when they were dislodged by the advancing troops"[41]

A Confederate soldier, lying in the woods waiting on the Federal troops, reportedly fired the first rifle shot on the flank. He reported a Yankee "shoved his face over a fence rail and [I] shot him square between the eyes and he squealed like a pig." It was claimed this was the first soldier killed in the Droop battle, but that is highly unlikely. 3rd Sgt. Andrew Mathews McLaughlin, Company I, 19th Virginia Cavalry, recalled he was among the Confederates who "first reached the top up to which the Union troops were advancing and were fire[d] upon, whereupon the battle began." Citizen George Hill, who lived on the west side of Droop Mountain, had been visiting friends in the Confederate lines. When he started for home he came to the glades and "saw the woods blue with approaching Union soldiers." He sensed he would definitely be killed and preferring to be among friends, dashed back to the Confederate lines as "bullets rained around him, but he made it back safely."[42]

Sporadic firing increased as Col. Moor advanced and had just increased his skirmish line with three companies of the 28th Ohio, led most ably by Capt. Edwin Frey, Company E, 28th Ohio. Through Frey's leadership the Federal skirmishers were able to locate "the exact position of the enemy." One correspondent said, "Colonel Moor, with great caution, threw out a double line of skirmishers and found the enemy heavily posted to receive him." Sometime between 1:45 P.M. and 2 o'clock, Col. Moor's infantry appeared within approximately 200 yards of the present-day park boundary along Lobelia road. Moor claimed that it was at approximately 1:45 P.M. that his column "arrived in front of" the enemy left, which "was covered by a kind of hedge constructed of logs and brush." M.A. Dunlap later said this was actually "an old log fence around one of the clearings on top of the mountain." Moor would later praise Lt. Col. Gottfried Becker, 28th Ohio, "in guiding and maneuvering the regiment [28th Ohio] in unbroken lines over the most difficult ground, through ravines, rocks, thick undergrowth, and fallen trees." A correspondent said Lt. Col. Becker "was among the soldiers, cheering them on, and, by the exposure of his own person, aiding that which all seemed so anxious to secure."[43]

Captain Frey "met the yells and bullets of the enemy, showing great bravery." Realizing he was in for a hard fight, Moor ordered the 28th Ohio into line and sent orders back to the 10th West Virginia to "move up in double-quick." Capt.

Modern-day Lobelia Road on Viney Mountain on the Confederate left flank. Colonel Moor's Federal infantry advanced through these woods and arrived at a clearing just to the left of this scene. BRIAN ABBOTT

Colonel Moor's infantry brigade of the 28th Ohio and 10th West Virginia burst out of this wood line at about 1:45 P.M., November 6, 1863, to strike the Confederate left. BRIAN ABBOTT

According to this 1928 view, this is the site of the "first formation of [the] 10th West Virginia [Infantry] and [the] 28th Ohio [Infantry] at edge of field [at] 12:45 P.M." The numbers 10 and 28 obviously designate the positions of the two regiments. DROOP MOUNTAIN COMMISSION BOOKLET

This park sign denotes Confederate trenches on the Confederate left flank. GARY BAYS

Denicke, of the center signal station, waiting for Moor to make contact, said, "I watched for Lieutenant Merritt's flag myself, and had a man continually on the lookout, but at no time during the engagement was I enabled to communicate with him." Allegedly, Lt. Merritt climbed a tree and waved his red underwear in a futile attempt to signal the center. Capt. Denicke recalled, "When after the engagement, I inquired the reason of this [no communication from Lt. Merritt], he stated in explanation that only at one time had he seen my flag (center station), and that at that time it had been impractical to open the desired communication, as some trees interfered with the view."

Probably defending the Confederate left at this time was Capt. Jacob W. Marshall's 100 dismounted troopers and about 28 additional men, although 1st Lt. David Poe, Company A, 20th Virginia Cavalry claimed, "...a report came to us to the effect that the enemy was going around us on our left and then on top of the mountain. [2nd] Lieut. [David B.] Burns, of Company 'A' was sent out with eighteen men to see after the matter. Burns met the 28th Ohio regiment and the firing commenced at once...Lieut. Burns held the 28th Ohio in check for quite a while. The whole regiment formed and fired on Burns and his men, who were behind trees and other things, without dislodging them. Burns was wounded in the hand...the rest of our company was stationed at the summit of Droop Mountain." The minute the Federal skirmishers made their appearance on the left "sharp firing" had ensued and Echols ordered Col. Jackson to send a force to the extreme left. Col. Jackson detailed Lt. Col. William P. Thompson, 19th Virginia Cavalry, for this assignment, but Thompson was misled by a dispatch from 1st Lt. William W. Boggs, Company I, 20th Virginia Cavalry [there was no Lt. Boggs in the 19th Cavalry] which indicated the Federals were attempting to gain Echols' rear by a "more circuitous" route than actually taken. As a result, Thompson had gone too far to his left. When he discovered the error Thompson left Capt. John S. Spriggs and his Company B of the 19th Virginia Cavalry on the threatened flank and "returned to the pike with his dismounted" members of the 19th and 20th Virginia cavalry regiments, which he deployed at the road [Lobelia?] under Major Joseph R. Kessler, former captain of Company C, 19th Virginia Cavalry, and resident of Covington, Virginia who would later serve as lieutenant colonel of the 46th Battalion Virginia Cavalry.[44]

Iron relics found on the Confederate left flank at Droop Mountain (left to right): iron pin or hook (use unknown) and remains of a Jew's harp (musical instrument) which many soldiers carried in their pockets. GARY BAYS

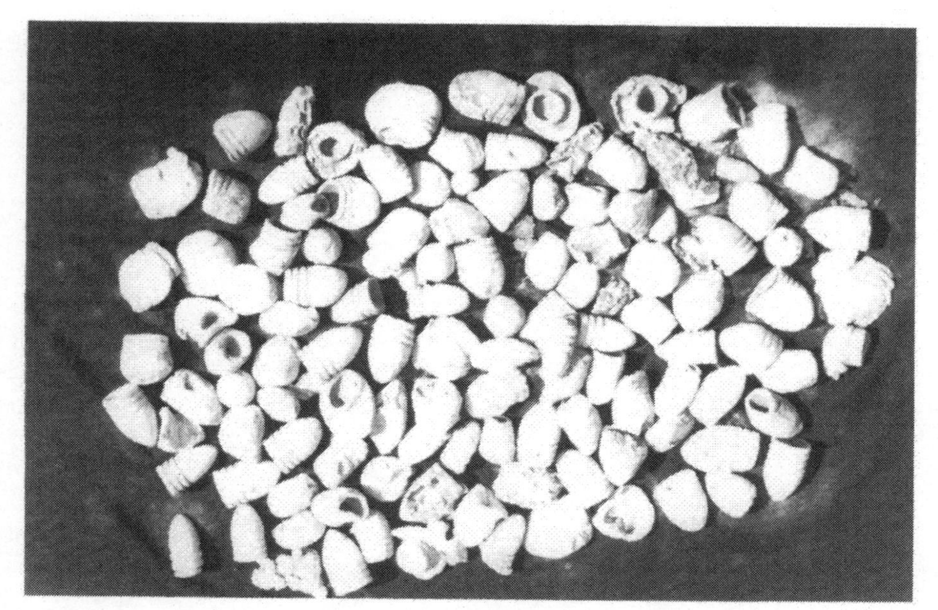

A grouping of fired and non-fired bullets recovered from the Confederate left flank at Droop Mountain. Note that the majority are .69 caliber three-ring bullets, which were probably the favored ammunition of the 28th Ohio Volunteer Infantry. GARY BAYS

Due to the trees and heavy undergrowth the men of the 28th Ohio could see no more than 25 to 30 yards to their front. Although we can only speculate, the rebel left now was apparently defended by Marshall's 100 men, some 28 additional troopers, Lt. Burns and his 18 men of the 20th Cavalry, and Major Joseph R. Kessler's 50 men, who had been sent to Marshall as a reinforcement. This added up to nearly 200 Confederates on the left at this time. In accordance with Lt. Col. Thompson's orders, Capt. Marshall and the other rebels permitted the Federals to advance to within yards of their position, when they suddenly stood up "yelling like Indians" and "poured a tremendous fire into the 28th Ohio," followed by a gallant and successful charge, driving back the 28th Ohio. Major Kessler, apparently in command of the rebels at this point, "shouted to his men to stand firm for two minutes," and felt the enemy was whipped.

Lieutenant Colonel Thompson, in charge of the Confederate left, realized his dismounted troopers were greatly outnumbered and sent a request to Col. Jackson and Gen. Echols for "heavy reinforcements." Thompson stated, "...my men were deployed [throughout the day] so as to keep the line extended to prevent flank movements, which were constantly attempted. Our men were sheltered by the timber while the enemy advanced in line of battle, and as our men shot with coolness and precision the enemy suffered considerably."

At about this time Echols, from his position (probably on the right) said it was very difficult to see the enemy, who were "so well masked and concealed" he found it near impossible to estimate Averell's strength, particularly as "a very large force was seen in front." Despite his uncertainty, Echols received Lt. Col. Thompson's plea for assistance and ordered Major Blessing's six companies of the 23rd Battalion Virginia Infantry to the left, along with two companies of the 14th Virginia Cavalry [Companies B and I].

Lt. Col. John Alexander Gibson of the 14th Virginia Cavalry, commanded four companies of his regiment on the Confederate extreme left at Droop Mountain. MRS. MARY LIPSCOMB, BROWNSBURG, VIRGINIA

Brass from a Henry repeating rifle bullet used by the cavalry. Found on the Confederate left. GARY BAYS

A .58 caliber Enfield bullet with numeral 57 in its base. This dropped ammunition was found on the Confederate left flank. GARY BAYS

Major Blessing received Echols' order, relayed through Col. Patton, and sent one company of his 23rd Battalion back to his original position on the right flank. The remaining companies of the battalion, numbering about 300 men, [although Micajah Woods gave a figure of 200] arrived to support Capt. Jacob Williamson Marshall on the left flank, who was being driven back. Blessing's group deployed in line to the right of Marshall's 125 dismounted cavalry, then charged and drove the Federals back "to his main body." At this point Blessing suddenly "met with a terrific fire and [was] forced back to the fence running parallel" with the rebel line. Here, at the fence, Blessing was reinforced by the two dismounted companies of the 14th Virginia Cavalry, Company B (Charlotte Cavalry) under Capt. Edwin Edmunds Bouldin, and Company I (Churchville Cavalry) of Capt. Joseph Alfred Wilson, which Lt. Col. Thompson had sent to the far left, where the fighting had become most fierce.

Lieutenant Colonel John Alexander Gibson, 14th Virginia Cavalry, would be in charge of these two companies as well as two others of the 14th Virginia Cavalry on the extreme left flank. Earlier in the war this squadron of the 14th Virginia Cavalry, led by Bouldin, had been designated the charging company by Col.

James A. Cochran. The "two companies were armed, for the most part, with sabers, pistols, and carbines. When[ever] advancing, this squadron preceded the regiment and on falling back its place was in the rear. When[ever] the squadron met the enemy it would fight at close quarters using pistols and sabers."[45]

This new alignment on the Confederate left made a stand for about 10 minutes, but was again forced back. Sgt. Major Robert Henry Gaines, Company B, 14th Virginia Cavalry, was among those who eventually went down wounded. The Federals continued to press the southern left, but as Lt. Col. Thompson had realized the importance of the position, he did his best to hold it against overwhelming odds. Lt. Col. Thompson wrote, "The enemy pressed our line persistently and with much impetuosity, and in despite of the gallant conduct of many officers the line gradually gave way before an overwhelming force. This being the left wing of our army, and as it protected the rear of the whole force, I made determined efforts to hold the position. At this time the men fought with great gallantry against overpowering odds, there being at the time at least 2,500 of the enemy. The enemy knew his advantage and pressed it with great vigor."

The 2nd and 3rd West Virginia

Averell said "intermittent reports of musketry heralded [Moor's] approach to Echols' left flank" and it was "evident from the sound of battle on the enemy's left and his disturbed appearance in front, that the time for the direct attack had arrived." Correspondent "M" for the *Wheeling Intelligencer*, apparently accompanying the dismounted West Virginia boys in the center, said "there was a crash—a volley—and then the rattling fire 'by file' and 'at will', [which] told that the time had come for us to act..." Lt. Col. Alexander Scott, 2nd West Virginia Mounted Infantry, said he heard a few reports from the rebel left "followed by volleys of musketry which were being hurled into the ranks in the rear of the unsuspecting" enemy.

The rounded hill in the center of photo denotes the center of the Confederate line, which was defended by the 14th and 20th Virginia cavalry regiments. Federal troops assaulted the face of this hill as Colonel Moor struck the rebel left flank. BRIAN ABBOTT

Modern view, facing north towards Hillsboro, of the Confederate center. To the immediate rear are faint remains of the rock breastworks. MIKE SMITH

Cave in the ravine where the 3rd West Virginia Mounted Infantry ascended Droop Mountain. The area has numerous such caves. MIKE SMITH, DROOP MOUNTAIN STATE PARK

Moor's contact with the Confederate left was the pre-arranged signal for all the Federal commands to come out of hiding and make a general assault upon the entire rebel line. Averell stated the three dismounted West Virginia regiments [had] "moved in a line obliquely to the right, up the face of the mountain, until their right [would eventually be] joined with Moor's left." Capt. Francis Mathers, Company I, 8th West Virginia Mounted Infantry, remembered, "the 2nd and 3rd West Virginia advanced, and just as the 8th [West Virginia] emerged from the woods the rebels began to waver, Colonel Oley moving his regiment up the mountain in the face of a murderous fire of shot and shell." As the West Virginia boys advanced, Battery G, 1st West Virginia Light Artillery (Ewing's Battery), situated on the right, opened in conjunction and with the same effectiveness as Keepers' guns. After hours of hard fighting Oley "arrived at the summit, immediately in front of the battery."

On the Federal left Col. Schoonmaker "hurried his right forward to assist in the general assault," but his main body was unable to give much aid due to the large distance to be covered. Schoonmaker apparently did advance two guns on the pike, which, according to Major McLaughlin, opened upon the 22nd Virginia Infantry, which was yet posted to the left of McLaughlin's front. This was probably a reference to 1st Lt. Howard Morton of Ewing's Battery, who, "At Droop Mountain...advanced the guns up the mountainside under the terrific fire of the enemies' batteries on the summit." A few well-directed shells from Chapman's and Lurty's howitzers quickly drove off the two Yankee pieces.[46]

As the fighting became more general along Echols' entire line, the 2nd West Virginia got about 10 to 15 yards from the crest of the mountain, where the rebels blazed away at their right and center. One member of the regiment said the "en-

View of the Confederate position from present-day Locust Creek Road. The park lookout tower is on the cleared hill on the left. It is believed the 14th Pennsylvania Cavalry used this approach. MIKE SMITH, DROOP MOUNTAIN STATE PARK

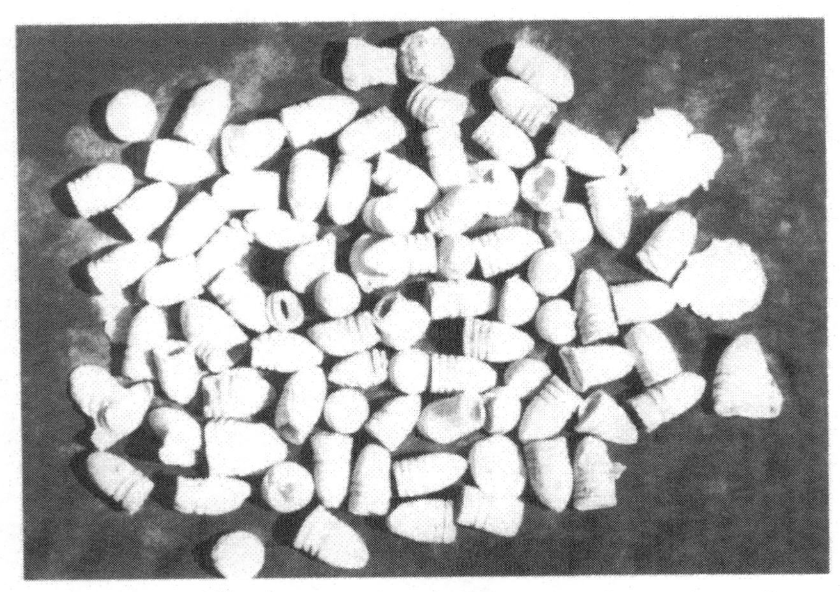

An assortment of fired and non-fired bullets recovered from the northern base of Droop Mountain. Most were probably fired at the Federal troops as they prepared to advance up the face of the mountain. GARY BAYS

emy opened upon us, and a sheet of flame issued from the mountaintop as the Confederates poured a terrific fire of musketry into the faces of our brave boys." The entire 2nd West Virginia pushed forward "with vigor, and never flinched or wavered," returning fire "with telling effect."

Reportedly, Lt. Col. Francis Thompson, 3rd West Virginia Mounted Infantry, signaled Averell for permission to charge, and once obtained, asked the Federal artillery to direct a few rounds at the breastworks to his front. Battery G [Ewing's] helped soften the rebel line by "throwing shot and shell" at the enemy when the dismounted West Virginia troops were "within 20 or 30 feet" of the rebels. This created a dangerous situation as the Federal batteries endangered their own comrades with their fire. When Company B, 2nd West Virginia emerged from the woods they were "saluted with shot and shell" from their own artillery. To correct the problem before any serious damage transpired, Pvt. Cyrus E. Ringler, Company B, stepped atop a huge rock and signaled the Yankee gunners to cease fire.

Private Ringler had never been shy about taking charge. Born at Johnstown, Pennsylvania September 30, 1835, he had moved to Maryland and learned the printing trade, growing "up a strong pro-slavery Democrat." He held a military commission under Gov. Henry A. Wise of Virginia and at the outbreak of the war stood against secession. Ringler was "probably the first man in Harrison County to procure names for the Union service." In May of 1861 he became a member of Capt. Latham's Company, and later, while posing as a spy, was the first member of his company to enter the Confederate works at Laurel Hill. During the battle of McDowell he was stunned by a musketball that shot the cord off his hat, and at

Pvt. Cyrus E. Ringler, Co. B, 2nd West Virginia Mounted Infantry. A printer by trade, by the time of the Droop Mountain battle, Ringler had been stunned by a musketball at McDowell, stunned by a Confederate shell at Cross Keys, and shot in the stomach and right hand at 2nd Bull Run. During the fighting at Droop Mountain he helped to carry the Confederate center and right.
FRANCIS READER COLLECTION

the battle of Cross Keys he was also stunned by a Confederate shell. If this were not enough, at 2nd Bull Run he was shot in the stomach and right hand.

Soon after Ringler's move a Federal artillery shell fell nearby, which provoked Ringler to place a "white nether garment on the front of a tall sapling" and wave to the Federal artillerists. The gunners finally acknowledged the signal and re-directed their fire. Another source indicated Lt. Col. Francis Thompson requested the artillery cease fire. Fortunately, no Federal injuries were sustained and Battery B continued to hurl "effective shot" into the "now hard pressed" Confederate ranks.[47]

Members of the 2nd West Virginia said they were at an advantage as the enemy rifle fire would pass over their heads due to the steep slope, while they struggled up the mountain in an effort to avoid being "mowed down." Casualty reports after the battle, however, indicated many of the dismounted West Virginia boys were shot in the head and trunk.

The 3rd West Virginia "had fought their way to within a short distance of the Confederate fortifications and could do nothing further except to stand and be shot or charge the enemy's works." Pvt. William Marshall Dearing, Company F, 3rd West Virginia, recalled that as his company advanced up the steep mountainside, orders were received to stop and lie down to permit the rebel bullets to pass overtop of them. Dearing turned his head downhill, fearing he would witness the slaughter of his entire command, as the rebel firepower was so intense.[48]

Sergeant Benjamin F. Hughes, Company F, 3rd West Virginia Mounted Infantry, apparently suffered a hearing loss during the battle. On December 9, 1863 the 39 year-old Hughes would write from New Creek, "what the real cause of my deafness is I can not tell. I was knocked down by a shell bursting near me at Droop Mountain but my hearing seemed to return after a few days, but began to fail and continued so for some time. I cannot hear the [railroad] cars at the depot a half mile from...here. Also, it is 400 yards from here to the fort and I can scarcely hear the cannon when she fires the daily shots. A 24 pounder at that."[49]

Bringing Up the 10th West Virginia

Caught under a withering fire on the Confederate left Col. Moor realized this was the critical moment of the day and ordered the 28th Ohio "to lie down and fire by file." This maneuver resulted in the Germans disappearing from view of the rebels, "and the increasing fire through the underbrush had an almost stunning effect upon the enemy." The Confederate defenders, slightly bewildered, hesitated in their defense, uncertain as to whether or not they had decimated the enemy. Moor had been awaiting arrival of the 10th West Virginia, which had encountered problems getting into position, plagued by "cavalry horses and other obstacles" blocking the trail.

Colonel Thomas M. Harris and his 10th West Virginia Infantry reached Moor's position "just in the nick of time." Sgt. Thomas R. Barnes, Company K, 10th West Virginia, said the column reached the rebel rear at 2:00 P.M. and was ordered by Moor "to front the regiment by inversion and form on the right of the 28th Ohio, which was promptly executed." 2nd Lt. David B. Burns, Company A, 20th Virginia Cavalry, claimed his detachment of 18 Confederates fell back upon the arrival of the 10th West Virginia. Col. Moor proceeded to detail one company from each of his two infantry regiments "to march in the rear as a reserve and to guard the flanks." 2nd Lt. Henry Bender, Company F, 10th West Virginia, added, "Our formation [10th West Virginia] was companies A, F, D and C, others which I am not sure, only Company A was on the extreme left. I do not remember any troops on the right." Once all the infantrymen were aligned and in position, Col. Moor advanced a frontal assault against the rebel left, his men cheering as they moved forward "completely drowning the hideous yells of the enemy." Correspondent "T.M." of the *Wheeling Intelligencer* reflected, "Never did a brigade go in battle with more gallantry and determination." Pvt. John D. Sutton, Company F, 10th West Virginia, recalled, "On the right the land is a broad flat and was covered with large timber. The ground fought over was something like a mile in width and was stubbornly contested." Pvt. Leman S. Clothier, Company A, 10th West Virginia, mentioned, "we were on the right wing of our forces and as soon as we located the enemy in thick timber we charged... without any casualties on our part, but with heavy loss to the enemy."[50]

Hamilton Griggs, reportedly of the 10th West Virginia [although the only man by this name on military records is Pvt. Hamilton Griggs, Company M, 1st West Virginia Cavalry, who did not enlist until March 18, 1865], recalled that as Col. Thomas M. Harris rode his horse, "Old Coaly," into battle, he encountered a particularly rugged section of terrain which forced him to dismount. As Col. Harris led his beloved horse on foot [during the final charge] an enemy bullet passed through the colonel's long, flowing, red beard. The projectile cut out a wisp of his beard, which prompted Harris to calmly strip out the scorched whiskers, drop them to the ground, and turn to Adjutant John Warnicke with the remark: "John, take my horse to the rear; I'm afraid he'll get shot." Harris, ever proud of both his beard and horse, reportedly never rode "Old Coaly" into battle again, and fought the remainder of Droop Mountain on foot.[51]

Lieutenant Colonel William P. Thompson, 19th Virginia Cavalry, related that soon after arrival of the two companies of the 14th Virginia Cavalry on the Confederate left "...we rallied the men and selected an admirable position, and when the enemy made his appearance he met with fearful loss, but our men, impressed with the belief that they were overpowered, gradually and in despite of the efforts of gallant officers, retired before the advancing line of the enemy." Lt. David B. Burns, Company A, 20th Virginia Cavalry, confirms that when Col. Harris of the 10th West Virginia arrived the rebels began to fall back.

The 10th West Virginia continued to advance successfully until arrival at an open space where the timber had been cut down. Lacking natural protection the 10th West Virginia came under a galling fire, a number of men were lost in one single enemy volley. Pvt. Leman S. Clothier of the 10th West Virginia said the regiment quickly sought shelter from the enemy bullets, hiding behind nearby logs and timber. At this point Sgt. Right Bird Curry, Company A, 10th West Virginia, then only 19-years-old, was killed as a bullet struck his head. He fell near the side of 2nd Lt. Henry Bender, Company F, 10th West Virginia, who remembered the incident, relating, "the enemy were stationed in line of battle when we advanced on them...I was near the head of the company where we joined Company A," and at which point Sgt. Curry fell.[52]

Others in the 10th West Virginia went down at about the same time, including Pvt. Milton Rollyson, Company F, who was wounded as a bullet passed through his left forearm, and Pvt. Edward B. Wheeler, also of Company F, who was wounded as a bullet struck his left shoulder. Caught in such an exposed position Company F of the 10th West Virginia accumulated numerous casualties, "many of its members being shot down and many wounded, and some of it's members began to fall back."

Private W. Wease [Nimrod Weese], Company H, 10th West Virginia, was wounded by a musketball in the abdomen. A surgeon of the 28th Ohio would later report the case of Wease as a "wound of the large intestine, with faecal fistula." In 1872 a Federal surgeon would examine Wease and say "the ball entered the right side of the abdomen, between the umbilicus and upper crest of the ilium, four inches from the umbilicus, and came out between the ilium and the spine, four inches from the spine, passing through the large intestine and upper part of the ilium. The contents of the bowel passed out of the posterior wound for some months [afterward], and portions of bone were also eliminated." The Federal surgeon continued his graphic medical description, writing, "The cicatrices of entrance and exit were depressed and about an inch in diameter. The abdominal muscles around the anterior cicatrix were contracted. Locomotion was difficult and the physical disability total and temporary." In April of 1864 a Federal surgeon would report of Weese, "the fistulous opening had then healed, and that the patient walked slowly about, and had nearly recovered."[53]

"T.M." of the *Wheeling Intelligencer* wrote, "The 10th had our right, and I am proud to say that never did officers and men behave better. We pushed the enemy steadily before us for over a mile, keeping up all the while a roar of musketry that was terrifically grand. The fight here lasted an hour and a quarter, and was all the way through brushy woods. The enemy was driven from one position to another, and so terrific was our fire that many in chosen positions, under the cover of a log or a thicket, awaited our approach were unable to leave their places in face of the storm of missiles to flee, and preferred rather to remain and fall into our hands."[54]

Colonel Moor knew that despite many valiant efforts by the rebel line his opponent could not hold against such overwhelming odds and continued to push his men forward. The vastly outnumbered Confederate defenders on the left flank were completely shocked to view the mass of Yankees moving against them, prompting many to begin tossing their weapons aside and flee in panic. The fighting would become so intense that this location at the fence was later nicknamed "The Bloody Angle."

The Bloody Angle

Since Capt. Nimrod McKyer of Company F, 10th West Virginia was absent, being in an enemy prison, and 1st Lt. Samuel A. Rollyson was on staff duty, 2nd Lt. Henry Bender assumed command of the company at Droop and moved it forward. Pvt. John D. Sutton noticed the company was greatly disorganized by the enemy fire, "some of the men holding back somewhat, mixed up with men of

John D. Sutton, below left, as he looked circa 1863. He served in Co. F, 10th West Virginia Volunteer Infantry and was a private when he fought at the "Bloody Angle" at Droop Mountain. At right, as he looked as Chairman of the Droop Mountain Battlefield Commission in 1928. Sutton was instrumental in preserving the battlefield site for future generations. DROOP MOUNTAIN STATE PARK

Ord. Sgt. John D. Baxter of Co. F, 10th West Virginia Volunteer Infantry. He was mortally wounded in the bowels as he led the last charge over the rail fence at the "Bloody Angle." DROOP MOUNTAIN STATE PARK

At right: Close-up of the monument to Sergeant Baxter, which reads: "John D. Baxter - This marks the spot where John D. Baxter, Orderly Sergeant, Co. F, 10th W.Va. Inf. fell inside the Confederate line leading the last charge November 6, 1863." GARY BAYS

other companies." Sutton went to so inform Orderly 1st Sgt. John D. Baxter, a 25-year-old officer from Sutton, Braxton County, West Virginia, of this development. He found Baxter in advance of the company, where Sutton requested Baxter move back and help line up the men. Baxter, who stood six foot three-and-a-half inches tall and had red-hair, was standing only a few steps from the rebel fence line. Baxter failed to answer Sutton and rushed forward, shooting two or three rebels off the fence. As the blue line surged forward Baxter kicked some of the top rails off the fence as he and Sutton jumped over the structure. Advancing a few steps, Baxter was hit in the bowels by rebel lead. Desperately sick from the wound, Baxter placed the butt of his rifle on the ground and lowered himself down. Sutton remained beside Baxter for a brief moment.[55]

Also near Baxter when he fell was Lt. Bender, who had part of his big toe shot off. It was said this was the only time Bender's men ever heard him swear. Bender was noted for gallantry and breveted Captain at the end of the war. Pvt. Wellington Fletcher Morrison of Company F would later say of Bender, "...[he] was a strict disciplinarian...generous, of a kindly disposition, and looked so carefully to the

Two wartime and one postwar view of the man who commanded Co. F, 10th West Virginia Volunteer Infantry at the "Bloody Angle" and witnessed the mortal wounding of Sgt. Right Bird Curry of Co. A of the regiment. Bender was promoted to Captain for gallantry and was reportedly wounded in the big toe. DROOP MOUNTAIN STATE PARK

Lieutenant Bender marker located a few feet from the marker of Sgt. Baxter, the plaque reads: "Lieut. Henry Bender commanded Co. F in the last charge that the 10th W.Va. Vol. Inft. made that broke the Confederate line at the Bloody Angle where so many brave men of both armies fell November 6, 1863." As with Baxter's marker, Lt. Bender's photo is encased at the top of the monument. GARY BAYS

welfare of those under his care that none was imposed upon and all shared alike under the circumstances. He was respectful toward prisoners. To say that he was a brave soldier is putting it mildly. He seemed to be fearless...no more gallant soldier ever unsheathed his sword in the face of an enemy," and "...He was a man of spotless character, gallant, courteous, offered his life in defense of his country's flag..."[56]

Privates Wellington Fletcher Morrison, William M. Barnett, and John A. Blagg, all from Company F of the 10th West Virginia, were next to cross the fence. Morrison got across safely but Blagg and Barnett were both wounded; Blagg suffering a serious gunshot wound of the right ankle involving the joint. Blagg later claimed he was wounded about 60 or 80 yards from the pike while he was attempting to disarm a wounded rebel who was resting on a log, and the incident took place near the close of the battle. Barnett suffered a gunshot wound of the right leg and knee joint. Pvt. John D. Sutton said Barnett's leg was shot off and the rebel who shot him was immediately killed. In the meantime Pvt. George H. Morrison, Sgt. Silas Carr, Company F, and Pvt. Morgan D. Shaver were the next to cross the fence. Morrison and Carr came up to Sutton and said they would tend to Baxter, who would expire the following day. Pvt. Levi Lockard, Company C, 10th West Virginia, was reportedly captured by the rebels but later escaped. "T.M."

Bloody Angle - Then and Now

Site where Sgt. John D. Baxter, Co. F, 10th West Virginia Infantry, fell mortally wounded. Two white spots in center of 1928 view (top) mark the spots where Baxter fell and Lt. Henry Bender, also of the 10th was wounded. Upper end of lake [bog] is at right. Bottom photo is modern view of same area. Two dark objects in center are monuments to Baxter and Bender replacing the white markers. TOP: DROOP MOUNTAIN COMMISSION BOOKLET, BOTTOM: GARY BAYS

of the *Wheeling Intelligencer* described the scene, writing, "Opposed to our two regiments were the 14th, 19th, 20th, and Derrick's Battalion, and finally four [three] companies of the 22nd, that [...would be sent...] to strengthen their line. The 19th you, are aware, is a cavalry regiment commanded by Bill Jackson; the 20th also is cavalry, commanded by Col. Arnett. Both are composed of rapscallions from our western counties. The 20th is a recent organization from those who left last winter and spring, and during the past summer. I think many of them are now satisfied that they have obtained all that they are likely ever to obtain of their rights." "Union," a correspondent for the *Cincinnati Daily Commercial*, described the scene on the rebel left: "...with a loud hurrah, the 10th and 28th paid their compliments in successive volleys of well told musketry, sounding as one continued crash; the rebel slain testifying to their determination in disputing Droop Mountain summit...While the 28th Ohio and 10th [West] Virginia were battling in close position, the sounds of musketry, the roaring of cannon, the cries of the wounded and dying, all seemed one sweeping hurricane. For a few moments the scale seemed poised, and it was difficult to tell which way the tide of fortune would turn..."[57]

While Col. Moor's two infantry regiments advanced, "the right of the dismounted men [3rd West Virginia] joined Moor's left, coming up through a ravine," creating "the wildest scene...in front," as the Yankee's raked the staunch rebel defenders with deadly firepower.

Attack on the Center

General Echols would later recall that as the Confederate left struggled to hold back Moor, the rebel center had also been struck and "fighting became general along [the] whole line," while the Confederate artillery reportedly served "with great rapidity and precision, and having succeeded in silencing the batteries of the enemy." In actuality, the Federal guns were neither injured nor silenced.

As earlier noted, beginning as early as 2:00 P.M. several vigorous assaults had been made at the rebel center against the dismounted cavalrymen of Col. James Addison Cochran (14th Virginia Cavalry) and Col. William Wiley Arnett (20th Virginia Cavalry), by the 2nd, 3rd, and 8th West Virginia regiments, as well as a portion of the 14th Pennsylvania Cavalry. While the outnumbered rebels on the left struggled to hold back the 28th Ohio and 10th West Virginia, Cochran and Arnett's men held strong against repeated attacks, not yielding an inch. However, following some two hours of heavy fighting the 2nd and 3rd West Virginia regiments moved against the rebel breastworks "with yells and cheers, loud and strong, [and] charged into the jaws of death and fire."

Lincoln S. Cochran remembered that 2nd Lt. John A. Cobb, Company K, 8th West Virginia, who had been mustered at Elk River, West Virginia at the age of 25, "took his company on foot across to the west of the pike to the woods before it was light, and when they heard the battle commence in the rear on top, they climbed the steep hill directly under the Confederate breastworks and charged. Cobb said the Confederates behind the breastworks were the most stubborn fighters that he encountered during the war. They stayed to their post and when their guns were empty they knocked his men off with their guns. He ordered his men to poke their pistols over the breastworks, and the Confederates [soon afterward] retreated."[58]

About 10 to 15 yards from the crest of the mountain the gray line opened upon the right and center of Lt. Col. Scott's 2nd West Virginia. Scott pushed his men forward vigorously and gained the crest "at which time the fighting was quite spirited for a few minutes." Arnett's men clung tenaciously to their position until "they could and did strike" the Yankees with their guns. It was said that at this time the fighting was "fierce and terrible, a battle to the death, the musketry fire

Various objects found at the Confederate center, including, a brass rivet with design; an unknown iron item with four holes; an iron knapsack hook; remains of an iron suspender clip; a brass rivet with some remaining leather and a brass percussion cap.
GARY BAYS

A colt revolver combination gun tool found at the Confederate center.
GARY BAYS

being very rapid." 1st Lt. Arthur J. Weaver, Company K, 2nd West Virginia Mounted Infantry, waved his sword aloft as he gallantly urged his men forward, only to be fatally struck by an enemy musketball. As Weaver fell he uttered his last words in a faint voice: "Tell Jimmie [the 15 year old bugler of Company F] to write Hattie [Weaver's fiancee in Baltimore]."

Weaver's loss was particularly tragic. Born in the south in 1837 and later a resident of Parkersburg, (West) Virginia for two or more years prior to the war, Weaver "was of a gĕnial, cheerful disposition," never married, and "in camp [he] freely mingled with the men, joining in their sports." He was never sick or absent from duty, participated in all the regiment's campaigns and engagements, and "on the march, he would cheer the men," as well as lighten their burden by helping to carry their gear. From the time of his enlistment he had a premonition of his death and was "often heard to say that should it be his lot to fall, he wanted

Front and back of brass U.S. cartridge boxplate found at the Confederate center, probably lost by a member of the 2nd West Virginia Mounted Infantry. The rear view shows the lead backing and iron wire attachments. GARY BAYS

the world to know that he gave his life for the best government in the world." Indeed, on the morning of the battle, while eating a meal, Weaver told his men to "eat hearty as they had hard work before them." He then turned to Sgt. G. A. Quimby's mess and, joining them, said to Quimby: "It may be 'Buddy' this will be the last meal you and I will eat together."[59]

About this time, "through some misunderstanding," the 2nd West Virginia fell back some distance. Perhaps the retrograde movement was prompted by Lt. Col. William P. Thompson of the 19th Virginia Cavalry, who ordered Capt. John W. Young, Company E, 20th Virginia Cavalry, "to lead his company in a charge," in which Young fell mortally wounded. Fighting beside Capt. Young was Pvt. John Y. Bassell [Company E] "scarcely 16 years old," [service records say he was 17] who was so severely wounded that Lt. Col. Thompson originally reported his wound as mortal. Bassell actually recovered and returned to duty later in the war. Pvt. Jacob S. Hall, Company E, 19th Virginia Cavalry, apparently stationed at the rebel center, recalled, "Our company was on the center and in a place where they would not like to come up on us and so they only left a few in our front and put the

Pvt. Israel B. Field, Co. C, 3rd West Virginia Mounted Infantry [6th West Virginia Cavalry]. Born in Preston County in 1835, Field enlisted June 22, 1861 and was wounded in the back March 19, 1862, at Elk Mountain in Pocahontas County. He survived the Droop Mountain battle only to be captured June 26, 1864, at Springfield, Virginia, sent to Andersonville prison camp and paroled in 1865. RICHARD A. WOLFE, QUANTICO, VIRGINIA

rest on our left." Despite Hall's relative safe position between the actual Confederate center and left he did not escape the fury of combat as he hid "...behind a tree and there was several balls struck the tree. It [the tree] was about 10 or 15 inches in diameter. The shells burst...clouds of smoke over us." Hall went on to describe "...the roar of musketry was so regular and volley after volley so we could not distinguish one volley from the other. It was so intense we could scarcely hear the booming of our artillery or the cracking of the vulgar [?] shell. The heavy small arms was on the left." He later said, "Our company had one man wounded slightly but is or nearly able for duty." Pvt. Abram Lowers, Company A, 19th Virginia Cavalry was shot at by U.S. sharpshooters. Abram hid behind a rock on which enemy bullets peppered and sizzled, sticking to the rock.[60]

Upon the given signal at the Confederate center, the 3rd West Virginia, with Lt. Col. Francis Thompson in the lead, had "Sprang forward with a wild huzza." The 20th Virginia Cavalry and portions of the 14th Virginia Cavalry fought stubbornly and continued to take casualties, including 2nd Lt. Augustus ("Gus") Liggott, Company K, 20th Virginia Cavalry, who was wounded, "shot through the head between the fourth and fifth pair of nerves." Pvt. Michael McGoldrick, Company I, 20th Virginia Cavalry, was wounded three times. McGoldrick, "a little pock-marked man...[who] was more afraid of a dose of castor oil than of a battle," was formerly of New Orleans and had been wounded twice in the Nicaraguan filibustering expeditions prior to the war, as well as three times in the Seven

Another view of Pvt. Israel B. Field, Co. C, 3rd West Virginia Mounted Infantry [6th West Virginia Cavalry]. Shown here with a loved one. RICHARD A. WOLFE, QUANTICO, VIRGINIA

Two cavalry spurs. Top, brass U.S. spur found near the Confederate center, bottom, Confederate brass spur (commonly known as a Richmond spur) found on the Confederate left flank to the rear of the park superintendent's residence. RICHARD ANDRE

An enlisted man's brass cavalry insignia, worn on the hat. Found at the Confederate center and probably lost by a member of one of the West Virginia Mounted Infantry regiments. GARY BAYS

Days battle in front of Richmond in 1862. He claimed that at Droop "a Dutchman [probably a member of the 28th Ohio] had shot him." At least three members of Company E, 20th Virginia Cavalry, were killed in action.[61]

Lieutenant David Poe, 20th Virginia Cavalry, said that while 2nd Lt. David B. Burns and his 18 men fought on the Confederate left, the remainder of Company A, 20th Virginia Cavalry was stationed "at the end of the mountain...[where] there was some very hard fighting and we lost a few men. I saw Lieut. Ulysses Morgan killed. I knew him to be a brave gallant soldier...Sergt. Wm. Straight, was wounded, Capt. [David M.] Camp's blanket was nearly shot off him, and had as many as eight or ten holes in it. [Pvt. Isaac] Milton Lake had a port hole in his breastworks. A blue coat crawled up to a tree...to Lake, peeped over a root and shot at Lake through the hole, filling his mouth and eyes with rotten wood and trash from his temporary works. Lake could see a small part of him above the root and arose from his hiding place, standing erect and fired. The ball passed through the bark of the root and the soldier rolled down the hill, while a dozen or more bellets [sic] whizzed around Lake's head. Gen. [Col.] Jackson was near by and told Lake to lie down or he would get his head shot off."[62]

Corporal Josiah Davis, Company A, 3rd West Virginia Mounted Infantry [later Company F, 6th West Virginia Cavalry] said, "The 3rd [West Virginia] fought in the center and was about 30 minutes in driving the enemy out of his breastworks on the brow of the mountain."[63]

Private George H.C. Alderson, Company A, 14th Virginia Cavalry, noted the 14th Pennsylvania Cavalry [probably Company A of the Pennsylvania troopers] assaulted the right of the breastworks "and came within 15 to 20 yards of us." Sgt. James Z. McChesney, Company H, 14th Virginia Cavalry, remembered, "the Yankees shot one of the 14th [Virginia Cavalry] named Myers [probably Pvt. John Aquilla Myles, Company A (2nd)] as he took out his pencil and memorandum book to write something, but died before he had time, and a Yankee got his

Weapons carried by Royal Paris Kershner, 14th Virginia Cavalry, include: top, two-band Enfield rifle with lockplate inscribed "Potts & Hunt-London; bottom, left, Kerr revolver and holster with belt. Octagon barrel says "London Armoury" and frame lists "Kerr's Patent 3647" and "London Arms W1;" bottom, right, three bayonets found on rocks behind the home of George Ralph Goode (Kershner's ancestor) on the brow of Droop about 50 years ago. MIKE SMITH

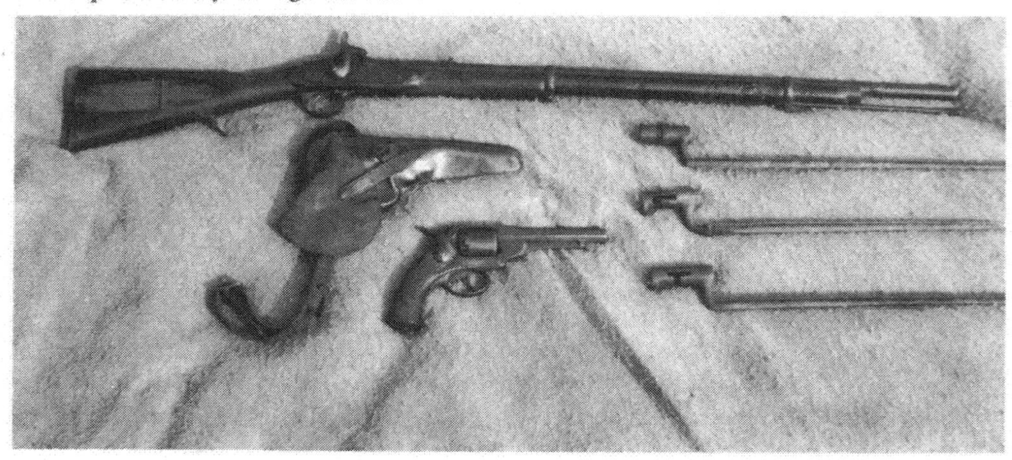

book and pencil." McChesney would later discover the Yankee would considerately send a letter and some personal effects to the dead Confederate's sister. The Federals captured some of the 14th Virginia Cavalry and would later inform a lady living near the battlefield "that the 14th fought like young devils."[64]

Colonel Cochran said that as the Yankees came up he was prepared to surrender, but when his potential captor exclaimed, "Stop, you damned, red-haired son-of-a-bitch," he became enraged and refused to surrender to one who addressed him in such a crude manner. His escape proved successful, although in the initial reports after the battle he was falsely reported as captured. Capt. William Thomas ("Billy") Smith, Company F, 14th Virginia Cavalry had three horses shot out from under him and received an "ugly wound." Pvt. John Andrew Fleshman, Company A (2nd), 14th Virginia Cavalry, had a horse valued at $700, which was killed. Capt. Alpheus Paris ("Captain Dod") McClung, Company K of the 14th, fell wounded in action. Pvt. Samuel Henry Lucass, Company F, 14th Virginia Cavalry, would later tell his father, "There was only about eight of our Co. engaged in it [Droop Mountain] and I was among them. The balls cut me pretty close and the Yankees pushed me tolerable tight and I was quite thankful to get out safely."[65]

Lieutenant Colonel Scott rallied the 2nd West Virginia and "advanced inside" the enemy breastworks, where it would later be determined the 2nd West Virginia took its highest rate of casualties, "near the breastworks on the crest of the hill." Lt. James B. Smith, Company H, 2nd West Virginia Mounted Infantry, the youngest officer in the regiment at 21, was the first to get inside the enemy breastworks at the Confederate center, accompanied by a portion of his men. Frank Reader of the 2nd West Virginia recalled, "After about two hours of fighting the 2nd & 3rd West Virginia regiments charged the enemy on foot and carried their posi-

Lt. James B. Smith of Co. H., 2nd West Virginia Mounted Infantry. The youngest officer in his regiment, he was credited as the first to enter the breastworks at the Confederate center. FRANK READER'S *HISTORY OF THE 5TH WEST VIRGINIA CAVALRY*

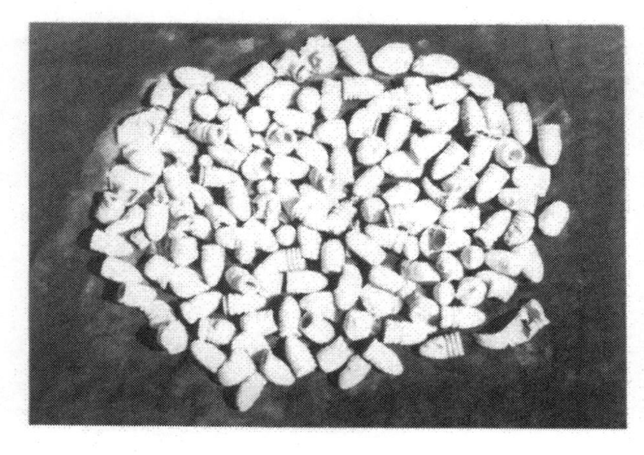

A hodgepodge of fired and non-fired bullets found on the Confederate center and right at Droop Mountain. GARY BAYS

tion." While assaulting the fortifications, Corp. Isaac Wilt, Company K, 2nd West Virginia, confronted a Confederate lieutenant [probably 1st Lt. N.B. Holland, Company K, 20th Virginia Cavalry] who refused to surrender while urging his men to hold the works. Wilt proceeded to thrust his bayonet through the rebel lieutenant, "who died with oaths on his lips, cursing the Yankees." This may be the same lieutenant referred to by John Henry Cammack of the regiment, who was recruiting in Richmond at the time of the Droop battle. Cammack, who had been elected 2nd lieutenant early in the war, recalled, "A young man from Monroe County, who was a special friend of some of the officers and men of the regiment, had been very active in forming a command...Before the elections he had boasted to family and friends he would be an officer, but was not elected." He was embarrassed by this so the colonel asked Cammack to step aside, which he did. Cammack added, "I think this young man was the most grateful soul I ever saw. He overwhelmed me with thanks and promised me undying friendship. Poor fellow, he was dreadfully wounded at Droop Mountain and incapacitated for further service"[66]

In a later account of the Federal attack on the center a Union soldier reported that the first dead man he and his perspiring, dirt-encrusted comrades observed after attacking the works was a "Negro with gun and cartridge box." Although some writers claim this is not possible, in actuality, this would not have been that unusual, as it is known the Negro cook of the 22nd Virginia fought with them at New Market in 1864. As the Federals continued their drive against the rebel center Sgt. Michael B. Keeny, Company H, 2nd West Virginia, was "stunned by a shot" which struck a tree near him. He said the shock caused him to lose "at least one-half hour of his life." The Federal line would advance some distance before he regained consciousness. Sgt. James L. Workman, Company B, 8th West Virginia, "was firing from behind a log and this puff of smoke revealed his position. When he heard a bullet from the enemy pass his ear rather closely, he changed positions." On the right of the breastworks Pvt. Thomas Swinburn, Company C, 8th West Virginia, entered the defenses and fell wounded in the neck and shoulder. Initially his wound was thought to be fatal but he recovered and returned to tour the battlefield in 1901. Lt. Col. Scott praised Adjutant John W. Combs and Lt. Alexander J. Pentecost, regimental quartermaster, who "exhibited great coolness and daring, and rendered important services throughout the entire fight." Scott also noted the performances of Capt. Douglas D. Barclay (Company D), lieutenants John R. Frisbee (Company D), Louis B. Salterbach (Company H), Andrew P. Russell (Company H), Charles H. Day (Company I), and Felix H. Hughes (Company B), "as being actively engaged during the entire engagement." A correspondent, believed to be a member of the 2nd West Virginia, described the fighting at the center, writing: "As soon as Col. Moor had fairly engaged the enemy's left flank, the 2nd and 3rd [West] Virginia being now within two or three rods of the works—under cover of the mountain—charged and took their strongest position, the 2nd [West] Virginia being the first to gain the works and hold them. The 3rd [West] Virginia was soon with us, and with the 8th [West] Virginia supporting our left under a severe fire of artillery and musketry, we soon succeeded in driving the enemy..."[67]

Colonel Jackson realized the Confederate center was crumbling and sent word

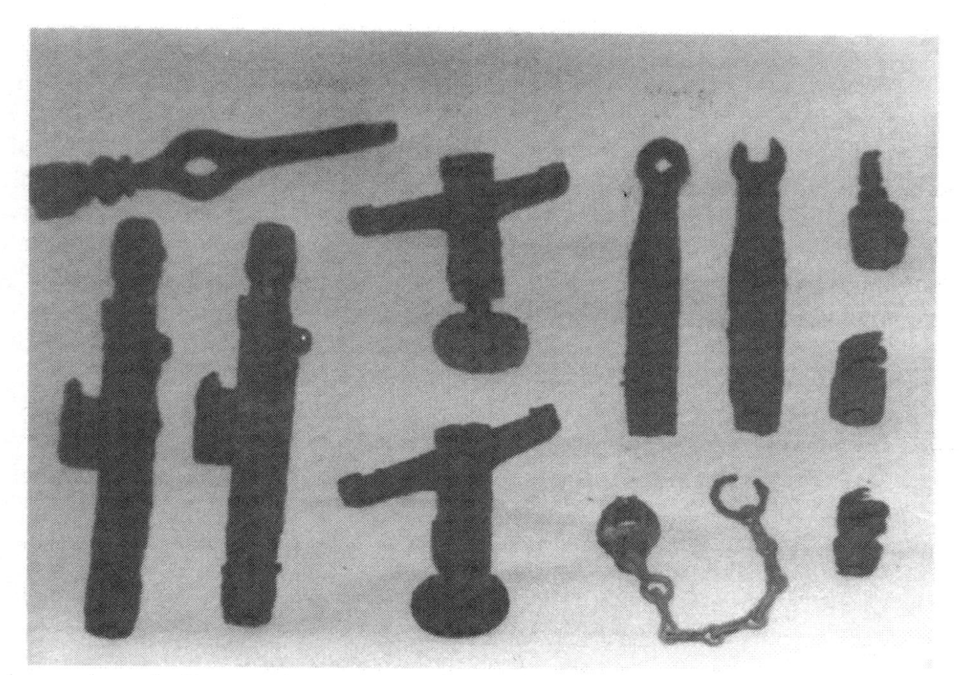

An assortment of gun tools found at the Confederate center. Top row, left to right: part of an Enfield pattern 1855 "T" headed combination tool; U.S. main spring musket vise; U.S. combination gun tools (closed and open ends) often referred to as Springfield gun tools and two musket worms. Second row, left to right: Two variants of part of an Enfield pattern "Y" combination tool, which consisted of a bullet punch, bullet screw, nipple wrench, screw driver, main spring clamp and nipple pick; U.S. main spring musket vise; musket nipple protector and musket worm (wiper). BRIAN ABBOTT

Iron objects recovered from the Confederate center at Droop Mountain. Left to right: unknown item (believed section of a musket interior) with roller buckle underneath; combination gun tool; remains of two Enfield pattern 1855 T-headed combination gun tools. GARY BAYS

of the situation to Echols, suggesting he move all or part of the artillery to the rear. Additionally, Major McLaughlin had also come to the realization the rebel left was "steadily falling back and the center faltering." As Lurty's Battery had nearly used up their ammunition, McLaughlin ordered Capt. Lurty to report to the rear along with Chapman's 24 pounder howitzer and the caissons. The remaining Confederate guns opened upon the advancing Federal line to the front. Having received instructions from Echols, Major McLaughlin ordered Lt. Randolph Blain of Jackson's Battery to place Jackson's Parrott gun and Chapman's 12 pounder howitzer "on the hill in rear, so as to cover the retreat should that be necessary." In the meantime, Captains Jackson and Chapman stayed with the two remaining pieces, Woods' three-inch from Jackson's Battery and one gun from Chapman's, which continued to pour a deadly fire into the advancing Yankees "with shell and canister, driving them back and preventing their advance" on the rebel left, front, as well as the pike.

Final Stand

Only an hour after being sent to the Confederate left, the 23rd Battalion Virginia Infantry, along with elements of the 14th and 19th Virginia cavalry regiments, had assisted Lt. Col. William P. Thompson stall the Yankee breakthrough, "alternately driving them back and being driven" back in return. Blessing noted Adjutant James A. Harden for "very gallantly, exposing himself to the hottest of the fire and doing all in his power to preserve order and win a victory." But with the intense fighting and the eventual breakthrough at the fence, Major Blessing began to retire his men slowly, "making several desperate stands." As the firing on the rebel left grew in ferocity the Confederate line began to fold back towards the center creating an angle. Thompson sent word to Echols that he was being overwhelmed by superior numbers, which prompted Gen. Echols to order Col. George S. Patton to move to the left and assume command. Patton was to take with him three companies of the 22nd Virginia Infantry; for this assignment Patton detached companies A, E, and I, comprising 125 men, and placed them under the immediate command of Capt. John Koontz Thompson of Company A.

A better choice could not have been made. Known to have been "brave to the point of recklessness," Thompson was a student at V.M.I. when the war broke out and left that institution in order to come to western Virginia to serve as drill master for the gathering Virginia volunteers in the Kanawha Valley. He soon afterward be-

Capt. John Koontz Thompson of Co. A (Putnam Border Rifles), 22nd Virginia Infantry. Received his fourth wound of the war while commanding three companies of the 22nd Virginia Infantry in a valiant stand on the Confederate left flank at Droop Mountain. CHARLESTON NEWSPAPERS, INC.

came Captain of Company A [Putnam County Border Rifles], 22nd Virginia Infantry. Already he had received three wounds in battle, including an 1861 injury when a musketball bounced off his head at Carnifex Ferry; his right eye had been shot out at Lewisburg in 1862; and he had been struck by a bullet in the face at White Sulphur Springs just three months previous to Droop Mountain.[68]

On reaching the left Patton surmised, "...it was evident that our little force was largely outnumbered and the enemy were entirely beyond both flanks. Our forces were retiring from the field in spite of the earnest and gallant efforts of Lieutenant Colonel Thompson and other officers to rally them."

Captain Thompson's three companies of the 22nd Virginia Infantry, along with one company of dismounted men from the 14th Virginia Cavalry, met the Confederate left about 300 yards from the pike. Capt. Thompson reported he found the rebel line at "the top of a spur of Droop Mountain, which was densely covered with laurel and underbrush." He also mentioned the Federals had gained the top, deployed across it, and were driving the rebels back. Under a galling fire Capt. Thompson deployed his 125 soldiers from the 22nd Virginia "in line parallel to and about 20 yards" from the enemy. Two companies of the 22nd Virginia and one-half of another were sent to Lt. Col. William P. Thompson's right in order to sustain the wavering 23rd Battalion, while the other half company was moved to Lt. Col. Thompson's left. This new alignment "immediately engaged" the enemy and temporarily checked the Federal advance, actually driving the bluecoats back some distance. It was at a high price, though, as Capt. Thompson sustained severe casualties. Historian Roy Bird Cook said that "in a space of one acre 13 were killed and 47 wounded." Thompson reported nine killed, 30 wounded, and 12 missing, for a total loss of 51 out of the 125 members of the three companies of the 22nd Virginia [although Confederate service records show only 11 killed, 17 wounded, and 12 captured, for a total loss of 40].[69]

Among Company A, Capt. John Thompson himself received his fourth wound of the war, as he was shot in the neck. Lt. Col. Andrew Russell Barbee of the 22nd Virginia Infantry, who was away at the time of the Droop Mountain battle recovering from his White Sulphur Springs wound, later described Capt. Thompson's wound by utilizing his medical knowledge. Barbee wrote, "Bro. John recd. a slight flesh wound about the neck...the Ball entered, just above the clavicle, & inclined to the right oblique upwards, came out on top of shoulder. He has suffered very little from it." Postwar historian Calvin Price would later incorrectly believe Thompson's eye wound was received at Droop Mountain instead of Lewisburg, and reported, "the fire was the hottest he [Thompson] had ever experienced...a bullet came so close to his face without touching him that the eye was drawn from the socket." Barbee, who was related to Thompson, pointed out Price's mistake when he also made reference to the Lewisburg injury stating, "His [Capt. Thompson's] *old* wound in right eye is getting along well. At times when exposed to great fatigue in cold & wind there is considerable discharge of thick matter from the orbit, but a little rest & the use of cold water & syringe to work out the paste & all is well again...the discharge is from orbit & not from disease of Lachrymal sack or duct. The free lachrymal discharge from which he once was so much annoyed, has well nigh entirely passed off."[70]

Other casualties in Company A, 22nd Virginia Infantry, included Pvt. Paul

Dickerson, who received a gunshot wound of the left side and pelvis; Pete [Dick?] Cartwell, '...who gets wounded every fight," was shot in the foot; Pvt. Stephen N. Burford, wounded in the leg; and 2nd Lt. William S. McClanahan, who fell mortally wounded. Lt. Col. Barbee, the original Captain of Company A, would say of the 2nd lieutenant: "...my lamented McClanahan...died as he lived the true old Virginia gentleman & brave Confederate soldier." Col. Patton called McClanahan "a gallant soldier." Within Company A at least four were killed, 11 wounded, and one captured. Barbee, torn with grief, would later write, "Poor me, with shock, [Corp. William S. S.] Morris & [2nd Corp. William H.] Hubbard, of our old company, together with other true friends & gallant soldiers now 'sleep their last sleep' on Droop Mountain. My loss is great. I mourn sincerely"[71]

Company E [Elk River Tigers] of the 22nd Virginia, composed primarily of men from Kanawha County, West Virginia, saw at least six killed, five wounded, and four captured, including Capt. George Steptoe Chilton, who was wounded by a bullet through the right jaw. Pvt. Jessee Shamblin was wounded in the throat and later captured in the retreat as he tended to the wounded.

Of the three companies of the 22nd Virginia, Company I [Boone County Company] took the lightest casualties, with at least one killed, one wounded and seven captured. Among the prisoners from Company I was Capt. John P. Toney. Pvt. James Henry Allen was wounded by a saber to the shoulder and his brother, Pvt. William Perry Allen, was killed. "A Mr. Powell from Greenview" was with William when he died. "T.M." of the *Wheeling Intelligencer* boasted, "The 22nd, Col. Patton, has frequently expressed a desire to have a chance at the 10th [West Virginia]. All I regret is that the other six companies of that regiment were not present to have participated in the sound thrashing which we were abundantly able to have given them."[72]

Colonel Patton and Lt. Col. Thompson made an unsuccessful attempt to rally the 14th and 19th Virginia cavalry regiments to aid Capt. Thompson. Patton wrote, "I now endeavored to rally men to his [Capt. John Thompson's] support, seconded most gallantly by [Lt.] Colonel Thompson, and was successful in collecting a considerable number together...when I received a communication from General Echols informing me the whole right had given way, and ordering me to fall back to the main road and join him." Capt. John Thompson said an effort was made to rally the 23rd Battalion and the detachments of 14th and 19th cavalry regiments "which were giving way in some confusion, but without success."

While the Confederate left struggled to hold position, Gen. Echols arrived at Col. Jackson's position at the center of the rebel line, which was also giving way. Echols helped rally many of the soldiers and hold the line while Major McLaughlin's two artillery pieces "rained canister and grape upon the enemy, checking their advance on the center and right" for about 20 minutes.

Despite the vigorous and gallant efforts of the Confederate officers and their commands, by 3:00 P.M. the rebel line was in shambles. Lt. Col. Thompson reported the reinforcements sent to aid him "fought with great coolness and gallantry, but they, too, with the rest, gave way." Patton sent word to Echols that superior numbers were forcing the left back and his force was not strong enough to provide resistance. He also said the "left was being driven and bent back [angle] to the rear."

Breakthrough on the Center and Right

The Confederate center and right also began to crack at about this time, as evidenced by correspondent "Irwin," believed with the 8th West Virginia, who said, "...with a cheer we charged up the steep mountainside and over the breast-works, officers and men mingling in confusion, covered with perspiration, dirt, and their clothes covered with burs. Just at this time Ewing's battery found a position and opened fire on the rebels" and "the rebel battery swept the point with grape and canister, but our men fought from behind stumps, trees, and logs...and the rebels became terror stricken and began to retreat." Samuel D. Edmonds, Company B, 22nd Virginia Infantry, recalled: "We seen the Yankees fixing to use a cannon on us & about 400 yards & Company H was ordered to make ready with the Enfields and fire at them & I noticed they didn't stop & then

The ridge to the right of the park lookout tower is believed to have been the first position of the 23rd Battalion Virginia Infantry on the Confederate right. GARY BAYS

we was ordered with Austrian rifles to fire on them & then they left there...in a few minutes here comes 'Mudwall's' men [retreating] & hollering at our men that the Yankees was coming & some of their officers begging them to stop & some a drawing their swords before them & but none stop; there wasn't anything to hinder them & as they moved right on & no stop." "Irwin" recalled the scene as the West Virginia troops moved up the face of the mountain writing, "The skirmishers kept a constant fire, while the heavy roll of musketry on the right, as it curved around the mountain, was as steady as the fire fanned by the wind advances through the leaves on the mountainside. The keen crack of the Enfields of the Tenth, and the deeper bass of the big bores of the Germans, could be readily distinguished, while overhead a strong wind made a deep, steady roar in the naked branches of

Pvt. Samuel D. Edmonds of Co. B (Jackson County Border Rifles), 22nd Virginia Infantry. Fought on the Confederate right flank at Droop Mountain and during his later years wrote a vivid account of his wartime experiences. DROOP MOUNTAIN STATE PARK

the forest, and to heighten the grand battle picture, the woods were on fire, the branches of the trees crossing to the ground, under the efforts of the shot and shell, accompanied by the heavy roar of the artillery, and music and bursting of shells, and the constant roll of the musketry."[73]

Lieutenant William Bahlmann, commanding Company K, 22nd Virginia, noted the flight of Col. Jackson's troopers from the enemy, and wrote, "...while standing in the woods some men of another brigade [Jackson's] broke right through our company." Bahlmann's men, however, stood firm, convinced the dismounted cavalrymen would not stand fast. Bahlmann said Col. Jackson's officers quickly rallied the men, but one of the officers was shot, possibly by his own men. "I saw him turn pale and just then I caught something. An honest, old-fashioned ounce minie ball struck me in the left knee," countered Bahlmann. As Bahlmann fell to his knees wounded, Pvt. Augustus ("Gus") Tyree, Company C, 22nd Virginia, and Pvt. Erastus P. Starkes and Sgt. Napoleon ("Nip") Baker, both of Company K, 22nd Virginia, picked up all 190 pounds of Bahlmann and began to carry him to the rear. The musketry and cannonading grew more intense, which provoked Bahlmann to worry that all four of the men would soon be killed.

Lt. William Frederick Bahlmann of Co. K (Fayette Rifles), 22nd Virginia Infantry. While commanding his company at Droop Mountain, he was shot in the knee and captured on the Confederate right. LARRY LEGGE, BARBOURSVILLE, WEST VIRGINIA

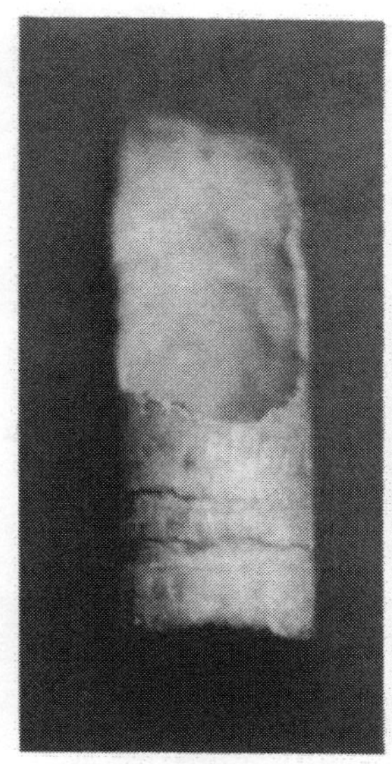

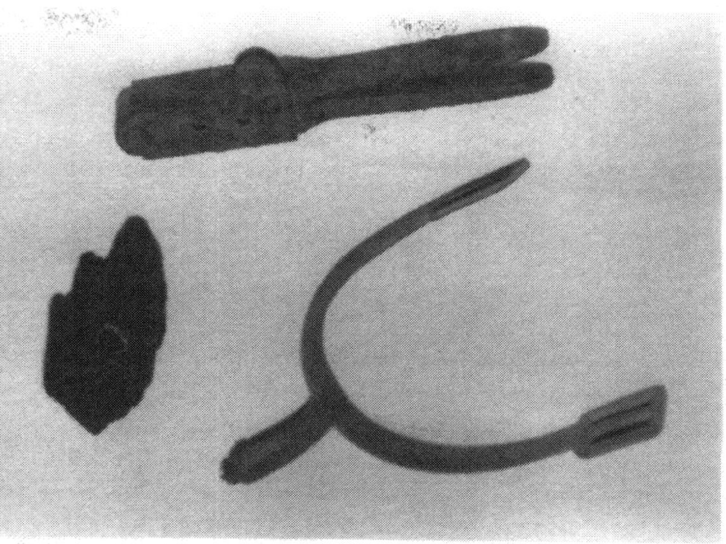

At left, a .58 caliber Confederate Gardiner bullet. When fired the nose area "blew out" due to the poor grain of lead. Found on the Confederate right. GARY BAYS
Right: three items found on the Confederate right. At top left, is a brass cavalry spur and a bullet lodged in wood. At bottom is an iron bullet mold. BRIAN ABBOTT

Bahlmann instructed the men to put him down and save themselves but Tyree offered to "carry him out on his back." Lt. Bahlmann refused the offer, stating, "No, Gus, we shall both be killed." Tyree and Starkes left but Baker remained despite Bahlmann's plea, "Nip, you got to go." Baker, not wishing to abandon his commanding officer, burst out in tears and replied, "I don't want to leave you." Forced to be stern Bahlmann demanded, "Nip, you can't do me no good, I order you to go." Baker finally departed, reluctantly, leaving Bahlmann in a small depression in the ground, where the wounded officer said he braced himself for the next bullet, convinced that if he was not struck in the abdomen or heart he would survive.[74]

The oncoming Yankees and the fleeing rebel troopers caused mass confusion on the Confederate right flank, particularly endangering the 22nd Virginia, which made a number of valiant stands. Privates George H.C. Alderson, John W. Legg, and Jehu Holley, all from Company K, 14th Virginia Cavalry, defending the right and center, fell back and "stopped on the ridge...[took a] shot each, then retreating some 400 yards...found a number of men lying behind a log and someone hollering 'wait until they come up.' When they came up we fired a volley and retreated back into the field" where Major Robert Augustus Bailey of the 22nd Virginia was posted. At this point the 22nd Virginia suffered heavily, as Capt. Richard Quarrier Laidley, Company H, fell wounded "while gallantly leading his men." Pvt. Pete Stribling of the same company was struck in the chest by a minie ball which "entered his left side and came out on the right of his spine," contributing to his death about seven months later. 2nd Lt. John P. Donaldson, also of Company H, was hit in the side and shoulder, while Pvt. Archibald ("Arch") P. Young, of Company H as well, was wounded as a bullet ploughed away the bone of his skull from front to rear of his head. Pvt. Mason ("Mase") V. Helms, Company K [Fayetteville Riflemen], who had entered the battle sick, was mortally

A 1928 view showing McCarty Farm (Gore's Grove and State Rt. 24). DROOP MOUNTAIN COMMISSION BOOKLET

wounded, shot "through the head and body and...a finger shot off." He would pass away the following day. Pvt. Hamilton B. Caldwell, Company K, was killed, leaving behind "a young family." 1st Lt. Woodson A. Tyree, Company C [Mountain Cove Guard], a "good fellow...[and] brave soldier," went down with a severe wound in the nape of the neck which "bled a great deal." Pvt. Henry H. Skaggs, age 24, Company C, 22nd Virginia "lost a leg at Droop" due to his wound. Pvt. Clarence L. Jackson, Company H, 22nd Virginia Infantry, son of Capt. Thomas E. Jackson of Jackson's Battery, was "just 17 years old [and] was very severely wounded–his hand being shattered by a minie."[75]

James H. Mays, Company F, 22nd Virginia, had earlier in the day been sent with 10 men "to hold a particular point" of threatened territory. Just prior to the Yankee army reaching his post and cutting off his route of retreat, a man crawled to the position and took five of his best men, leaving Mays with only five. Now, with the Federals closing in, Mays and his men had no choice but to either surrender or attempt to avoid capture. Of the group only Mays got away, while the "others were all captured without making much effort to escape." Mays said the choice of his comrades was "a fate hardly better than being killed outright, for most of them died in prison."[76]

James Henry Mays of Co. F (Rocky Point Grays), 22nd Virginia Infantry, circa 1860. During the early part of the Droop Mountain battle, Mays was sent with 10 men to hold a threatened position and was nearly captured. He also witnessed the mortal wounding of Major Robert A. Bailey. LEE MAYS, RIPLEY, WEST VIRGINIA

Withdrawal and Retreat

During the latter part of the battle Lt. Denicke, 68th New York Infantry, signal corp detachment, had been ordered by Capt. Ernst Denicke "to change his station to the new position taken by our artillery. From this station he also communicated with [the] center station, sending and receiving messages and observing the enemy's movement." From his position with the Federal artillery Lt. Denicke was the first to spot the collapse of the Confederate line and so communicated the information to Gen. Averell. Having received Denicke's communication of the Confederate withdrawal at approximately 3:00 P.M. Averell sent a verbal message, via an orderly, to Major Thomas Gibson to advance his Independent Cavalry Battalion as quickly as possible. Gibson's men had been posted three and three-fourths miles "in a direct line from the enemy's batteries" when the battle had opened and had remained there until receipt of Averell's message. Gibson immediately ordered them out at a trot [a fast gait] for the mountaintop. One section of Ewing's Battery was also sent in pursuit as Col. Schoonmaker's advance also arrived "and went forward with the troops that [soon afterward] carried the summit."

At the unofficial termination of the battle, at about 4:00 P.M., Echols received information that at 2:00 P.M. Gen. Duffié had advanced to the top of Little Sewell Mountain, 18 miles west of Lewisburg, advancing rapidly upon Lewisburg with 2,500 men and five pieces of artillery. At this time Gen. Echols knew it was useless to continue the unequal contest and issued orders for his army, now greatly confused and with many already in flight, to "fall back slowly," followed by the withdrawal of the artillery. Echols informed Col. Jackson the Federals had breached the left flank and were about to gain the rear. He also ordered Jackson to fall back, which he did in about as good order as could be expected "under severe shelling and enfilading fire of musketry." Pvt. James Steel McClung, Company K, 14th Virginia Cavalry, who had served on the skirmish line on the face of the mountain during the morning, wrote, "...while eating, orders came to [Pvt. David C.] Groves, and me to go down next to Lewisburg and take some cattle into Irish Corner, that we were retreating. The company got together after being scattered for awhile." Col. Patton, who had been with Lt. Col. Thompson on the left flank not more than six or eight minutes, received a message from Echols informing him that the entire right flank had crumbled and was ordered to fall back to the main road and join Echols. Lt. Col. Thompson also recalled that a few moments after Patton's arrival on the left "we received an order to march in retreat." Patton complied by sending orders out to all field commanders to fall back slowly and march to the road. Pvt. Milton Butcher, Company B, 20th Virginia Cavalry, [later of Company B, 46th Battalion Virginia Cavalry], claimed he was a courier for Col. Jackson during the battle and carried the last order that "his chief" gave, which was for Lt. Col. Clarence Derrick [apparently Butcher was confused, as Major Blessing, not Lt. Col. Derrick, commanded the 23rd Battalion Virginia Infantry in the affair] to take his 23rd Battalion and "fall back to the pike, west of artillery." After delivering this message to Derrick [Blessing], Butcher was caught in the enemy assault, lost his horse and was captured, as were some forty other rebels at the same location. However, Butcher managed to escape from his captors before getting off the mountaintop, and it is reported that years after the battle the bones of his dead horse were located on the battlefield.[77]

Captain John K. Thompson, Company A, 22nd Virginia, yet suffering from his latest wound, said the Federal forces had reformed and the superior numbers had caused the rebel line to break, with Thompson retiring slowly, "contesting every inch of ground...yielded." Having received Patton's order, Capt. Thompson fell back to the road [pike] in good order. Major Blessing's 23rd Battalion made one last unsuccessful effort to hold the enemy in check, then fell back to the road per Patton's instructions. Capt. James Barnes, Company H, 23rd Battalion, would write, "...I don't think he [Pvt. Nathaniel Davis Hill, Company H] is kiled [sic] though I believe that he is a prisoner...the last time that I saw him was after we had retreted [sic] some 5 or 6 hundred yards–he was close to me and was loding [sic] and shooting and we got seperated their [sic] and I saw him no more though he might of got wounded...I don't think...[he]... is killed. Though if he is[,] he dide [sic] fighting for his country and he was as goo[d] a soldier as ever Soldered [sic] a musket and Stood his groun[d] like a faitful [sic] soldier and was as little excited as any man I ever Saw fight." Pvt. Hill was actually captured as were some 30 other members of the battalion.[78]

Colonel Moor's two infantry regiments continued to advance forward until his left reached the cleared hill where the Confederate artillery was posted. Correspondent "Irwin" wrote, "...the gallant Tenth and glorious old Twenty-eighth closed in, and the rebels became terror stricken and began to retreat, and then the retreat became a rout, while from our boys went up one prolonged cheer that was kept up, and the pursuit began immediately–it was a hard day's work, but officers and men worked with a will, and did their whole duty–no flinching, no shirking." Major McLaughlin realized his two guns were in a precarious situation, as the infantry supports were quickly retiring and his guns had come under flank fire from Moor's soldiers on the left. This movement by the Federals from the left to right flank was witnessed by Pvt. Alexander Hall, Company B, 22nd Virginia, who "turned up a rock and layed down on [the] grass on [his] belly. Rock in front, put cartridges by side, and soon saw Yankees passing to our left across a hollow &...Hall cracking at them. We all got busy..." Orderly 1st Sgt. August ["Gus"] Hess, Company A, 28th Ohio Infantry, had "a bullet cut across his breast and his daybook, which he carried in his pocket, in halves." Col. Moor later praised 2nd Lt. Jacob H. Mork [Co. B, 28th Ohio, who had been made Acting Assistant Adjutant General, 1st Brigade, 4th Division on May 15, 1863] "by carrying orders and even executing some in most exposed places with coolness and judgment." But all in the 28th Ohio did not perform well, as evidenced by Sgt. Christian Hahn, Company E, a 37-year-old stonecutter, who was "reduced in rank to a Private the following day for cowardice at Droop Mt." Lt. Micajah Woods of Capt. Jackon's Battery remembered, "all eight [actually seven] of our guns, Chapman's, Lurty's and ours opened on the infantry which was advancing rapidly on our left. All did well–the enemy was evidently hesitating on further advance under such fire of shot & shell."[79]

Just as Major McLaughlin gave the order for his two remaining guns "to limber to the rear" he received an order from Gen. Echols, relayed through Col.

Jackson, to move all of his guns as quickly as possible to the rear in order to cut off a retreat. McLaughlin had these remaining two guns limber up just as Ewing's Federal piece got in position and opened a heavy fire on the two rebel guns, along with flank fire from the Yankee infantry. A portion of Col. Jackson's men made one brief stand in order to cover the retreating Confederate guns, although following the retirement of most of the troopers, the rebel pieces continued to hold the Federals in check with "well-directed discharges of grape and canister." Lt. Micajah Woods of Jackson's Battery wrote, "...the supports to our guns ran ignominiously, leaving us entirely to the mercy of the enemy, who were nearly in our rear, thus we stood depending on our own resources. The enemy ran another Battery right below us & opened furiously—their infantry approached rapidly. I shelled them to the best of my ability—till within 450 yards when I gave them 12 rounds of canister well directed which threw them into great confusion. At this juncture I withdrew my gun under a heavy fire of infantry & artillery which however only killed one man near us." Woods added, "The enemy, still came—our supports everywhere were yielding in confusion until our two guns (Chapman's & mine)—were left entirely in rear & unsupported. The victorious foe advanced eager for their prey but when at a distance of about 400 yds., we opened rapidly with canister & so effectively that 12 rounds of double charges threw them into confusion & compelled them to retire behind a little eminence." This is further confirmed by Col. John H. Oley, 8th West Virginia, who said that after an hour's fight to gain the summit his men got within 50 yards of the enemy battery but were unable to capture it due to heavy fire from the artillery and it's supports, and "the sudden withdrawal of the battery."[80]

Lieutenant Randolph Blain, also of Capt. Thomas E. Jackson's Battery, claimed that when the infantry in his front was struck by the oncoming Federals, to his "surprise & shame they turned and ran like dogs." Blain, Echols, and several others attempted to rally the troops to protect the last piece on the field, but all to no avail. Apparently this piece belonged to Micajah Woods, who said, "my gun was the last to leave the field." Blain's gun was then "limbered & brought off at a gallop, without an infantryman around it." This contradicted Col. Jackson, who said his men remained with the battery until he felt confident the guns were safe, at which time his men moved to the pike. There may be some truth to this, as Col. William W. Arnett, 20th Virginia Cavalry, said that as his men were afraid of being cut off from behind they fell back to the rear of the battery until Col. Jackson ordered them [further] back.[81]

The remainder of the Confederate artillery preceded Major McLaughlin off the battlefield area, having received the same order directing them to retire, which was accomplished in good order. However, Captains Jackson and Chapman remained on the field "until the last pieces were withdrawn, directing their fire and assisting in bringing them off." Micajah Woods remembered, "...we limbered up & moved rapidly to overtake the retreating command." Major McLaughlin later praised Sgt. Cephalus Black of Chapman's Battery, who was in charge of the removal of the artillery caissons.[82]

Samuel D. Edmonds, Company B, 22nd Virginia, recalled the rebel artillery had been "...down on [the] third bench from top" of the mountain and the Confederates had a difficult time saving the guns as the Yankees mixed in with them

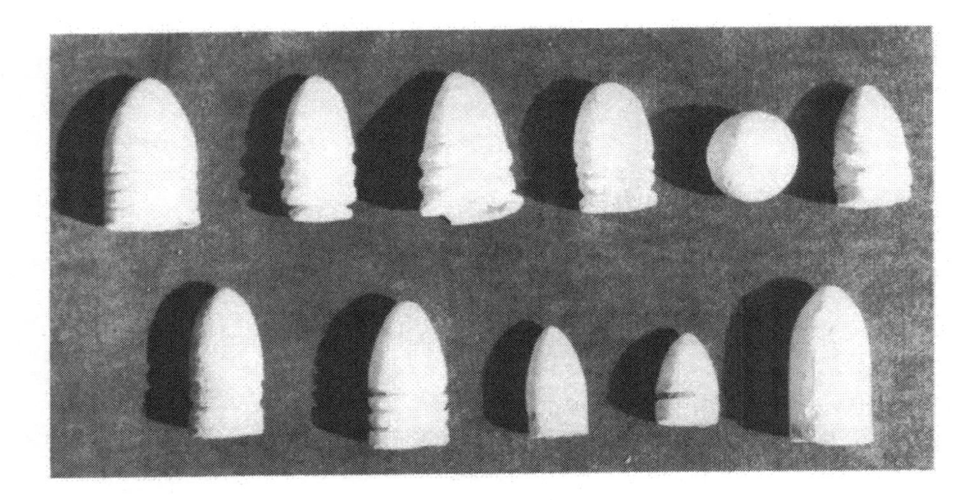

Top: Various fired and non-fired bullets found on the Droop Mountain battlefield. Top row (left to right): .69 caliber three-ring bullet; .54 caliber Sharps carbine bullet (known as a ring-tailed Sharps); .69 caliber Italian Garabaldi bullet; .57 caliber Lindner carbine bullet; musketball; .54 caliber Sharps carbine bullet. Second row (left to right): .58 caliber Gardner bullet (Confederate); .58 caliber three-ring bullet; .44 caliber Colt dragoon bullet; .44 caliber Colt Army revolver bullet and .58 caliber Enfield bullet. GARY BAYS
Middle: Four iron objects found in various locations on the Droop Mountain battlefield, including remains of a pocket knife, two roller buckles and a large iron buckle. GARY BAYS
Bottom: Rifle and ammunition pouch or haversack reportedly carried by a Confederate soldier at Droop Mountain. VERNON C. WILLIAMS, BLUE ISLAND, ILLINOIS

Civil War drum found on the Droop Mountain battlefield by a citizen the day after the battle. The tension ropes and drumheads are not original parts. MIKE SMITH, DROOP MOUNTAIN STATE PARK

continually. Edmonds added, "Then here come the Yankees, nothing to hinder them & we make a general move & Yanks firing right after us." Pvt. Polk Herdman, Company B, 22nd Virginia Infantry, paused behind some protective cover and noticed that Pvt. Alexander Hall, also of Company B, 22nd Virginia, had taken refuge behind a tree when a "Dutchman" [probably a member of the 28th Ohio] came up and cursed at Hall, "Surrender, damned rebel." In response, Hall "leveled his gun & fired & he jumped & run." Edmonds and his comrades then ran to the top of the mountain, then stopped and looked back at the middle bench where "there was a lot of our men holding them back." Edmonds and Pvt. Columbus ("Lump") Lewis, Company B, 22nd Virginia, made a plea for assistance for the defenders on the middle bench, and heard Gen. Echols, who was mounted nearby, respond, "Yes, boys. We better go & do all we can but we're about whipped." Edmonds and the other soldiers ran back down to the bench, where they fired one volley, but as the Federals outnumbered them three to one and were within 10 steps of them, Edmonds and the group opted to fall back. A member of Company B, 22nd Virginia, said that Edmonds ran just ahead of him through "the swag—said he thought to himself if he could just run as fast as...[Edmonds]...he would soon be safe." When the men again arrived at the top Echols was gone and there were three officers and Major Robert Augustus Bailey attempting to rally the panic-stricken men.[83]

Echols and Col. Milton Jameson Ferguson said confusion prevailed due to the uneven terrain and an alarm which had earlier spread among the horse-holders of the dismounted cavalry that the enemy had turned the left and was going to strike the rear. Sgt. James Z. McChesney, 14th Virginia Cavalry, complained that the retreating infantry "got on some of the 14th [Virginia Cavalry] horses and run over their own men, that's the reason so many of our Regt. had to foot it out and lost their horses." Lt. Randolph Blain of Jackson's Battery also noted his surprise that the 22nd Virginia ran as soon as they did, although this later proved to be a blessing in disguise. Pvt. Samuel Baldwin Hannah, Company B, 14th Virginia

1st Lt. Samuel B. Hannah of Co. B, 14th Virginia Cavalry, lost his horse, had his hat shot off and was nearly captured at Droop Mountain. ROBERT DRIVER, BROWNSBURG, VIRGINIA

Cavalry, reflected, "I must confess that we were badly whipped & suffered one of the most disgraceful routs to happen ever conceived of since the war. I came near being taken in consequence of being dismounted to fight & when the retreat began it soon became a stampede. I lost my horse and if it had not been for ------ who took me up behind him I would certainly have gone to the bushes to make my escape. None of our boys were hurt...I lost my hat after being shot through it, we were totally overpowered...I lost my bay horse in the fracas..." The *Richmond Examiner* was much more harsh, describing the retreat and battle as "a shameful, unmitigated disgrace."[84]

When Col. Patton arrived back at the pike with his command from the left flank, Gen. Echols ordered him to return to the front, the Confederate right, to try and rally the men, but it was already too late and Patton was unable to restore order. Moving back to the pike Col. Patton found the Confederate center and right had also fallen back, creating a massive scene of confusion, the roads clogged with "artillery, caissons, wagons, and horses." Capt. John K. Thompson, Company A, 22nd Virginia, said total confusion reigned as the pike was blocked by "artillery, baggage wagons, and fleeing cavalry...for some distance" to the Confederate rear. Col. Jackson stated, "A portion of my command with others were cut off, there being but the one road to retreat upon, and that at one time somewhat jammed by horsemen, infantry, and trains..." He later reported, "...My train and artillery were all brought out safely, except one wagon loaded with corn, which broke down. The horses are safe." 4th Corp. James W. Sisler [Shisler/Schisler], Company F, 19th Virginia Cavalry, who claimed he was quartermaster of Jackson's Brigade, later stated he ran to Col. Jackson and inquired as to what he should do with the wagons, to which Jackson reportedly replied, "damned if I know." Taking the initiative, Sisler ordered his teamsters to reverse the wagons and retreat down the pike. In the ensuing confusion the team of Echols' ordnance wagon "became frightened, whirled around, and the tongue broke on the wagon." Unable to move the vehicle, the teamsters threw some fence rails in the wagon and set it ablaze, igniting the ordnance stores. As a result of the ensuing explosion, Sisler claimed, for several years after the war people would visit the spot and gather up lead from throughout the field. The Federals later boasted of killing a Major McMahon of the rebel army, but the only officer with Echols who possessed this name was Major Edward McMahon, Quartermaster for Echols, who is known to have survived the battle. Federal correspondent "Irwin" added, "Immediately in rear of the battlefield was the rebel commissary building, and they had tumbled out barrels of flour and provisions, with arms, ammunition, accouterments, clothing, etc., thrown away in their flight." Echols said that despite the chaos Patton did everything possible "and was as usual, conspicuous on the field in the thickest of the fight."[85]

Finally, disorganization caused many of the remaining Confederates to also panic and flee to the woods for safety. However, Capt. John K. Thompson's three companies of the 22nd Virginia, though aware of being vastly outnumbered, "retired fighting obstinately, maintaining their alignment until run over and dispersed" by their own fleeing cavalry. A portion of Col. Jackson's men were cut off as well, although most would eventually find their way back to the command. Lt. Col. William P. Thompson reported that after he received the order to retreat he fell back to the pike, where he found the "main command marching in retreat." He also noted that Company E, 19th Virginia Cavalry, under Capt. James W. Ball, was never in disorder during the fight or subsequent retreat. Pvt. Lanty Alexander Hefner, Company E, 26th Battalion Virginia Infantry, lamented, "our retreat over Droop Mt. and across the Greenbrier River was so rapid [we] crossed the river and never got [our] feet wet."[86]

Colonel Augustus Moor's Federal infantry eventually arrived at the point of mass confusion and poured deadly rounds into "the moving rebels, killing and wounding artillery horses, rebel officers urging to make another stand, others cutting loose fallen horses, driving and pushing on cannon and caissons through their infantry."

Death of Major Bailey

It was reported by Samuel D. Edmonds of the 22nd Virginia that as the regiment fell back the pike was littered for a half-mile with guns and haversacks, until the regiment arrived at a point where the officers had stopped to make a stand. Company B, 22nd Virginia, then numbering about 25 men, was ordered to fall in beside the road. A number of unsuccessful efforts were then made by the rebel officers to rally the demoralized troops. Conspicuous among this group was Major Robert Augustus ("Gus") Bailey, 22nd Virginia Infantry, although various accounts of his deeds exist. One version states Bailey grabbed the regimental flag from a wounded Ensign Elijah D. Dotson [he was probably a color sergeant, as he did not become an ensign until April 30, 1864], Company E, 22nd Virginia, and

A 1928 view of the spot where Major Robert A. Bailey, 22nd Virginia Infantry, fell mortally wounded. The small white marker just to the right of "State Road #24" indicated the exact location. The marker "vanished" years ago and today the exact location where Bailey fell remains uncertain. DROOP MOUNTAIN COMMISSION BOOKLET

waved the banner to rally the men. Pvt. George Henry Clay Alderson, Company A, 14th Virginia Cavalry, claimed as he fell back during the battle to the field where Bailey was located the Major was seen waving the flag and yelling, "Rally 'round the flag, boys." About 100 to 125 men stopped to assist Bailey, who was supporting a rebel artillery piece. Alderson, standing to the immediate right of Bailey, watched in horror as Bailey fell mortally wounded. George attempted to grab the flag but someone else got to it first. Alderson claimed Bailey had the flag of the 14th Virginia Cavalry, which was possible, as Bailey had briefly served as lieutenant colonel of that regiment, but it was more than likely that of the 22nd Virginia, as Patton said when Bailey was "...struck he was bearing the colors of <u>his</u> regiment and rallying his men by his voice and example." Alderson also recalled that after the war he had always heard Bailey "fell close to the place where the road to Jacox leaves the Seneca Trail [Rt. 219]." Samuel D. Edmonds of the 22nd Virginia recalled the incident in more detail, and related that the regimental colorbearer [Dotson] had passed the flag to Bailey in order to aid Pvt. John Bennett, Company B, 22nd Virginia, who had a minie ball enter the front of his thigh, which "left four inches above the knee joint,...[then] came out one inch higher, passing to [the] inside of the bone." Two of Company B made an unsuccessful effort to retrieve Bennett, who was captured. Edmonds added, the rebels were then forced to make "a general run about half mile & last I seen of Bailey the enemy was in a few steps of him. He didn't stay there long as the pike was level for half mile & as I got started on pike I seen the Yanks coming thick through brush & if we'd been a few minutes later we'd had to a fought our way out." Edmonds went on to relate that as Bailey reached them with the flag, "Here came the enemy firing in to us with cannons & small arms. Bailey was shot there & died in four hours [Bailey actually did not pass away until November 11]." Another witness to the mortal wounding of Major Bailey was James H. Mays, Company F, 22nd Virginia, who claimed that as he fell back he ran past Bailey, who "was almost alone and vigorously waving the Stars and Bars, and yelling at the top of

Modern view of the wooded area near the monument base marking the spot where Major Robert A. Bailey, 22nd Virginia Infantry, was mortally wounded. GARY BAYS

his voice." Mays said as he approached Bailey the Major ordered him to stop and help, but sensing Bailey was doomed, ignored the command and continued his flight just as Bailey went down. Lt. Micajah Woods of Jackon's Battery added, "A few brave men were rallied by gallant officers & came to the rescue [of Woods' artillery] when we had gone a short distance. Here Major Bailey of the 22nd fell, severely, if not mortally wounded. Lieut. [John P.] Donaldson [Company H, 22nd Virginia] and a large number of their wounded, immediately around us, besides several killed—mostly belonging to the 22nd. Fortunately, almost miraculously, we escaped and our horses were not killed." Possibly as Bailey fell Pvt. John Henry McDevitte, Company G [Wyoming Riflemen], 22nd Virginia, managed to save a flag and was later promoted for his brave action. Patton praised Bailey, writing, "In him the cause has lost a brave and dedicated officer, whose cool courage and excellent judgment had been tested on many fields." Gen. Echols said, "No soldier ever displayed more dauntless courage than did he upon this, his last battlefield." Major Bailey's body, which fell into Federal hands, would eventually be brought back to Lewisburg and buried in the Old Stone Presbyterian Church Cemetery, about "30 steps east of the back" of the building.[87]

Following Major Bailey's mortal wounding the Confederate retreat became pell-mell. The commanding officer of Company B, 22nd Virginia, barked out, "Company B, get out & save yourselves. We can't do anything here." A few of Company B remained but about 10, including Samuel D. Edmonds, ran about 100 yards to a level flat, then over a steep break down to the Greenbrier River. Many tossed their weapons over the mountainside although Edmonds kept his "till war's end." Edmonds said that as soon as they were over the break they were "out of sight of the enemy." James H. Mays, Company F, 22nd Virginia, claimed the firing almost ceased after Bailey fell and the bulk of the remaining defenders

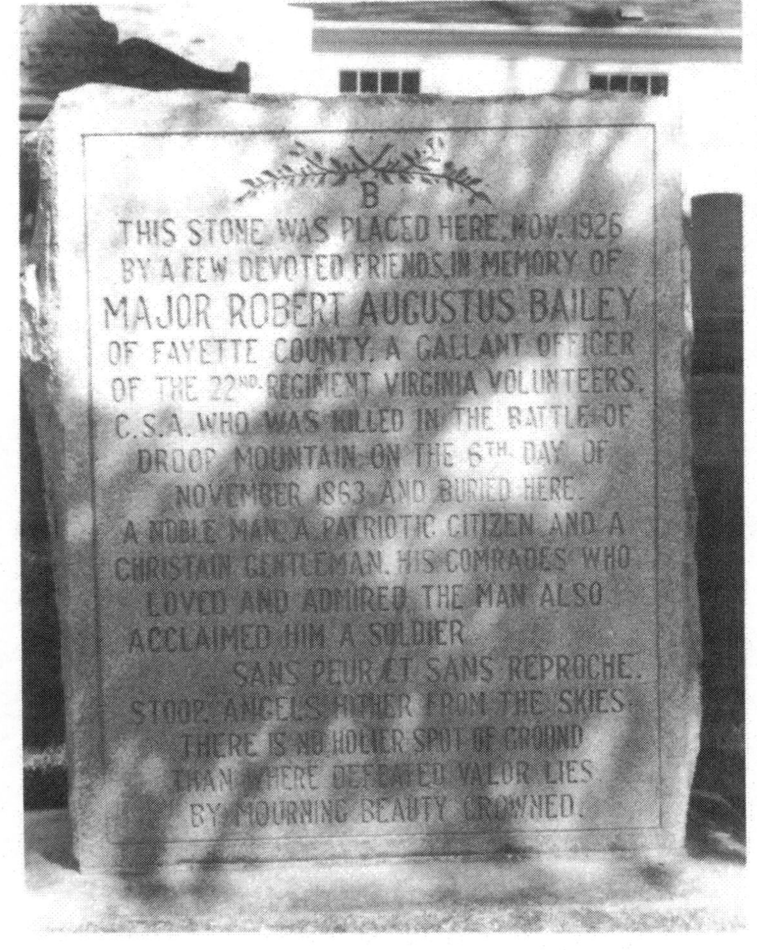

Top: This monument base is believed to possibly be the uncompleted monument marking the site where Major Robert A. Bailey was mortally wounded. It is located in the park woods just off Rt. 219 and in line with the state highway historical marker describing the battle. The base is of the same measurements as those of Baxter and Bender. Speculation is the remainder of the monument is on Bailey's grave at Lewisburg and never made it to the park. GARY BAYS

Bottom: Grave of Major Robert Augustus Bailey, 22nd Virginia Infantry in the Old Stone Presbyterian Church Cemetery at Lewisburg, West Virginia. Bailey was mortally wounded in action in Droop Mountain. BRIAN ABBOTT

was widely scattered. Mays continued his flight "running down a steep river ridge and at the bottom wading and swimming across the Greenbrier River." After crossing the waterway Mays eventually overtook part of his company in the general retreat. Another Confederate veteran, possessed with fear, supposedly actually leapt across the Greenbrier River, which was about 100 feet across at the narrowest point at the base of Droop Mountain. Finding himself on the east bank of the river with dry feet his only explanation was that he unknowingly, in the midst of his fearful flight, "had jumped the stream."[88]

Sergeant Andrew Mathews McLaughlin, Company I, 19th Virginia Cavalry, related that during his escape from the field he came upon a wounded Federal soldier sitting with his back against a log. Whenever the Yankee lifted his rifle to aim at McLaughlin the weight of the gun would cause him to fall forward and the gun would point to the ground, whereupon the soldier would use the rifle to push himself back into a sitting position, and repeat the process. McLaughlin sprang upon him and took the weapon, bent the rifle barrel around a tree, and went on. Years after the war McLaughlin returned to the battlefield and purportedly found the bent gun barrel. Although "the wood part had disappeared...his story was substantiated."[89]

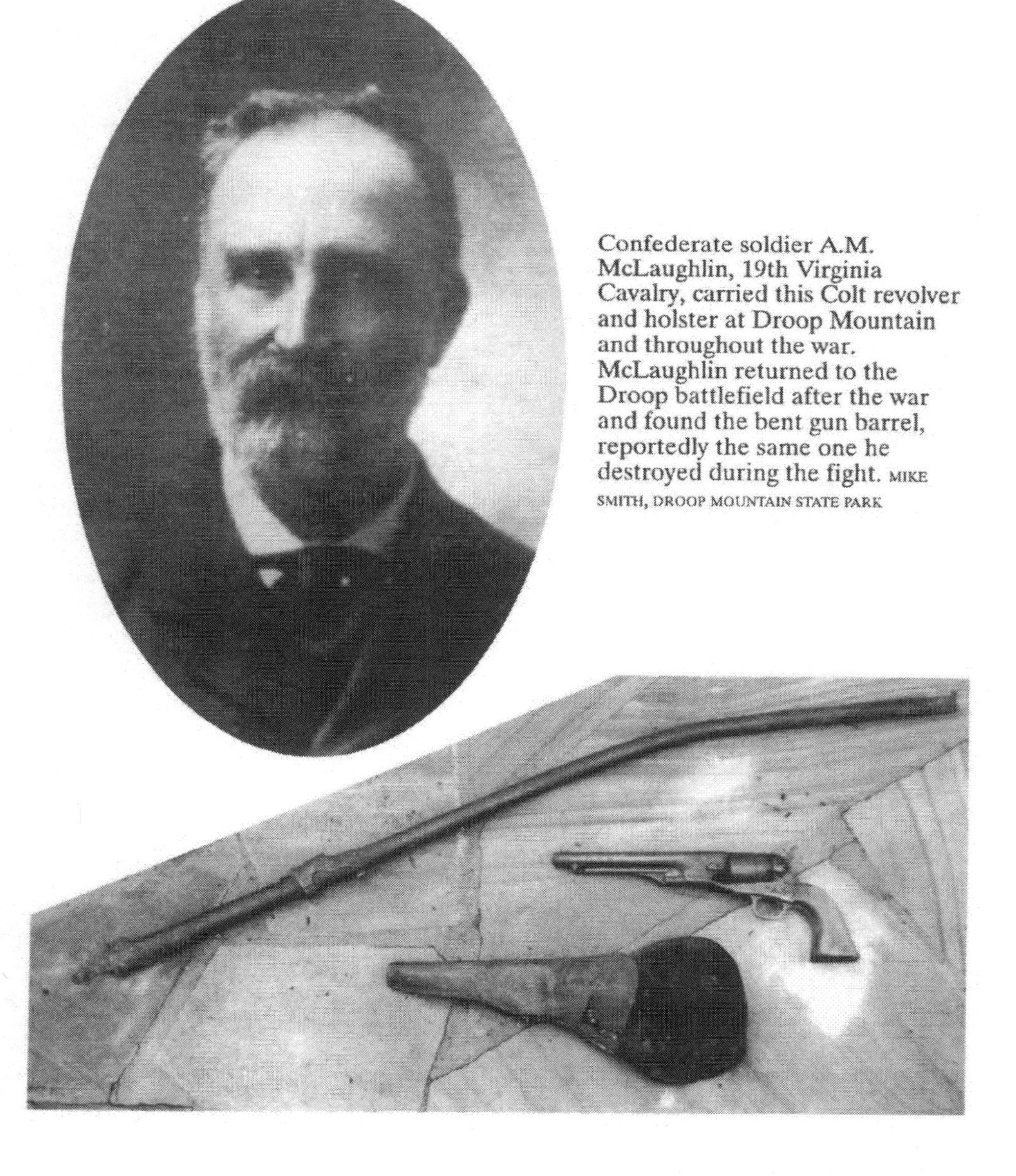

Confederate soldier A.M. McLaughlin, 19th Virginia Cavalry, carried this Colt revolver and holster at Droop Mountain and throughout the war. McLaughlin returned to the Droop battlefield after the war and found the bent gun barrel, reportedly the same one he destroyed during the fight. MIKE SMITH, DROOP MOUNTAIN STATE PARK

Another amusing incident was that of Capt. James ("Jim") Monroe McNeil, Company D (Nicholas Blues), 22nd Virginia, who was captured while reloading behind a log. McNeil's half-brother, Pvt. Alfred L. McKeever, Company C, 3rd West Virginia Mounted Infantry, [6th West Virginia Cavalry], had been looking for James throughout the day and found the dejected Confederate prisoner sitting beside the road with his captors. McKeever, who had not seen his half-brother for some time as they had chosen to fight on opposite sides, rushed forward with outstretched hands and said, "Hello Jim." McNeil, however, folded his arms and said, "I am glad to know, Alfred, that you are alive and well, but Alfred, we are not shaking hands today." Another source says McKeever reportedly captured McNeil while the captain was behind a log reloading. Reportedly Capt. John K. Thompson assumed command of Company D upon McNeil's capture. A similar incident was recalled by a soldier of the 3rd West Virginia Mounted Infantry by the last name of Fortney, who told his brother, "there were some Clarksburg men on both sides...and it so happened that a Clarksburg boy on the Union side was detailed to guard a Confederate prisoner who chanced to be the boy's own father. The father had enlisted in the cause of the Confederacy. A Union soldier who knew them both said, 'that old rebel is the guard's father and he may let him get away.' The son hearing the remark said 'father knows better than to try and get away, as he knows that I would have to do my duty in that case.'" [90]

During the Confederate retreat Pvt. Henry H. Brookhart, Company C, 19th Virginia Cavalry, suffered from a shoulder wound and sat down to rest. He would have been captured if not for a comrade who dismounted and placed Brookhart on his horse. Henry proceeded to ride 30 miles before getting his wounds dressed. It would later be written of Brookhart, "He was a fearless soldier and was first a dispatch carrier in battle. He carried water to the fighting lines in canteens, but was in the line of battle when wounded. The horse on which he rode had seven bullets in him."[91]

Although the Dye brothers apparently did not meet during the affair at Droop Mountain, it is worth mentioning, to further illustrate the fratricidal nature of the battle, Frank Dye, [the only Dye on official records is Benjamin F. Dye of Company C, 2nd West Virginia Cavalry], of Wood County, West Virginia, marched up the mountain that day and fought against his brother, Pvt. Cornelius Harrison Dye, Company B, 22nd Virginia Infantry, on the left flank.

Lieutenant Randolph Blain said Capt. Jackson's Battery was in the rear and "came within a hundred yards of being cut off by the flankers on the left." When the retreat became universal "everything went pell-mell for at least ten miles." Pvt. George Henry Clay Alderson, 14th Virginia Cavalry, said, "We had cut a road through the woods for this cannon some half mile around and very crooked. I caught hold of the sight of the cannon and with difficulty held on to it for a half-mile, till we came to the main road, where Bob Wallace [Pvt. Robert Bruce Wallace, Company A, 14th Virginia Cavalry]...[was] holding my horse, [and] stayed until I got to him." Micajah Woods of Jackson's Battery charged that "on reaching the main road we found everything in utter confusion—infantry, cavalry & all dashing pell mell to the rear—crowding together, running over each other. Genl. Echols' seeing us return in perfect order rode up to Capt. Jackson who was near & exclaimed, 'Capt. I saw your little band of heroes standing alone and can [be] proud that such men are on the field in all this confusion.'" 5th Sgt. John Waring Hampton of Jackson's Battery later said he "covered the retreat from Droop Mountain with his light artillery."[92]

1st Lieutenant David Poe, 20th Virginia Cavalry recalled, "Company A, was among the last to leave the field...when we fell back to the road we found that our retreat by that route was cut off. Then we filed to our left leaving the road, and went in the direction of Bonsacks Depot."[93]

The 8th West Virginia, exhausted from climbing the steep mountainside, received orders from Col. Oley to halt and rest while their horses were brought up, although a few squads of the regiment continued pursuit on foot. Oley's own squadron, which had been held in reserve and not engaged, was also brought forward as quickly as possible to join in the pursuit. Averell said the "horses of the 2nd, 3rd [and] 8th [West Virginia] and the 14th [Pennsylvania Cavalry]...[were] brought up the mountain as soon as possible," although Col. Schoonmaker said his horses were on the right of the pike and "it was some half-an-hour after the entire command had passed before" the 14th Pennsylvania Cavalry was mounted. At least one company of the Pennsylvania cavalry participated in the infantry fight, though, as evidenced by 1st Sgt. George H. Mowrer, Company A, 14th Pennsylvania Cavalry, who said, "...in the final charge Company A was brought to the center in the woods." M.A. Dunlap would also write that when the Confederate line broke [Dr.] Pvt. William O. Hartshorne, Capt. Chatham T. Ewing's nephew and aide, took his own saddle and bridle, caught the sorrell of area resident Franklin Andrew Renick [who would join the 14th Virginia Cavalry in 1864], and advanced up the mountain with the command. In the process a rebel Captain and 10 or 12 of his men were confronted and captured. The Confederate officer "had a gun in his hand fighting as a common soldier," and scolded the Yankees for cowardice in "sneaking up behind him while he was shooting at those in front."[94]

Scattering throughout the woods south of the pike Echols' men were nearly all gone when Col. Moor's right wing came to the pike, and all that remained were the dead and wounded. Correspondent "Irwin" described the scene, writing, "The rebel dead and wounded lay on top of the mountain, and almost the first one we saw was a dead negro, with a gun in hand and cartridge box buckled on; while prisoners were being taken every moment, the men in their eagerness were following on, but the Tenth and Twenty-eighth were resting from sheer exhaustion." Thomas R. Barnes, 10th West Virginia, recalled, "...we whipped the combined forces of Eckles [sic], Jenkins [Ferguson], and Jackson. We ran them through their camps where their dinner were over the fires cooking, they had not time to take them off the fire. They threw guns, haversacks, knapsacks, blankets, in fact most everything they owned was either lost or fell into our hands. Their army was completely demoralized. Our regt. whipped the Rebel 22 Va., they were called the star of the south." The disorganized Confederate army retreated "over a narrow and straight road along the top of the mountain" for about four miles. Micajah Woods wrote, "...the enemy [pressed] us hotly for several miles & killing a large number of our men & officers." A short distance further up the road a group of Moor's men accosted a rapidly moving spring wagon bearing some Confederate wounded. After opening fire and killing two horses, the wagon and it's passengers were captured. Another rebel ambulance, "drawn by two fleet little

Then and Now

At left: A 1928 view of the site "where the Federal infantry struck the Lewisburg pike at lower end of lake." DROOP MOUNTAIN COMMISSION BOOKLET

Below: A modern view showing Rt. 219 (Lewisburg Pike) and the southern entrance to Droop Mountain Battlefield State Park. GARY BAYS

mules," and carrying several wounded rebels, including Corp. Joseph Alline Brown, Company H, 22nd Virginia, who had been wounded near the heart, was also approached by the Federals. Luckily, whenever the Federals got near this ambulance, the wounded rebels would fire upon them and slow the enemy pursuit, eventually making good their own escape.[95]

Captain Ernst A. Denicke, 68th New York Infantry, signal corps detachment, having been informed of the retreat, called in his station on the left and quickly moved to the summit of Droop Mountain "with the view of obtaining some position from which to observe the valley beyond in which the enemy were moving." Averell also arrived at the top of the mountain and boasted, "The enemy came at us and in four hours we had them on the go. I never experienced a happier moment in my life than when we went into the works. And the troops felt the same way. I remember one Irishman jumping up and exuberantly yelling 'the thirty fourth fight and the first victory.'" Averell ordered Col. Moor to march the infantry in pursuit as far as possible. When Major Gibson's Independent Cavalry Battalion arrived at the summit Averell ordered them to give chase as well, and to attack the enemy rear "vigorously with the saber." Gibson moved forward quickly, passing Capt. Ewing's Battery, which was busy shelling the Confederate rear. Correspondent "Irwin" recalled the scene, writing, "In a short time the horses were brought up, we mounted, and the pursuit began, and Major Gibson, with his battalion, took the lead. In a few moments we came to two broken ambulances, with their contents lying by the roadside; here lay Major Bailey, of the Twenty-second; here some wounded; there, some dead; a little further on, another group; in the middle of the road, a broken wagon, and a large bay horse shot in the head; and a little further on, a burning caisson, with the terrified rebels flying and scattering through the woods, where cavalry could not pursue them, while the road was strewn with debris of a terror-stricken routed army."[96]

Following burial details and arrangements for the wounded the bulk of the Federal command joined in the pursuit, Col. Moor's infantry in the lead. Capt. Ernst A. Denicke, joined at the top of the mountain by Lt. Merritt, also moved with the command, but was "never able to find a desired position [to open signal communications], the views in all directions being obstructed by hills densely covered with brush and timber." When the dismounted Federal troopers came to the location of the old Murphy Post Office [where Wallace Kershner resided in 1935] they remounted their horses. There they spotted the toe of a boot sticking out from under a huge pile of fodder. Upon closer examination twelve rebels were found hiding and immediately captured. Pvt. George H.C. Alderson, 14th Virginia Cavalry, said as the Federal cavalry chased the retreating enemy several shots were fired at the rebels as they fell back through Beard's Lane, and "one ball passed over our heads and killed [Pvt.] George [W.] Lewis, [Company K, 14th Virginia Cavalry], brother of Harvey Lewis," also of the 14th Virginia Cavalry. Pvt. George Lewis died "bravely defending his native state." Many of the 2nd West Virginia hastily remounted at the Spice Post Office location and joined in the pursuit "in front of the chase." For about 12 miles they would observe "the ground was strewn with guns and accouterments, and the upturned faces of the poor victims, formed a ghastly picture on this terrible scene of carnage."[97]

General Echols had ordered Col. Jackson to assist the Confederate rear guard, which Jackson [who also remained with the rear] accomplished. He ordered Lt. Col. William P. Thompson to remain with the rear, which was commanded by Colonels Ferguson and Cochran from the beginning of the retreat until the army eventually reached Union in Monroe County. Col. Ferguson recalled that after the command remounted order was restored and the rear guard was formed, comprised of Major Benjamin Franklin Eakle, Capt. Edwin Edmunds Bouldin, 2nd Lt. John A. Feamster, all from the 14th Virginia Cavalry, and various privates. Micajah Woods wrote, "Gen. Echols, Col. Ferguson of Jenkin's Brigade & a few other intrepid spirits gathered around them a band of 30 or 40 in rear & saved the entire army from utter dispersion."[98]

Major Gibson's Independent Cavalry Battalion, having passed Capt. Ewing, whose artillery was shelling the Confederate rear, caught up with the demoralized rebel rear guard about five miles south of the battlefield. Gibson ordered Capt. Frank Smith, 3rd Independent Company Ohio Volunteer Cavalry, to "charge with the saber," the same order Averell had given Gibson upon reaching the mountain's summit. Although greatly outnumbered by the Confederate rear guard,

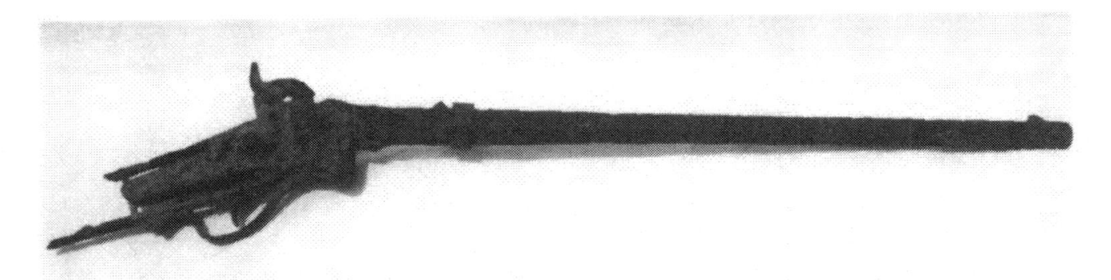

Confederate Sharps carbine found in 1986 by Greg Hively, Junior Goode and Jeff Hanna on a tributary of Locust Creek. A few days later a brass and gold wedding band was found about 18 feet from the gun by Tim McKinney. These items were probably lost during the Confederate retreat. MIKE SMITH

Smith attacked as ordered and managed to rout the enemy, killing one and wounding three others who were also captured. The remaining defenders fled in disorder. Capt. Smith, however, was incapacitated by a shoulder wound suffered in the clash. Unable to continue, this left Smith's company without a commissioned officer. Smith's 3rd Independent Company Ohio Volunteer Cavalry also lost Sgt. Henry Foplie, who was wounded, as well as Pvt. Frederick Donnerline, who was captured and died July 7, 1864 in the infamous prison camp at Andersonville, Georgia. Records indicate the last Confederate stand was at the "Old Schister Place," which was probably at the south brow of Droop Mountain, and this may be the fracas noted here.

Gibson continued to annoy the Confederate rear, gradually driving his opponent two more miles southward. But as Gibson's horses were exhausted from the rapid pursuit, many of his troopers dropped out of the chase. With his command dwindled down to some 50 horsemen Gibson continued to drive the enemy back, which was then composed of some 200 infantrymen and cavalry. This proceeded for another mile, where Gibson captured three rebels. Echols moved further south until Major Gibson "had driven their whole rear guard, together with a piece of artillery, about 200 infantry, and several wagons, in a mass of disorderly fugitives."

M.A. Dunlap said [Dr.] Pvt. William O. Hartshorne, Capt. Ewing's nephew and aide, accompanied the Federal advance and said the Confederate rear guard would fire at them at every turn and then gallop out of sight. This prompted the Federals to approach every bend in the road with extreme caution. As the Federals came to the open ground of Renick's Valley the rebel rear guard was spotted. The Yankees spread out in "fan fashion" and called upon a lone rebel stationed nearby to surrender, to which the enemy soldier refused and fired on the Federals, killing a horse and then running away. The entire Federal advance fired and killed him, but, according to M.A. Dunlap, this "single man had saved the [rebel] army by slowing the Yankee pursuit." It was believed the rebel resided at the "big brick house," where the Yankees notified the occupants of his death.[99]

In an attempt to cover the southern retreat the Confederate army located a strong defensive position and opened a severe infantry fire upon Major Gibson's men. Gibson attempted to form his troopers and charge the enemy artillery and train, but a "severe fire" from the rebels prevented this, wounding one Federal and three horses "in less than a minute."

The Renick House is possibly the residence of the one lone rebel who was killed nearby while stalling the Federal pursuit. Construction of the limestone section began in 1787 and was completed in 1792. The red brick portion (right side of photo) was added about 1825. James Henry Renick, resident and son of the owner during the war, was an officer in the Confederate Commissary Department under Gen. Robert E. Lee, "buying cattle and horses for the Confederate Army." "At the outbreak of the war the farm had 300 head of cattle, 1,000 sheep, 200 hogs and 125 head of fine horse flesh, together with a large number of slaves." BRIAN ABBOTT

While Gibson's men continued to press vigorously, Jr. 2nd/3rd Lt. John J. Beard, and Pvt. George B. Pollard, both of Capt. William McNeel's Company (Company F, 19th Virginia Cavalry), and two or three other cavalrymen, threw themselves in the rear of the Confederate command and charged Gibson's advancing troopers. The Confederates managed to capture one Yankee, who was at the head of the column, along with his horse and accouterments. Lt. Randolph Blain, Jackson's Battery, described the scene, writing that Col. Ferguson gathered together about *eight* men and stalled the Federal advance. Of this encounter Major Gibson said the enemy force, comprised of some 50 men, struck him before he was able to accumulate "over about 15 men."

Faced by superior numbers Gibson was forced to fall back with the loss of one man captured. 4th Sgt. Matthew John McNeel, Company F, 19th Virginia Cavalry, recalled that in addition to this, "On [the] day of [the] battle [he] was on his way by way of Anthony's Creek from Dunmore...met...[C.S.] army in retreat and captured a few [U.S.] men who had pressed forward too fast and far in pursuit."[100]

Later, after having managed to get five or six men in a field next to the road, Major Gibson was able to check the enemy cavalry. But as the skirmishing was protracted Gibson found himself unable to assault his opponent. About this time Ewing's Battery arrived and assisted by putting the rebels to flight. One source later claimed the last shots fired from Federal artillery came from the south brow of Droop Mountain as the Confederates retreated down Renick's Valley. Capt. Julius Jaehne, Company C, 16th Illinois Cavalry, also came up with his men and continued to advance with Gibson.

The Lost Cannon

Sometime during the Confederate retreat Capt. George B. Chapman's brass howitzer, previously damaged in the battle at White Sulphur Springs, had its carriage break down, although one Federal mistakenly thought a Yankee cannonball had taken the wheels off the brass cannon. The southern artillerists quickly rejoined the assembly by locking the carriage to the limber, but the old pintle-hook broke. After placing the pintle-hook in the limber chest the chest itself broke down, forcing the crew to abandon the gun and hide it.

Throughout the years various stories as to the location of this gun have circulated throughout the community. Pvt. John Montgomery Irvine, of Deerfield Valley, Virginia, a member of Company I of the 14th Virginia Cavalry, claimed he was with the gun when it was buried. He, too, mistakenly said the Federals had shot the wheels off after the gun had been engaged for awhile in the battle. Irvine believed the cannon was buried on the William McCoy place on Droop Mountain and "a rotten chestnut tree was rolled over the burial spot." Pvt. James C. Wiley, Chapman's Battery, a later resident of Pocahontas County, said he also helped bury the gun, although Confederate service records [which are not always accurate] state he did not enlist in the battery until 1864.[101]

A descendant of Pvt. John A. Morris of Company A, 22nd Virginia Infantry [although the family says he served in Chapman's Battery] claims that Morris, a teamster at the time of the battle, returned to the battlefield two weeks after the fight. Dressed as a civilian, he reportedly found the lost cannon and loaded it on a wagon. He then proceeded to bring it back with him to the vicinity of Covington, Virginia.[102]

Historian Calvin Price said, "It was a 12 pound brass howitzer...howitzer from a foreign word meaning sling...pride of the army...soldiers said that every time it was fired it called for the First-born!...it had been injured at White Sulphur Springs...hid in a morass on top of the mountain, such place generally referred to as bear wallows."[103]

Averell added to the confusion in his report, stating that "several of my command reported having seen and removed two other pieces of artillery, abandoned by the enemy and secreted by the wayside. Time was not had, however, to look after them."

Although stories continue to this day of the whereabouts of this artillery piece, some claiming it is still in the swamp (glades) on the battlefield, or in area caves (which are numerous), or elsewhere, the truth is the Federals found the gun and Col. Moor later returned to Beverly with it, clearly evidenced in his report of captured materials. "Irwin," a northern correspondent, wrote the Federals found the cannon the following day while descending the south side of Droop. "T.M." of the *Wheeling Intelligencer* wrote that "eight miles out [on the retreat...the rebels] abandoned a brass 12 pounder, which we brought in."[104]

Final Fire

Near dark the retreating Confederate army fired on Major Gibson from an ambuscade, but were driven off. A squadron of Federal mounted infantry arrived to assist Gibson, although it was "too dark to see much." Gibson moved on hoping for some cleared ground when Austrian born Acting Assistant Adjutant Gen. Lt. Leopold Markbreit, 28th Ohio Volunteer Infantry, who had been detached September 6, 1863 to assist Averell, came up and found "it was very dark." Major Gibson opted to end the pursuit, fell back one mile, and went into camp for the night. During all this rear guard activity Lt. Col. William P. Thompson, 19th Virginia Cavalry, said his men wounded two Federals and captured three, although he believed the prisoners managed to escape afterwards. At about 7:00 P.M. news of the Droop Mountain battle reached Lexington, Virginia and the Rockbridge Home Guard.

Darkness had also ended pursuit by the other Federal units. Earlier, Gen. Averell had ordered Col. Moor to march his infantry as far as possible, but after having covered six miles, darkness prevailed and Moor's men bivouacked on the side of the pike, "rather tired, but high in spirits." Capt. Will Rumsey sent a dispatch to Col. Moor from Renicks Valley, which read, "The lieut. cmdg. desires me to say that the enemy and one piece of artillery went down a side road to the left. He desires that you will picket that road and put out a strong picket in rear." Col. Moor would later file a report of captured Confederate weapons which fairly exemplified the armament of Echols' army. The list read: 54 cartridge boxes (most of them very bad); 54 bayonets (good); 46 Enfield rifles (16 unfit for service being badly broken); 15 Springfield muskets (12 unfit for service); three Mississippi rifles (two unfit for service); five Harpers Ferry muskets (three unfit for service); 18 Austrian rifles (four unfit for service); two Vermont rifles (two unfit for service); and one brass cannon.[105]

The 2nd West Virginia Mounted Infantry had also participated in the chase for some distance before going into camp. A correspondent believed to be attached to the 2nd West Virginia said the Federals captured, "One piece of artillery, two ambulances, and a wagon, loaded with hospital stores (stolen from Chambersburg, Pennsylvania), and several hundred stand of small arms were left on the field." The 8th West Virginia Mounted Infantry moved with the column as well until dark, at which time Col. Oley received orders to make camp. Correspondent "Irwin" reflected, "It was late in the day, and we kept up the pursuit for ten miles, until after dark, when we went into camp in a field around a 'sink-hole' that afforded water for our horses..."[106]

Colonel Schoonmaker said his 14th Pennsylvania Cavalry went into camp for the night at 8:00 P.M. Northern newspaper correspondent "C" said, "We pursued the enemy until 10 P.M., discovering two additional pieces of artillery hidden in the woods. A large number of small arms, equipment, wagons and several caissons partially destroyed. One wagon was left standing in the road, heavily loaded with ammunition, for small arms." Oddly, John A. Blagg, reportedly of the 10th West Virginia Infantry, claimed the battle ended between 12:00 midnight and

1:00 A.M. Postwar historian Calvin Price, writing in *The Pocahontas Times* in 1925, believed Averell ended his pursuit "on top of Spring Creek Mountain, overlooking the Big Levels of Greenbrier" County. Sgt. James Z. McChesney, 14th Virginia Cavalry, remembered that "One of our company was sick at a house where the Yankees camped on the night of the fight, and Genl. Averell and his staff and several Cols. and officers went into his room and talked with him. Genl. Averell asked him if he was in the fight, he told them he was not, that he had been sick for some weeks & so, they didn't disturb him, not even paroled him. He heard the officers say that they didn't gain anything by the fight..."[107]

While Averell's men settled in for the night many of Echols' soldiers remained scattered throughout the countryside, lost in the woods, searching for the main column and safety. Samuel D. Edmonds, 22nd Virginia, said when the retreat commenced he was afraid of being trapped and ran to the Greenbrier River, crossed by a canoe, and went to the top of a mountain, where he built a campfire and spent the night. George H.C. Alderson, 14th Virginia Cavalry, found refuge two miles above Frankford with some of his comrades. To make matters worse, when Echols fell back apparently no word of the retreat was sent to Edgar's Battalion, which meant the men of the 26th Battalion Virginia Infantry found themselves cut off from Echols, behind enemy lines, and would have to find their own way back to the Confederate command. This was contradicted by Lincoln S. Cochran, who said Joesph R. Perkins, 26th Battalion, told him "they received a dispatch to get to the pike at Renick's Valley. They drove their artillery into the lot of my grandmother's [Cochran's] and turned. When they got to Renick's Valley, they met the other army retreating," although this was probably just Chapman's artillery piece and not Edgar's Battalion.[108]

Skirmish on Little Sewell

While the armies of Averell and Echols fought at Droop Mountain, Gen. Duffié had marched his column 15 miles, from Tyree's to Meadow Bluff. At about 2:00 P.M. [the same time Col. Moor struck the Confederate left at Droop] he skirmished with rebel pickets on Little Sewell Mountain, drove them five miles back, and captured two of them. The prisoners informed Duffié that the entire Confederate force had departed Lewisburg on the Frankford road to fight Averell, except for one cavalry regiment [16th Virginia Cavalry] left behind to garrison Lewisburg. The historian of the 2nd West Virginia Cavalry recalled the day, writing, "our advance succeeded in capturing two of the enemies pickets on Little Sewell...marched to Meadow Bluff...about 18 miles from Lewisburg."[109]

Private James Ireland, Company A, 12th Ohio Infantry, vividly summed up Duffié's movements on November 6, writing, "We passed sleepless night. It rained until 2 o'clock A.M. making it very disagreeable having no shelter but our blankets. By daylight everything ready to move—cavalry in advance. Turns cool & clears up. Cross by Sewell Mt. Take dinner and move on crossing over little Sewell. View the position of Genl. Rosecrans & Lee two years ago. Camp at Meadow Bluff where Genl. Crook camped in the Spring of 1862. Boys take more liberty than I ever saw them before. Some houses are almost stripped of almost everything. Traveled about 18 miles. Captured some rebel pickets."[110]

In addition, as the Droop Mountain battle raged, Gen. John D. Imboden had marched from Buffalo Gap to Goshen Depot with about 600 of his best mounted men and a section of artillery, without baggage. He issued provisions to his soldiers and encamped for the night at Bratton's in Bath County.

Aftermath

Although the battle of Droop Mountain was over the most gruesome task remained—burying the dead and tending to the wounded. Averell, in possession of the field, held the responsibility of looking after the casualties of both sides. The 10th West Virginia apparently assumed much of this chore, while Major Gibson and other Federals had chased the retreating rebel army. Immediately, various structures in the vicinity were converted into makeshift hospitals, including the Stulting house, the Joe (Josiah) Beard home (where Lee McLaughlin resided in 1935), both at Hillsboro, and the Mountain House at the top of Droop. The home of Franklin Andrew Renick, later of Company A (2nd), 14th Virginia Cavalry, whose farm at the foot of Droop Mountain was part of the battlefield, was also used as a Union hospital. A log house in Renick's Valley near the Brownstown Road was probably used as well as "Hundreds of minie balls collected on the farm by Gladys [Jenkins] lends support to the story that the place was used as a field hospital during the Civil War." Thomas Swinburn, 8th West Virginia Mounted Infantry, noted he "stayed in [the] hospital established in [the] house occupied in 1901 by George W. Callison." Swinburn visited the house that year and noted "that the wall paper on one of the rooms still remained as fresh looking" as in 1863.[111]

While Lt. William Bahlmann, 22nd Virginia, lay wounded on the battlefield, an elderly Yankee approached him and asked, "Is the 22nd here?," to which Bahlmann replied, "Yes, the 22nd is always here." Soon afterward Col. Thomas M. Harris of the 10th West Virginia came by and asked Bahlmann, "How many men have you got here?" Bahlmann, naturally, did not wish to divulge such pertinent information, and doubled the figure to 3,000, hoping to fool Harris. Bahlmann, sitting with his back to a tree stump, said his next visitor was Capt. James S. Cassady, Company G, 8th West Virginia, a pre-war neighbor of Bahlmann's, who came up and shook hands with him. Next to arrive at Bahlmann's position was 1st Lt. Spragg [Sprange] Lawrence, Company C, 8th West Virginia, who spoke kindly to him. Soon afterward, Capt. Brown, of Averell's staff passed by and asked Bahlmann if he wished to be carried to a campfire, to which Bahlmann consented. Placed on the ground near a fire Bahlmann noticed a number of bloody Confederates lying nearby, one in particular was groaning loudly. Bahlmann asked Capt. Brown for the name of the wounded man, but when the wounded rebel replied Bahlmann immediately recognized the voice as belonging to Pvt. Pete E. Stribling, Company H, 22nd Virginia Infantry. "Is that you, Stribling?," Bahlmann inquired, to which Pete replied in the affirmative. Bahlmann told Stribling, "I am going to lose my leg" [Bahlmann had taken a minie ball in the left knee but apparently his leg was not amputated], while Pete told him, "I am shot through my body."

Then and Now

Above: A 1928 view of the house on Droop Mountain in which Major Bailey, 22nd Virginia Infantry, died on either November 7 or 11 [depending on source]. Descendants claim that, even in this early view, the only part of the house remaining from the Civil War period is a small portion of the interior. This house is located along Route 219 just south of the southern park entrance. DROOP MOUNTAIN COMMISSION BOOKLET Below: A modern view of "Mountain House". Not much has changed since 1928. GARY BAYS

[Stribling had been shot in the chest, the ball entering the left side and exiting on the right of his spine. He died in June of the following year of complications from the wound]. Capt. Brown returned and told Lt. Bahlmann, "It's not the men that are hurt the worst that make the most noise."Brown asked Bahlmann, "Have you any of our money?," to which Bahlmann replied, "No." Brown then proceeded to remove a $20 bill from his own pocketbook and placed it in Bahlmann's, which then only contained a Confederate five dollar bill.[112]

While Lt. William F. Bahlmann lay on the ground he had improvised a tourniquet by tying his handkerchief "above the knee and tightening or loosening it with a little stick." When the Federal surgeon, Dr. Lucius Comstock, came up to Bahlmann he asked whom had applied the tourniquet and seemed quite surprised when Bahlmann confessed to the achievement. Comstock probed the wound with his little finger but was unable to locate the bullet. After the war Dr. Comstock would publish an article in the medical journal of Washington, D.C. on the case.[113]

An ambulance bearing one additional wounded rebel soon arrived and as Bahlmann was placed in it he recognized the other passenger as Pvt. Mason V. ("Mase") Helms, Company K, 22nd Virginia, who would die from his wounds the following day, "leaving a widow and one or two young children." At about 8:00 P.M. the ambulance reached the Beard house at Hillsboro, where Bahlmann and Pete Cartwell, Company A, 22nd Virginia, were placed in a room with 15 wounded Federals. One observer noted that at the temporary hospital at the Beard house there was "blood on the floor until it ran out the doors." Bahlman recalled, "There were four of us in a large room on the second floor each in a double bed [Bahlmann, Michael McGoldrick, Stribling and Augustus C. Liggott]." Bahlmann would remain there "eight weeks and every day Miss Nannie Beard, a very pretty lady...," tended to the men. Bahlmann added, "Samuel Hudson was a member of the 3rd West Virginia Mounted Infantry [actually the 2nd West Virginia], I think. In the fight his scalp was grazed by a bullet. He volunteered to nurse some of the wounded Confederates. He took the night watch from midnight until midnight and was very good to us. Another generous man. But he took erysipelas in his wound and had to be sent to a larger hospital."[114]

Young C.L. Stulting, whose home was also used as a hospital, recalled the Federals "used spring wagons drawn by horses which were covered with white sheets to haul the wounded from the battlefield" to Beard's, where the men were placed on the lawn so physicians could tend to them. Several of the wounded died and were buried in a nearby field beside the pike, where their bodies remained for a few weeks. The corpses were later removed and taken away in ambulances.A correspondent for the 2nd West Virginia claimed, "We buried over 70 rebels, on the field, including several officers, and [we] captured about 250 prisoners, including the wounded." This may be a bit of exaggeration, though, as statistics show only 33 rebels killed in the fight.[115]

Among the burial detail on Droop Mountain was Pvt. Andrew ("Andy") Jackson Short, Company F, 10th West Virginia Infantry. Working in the darkness, Short dragged the lifeless body of an enemy soldier to the gathering point for the dead and wounded. While lifting the corpse he noticed the deceased possessed a crooked finger and was very similar in size and height to his own brother, who had joined the Confederate army at the beginning of the war. Requesting a light,

Then and Now

The Beard House, built in the 1840s is located on the southern edge of Hillsboro. The Josiah Beard residence was used as a hospital by Averell following the battle of Droop Mountain. The above photo is from 1928. DROOP MOUNTAIN COMMISSION BOOKLET
In the modern view below, you can see not much has changed. The back section was added sometime after the war. BRIAN ABBOTT

Short discovered the man was, indeed, his brother, Pvt. John J. Short, Company A, 22nd Virginia Infantry. Both men had been "raised near the Pocahontas and Greenbrier counties line, but had taken different courses at the outbreak of the war." Once again, the phrase "brother versus brother" rang horrendously true.[116]

While burial details worked throughout the night, a mysterious young lady was spotted passing among the dead and wounded, closely checking the face of each rebel she encountered. When the Federals asked as to her motive she replied she was a guest at the house of Col. McNeel and was searching for Capt. George I. Davison [probably Capt. George J. Davisson later of Company B, 46th Battalion Virginia Cavalry] of Weston, in Lewis County, whom she had recently

This small graveyard, reportedly containing some five or six soldiers killed in the battle of Droop Mountain, is located to the rear of the park superintendent's office.
GARY BAYS

married. It was later discovered Davisson had emerged from the battle unscathed and was then miles away. Such fears were not unusual, as some of the Confederate officers, such as Lt. Col. William P. Thompson and Col. James Cochran, were originally thought to have been caught or injured in the battle. But all were later found to have made good their escape with the Confederate army. The *Staunton Vindicator* later read, "In our city last Monday, [Lt. Col.] Thompson was reported among the killed and we learn that in the retrograde movement the Yankees took special pains to spread the same rumor..." Another report in the paper said, "...[the] gallant [Lt.] Colonel is not only not killed, but in excellent health and spirits and ready again to meet the Yankees." Lt. Col. Thompson, who commanded the left wing at Droop, would later visit the city "and reports Col. Cochran, 14th [Virginia] Cavalry, was near him and was conspicuous for his gallantry." Although Thompson and Cochran were both reported as missing and killed, Thompson said, "friend [Col.] James [Cochran] was uninjured and among the last to leave the ground." Pvt. George H.C. Alderson, 14th Virginia Cavalry, recalled, "My father hearing of the battle [later] went over to our command at Pickaway Plains, Monroe Co., and there heard that I was killed." Supposedly, a lady from Braxton County, with a wagon and one house slave, arrived after the battle. As her parents were fanatical about aiding Confederate soldiers, she took three of the wounded rebels back to Braxton County, nursed them back to health, and eventually married one of them by the name of "Cutlip" who had been shot in the arm near the shoulder. Although no Cutlip appears on the Confederate casualty lists, a number of Cutlips with Braxton County connections served in the 19th Virginia Cavalry.[117]

At right: Monument to two unknown Confederate soldiers, located on private property, just north of the park boundary at the modern cemetery. The Confederate artillery was posted at this location during the battle. The marker reads: "Dedicated July 4, 1931 by Capt. E.D. Camden Chapter U.D.C. Sutton, W.Va. to Confederate Soldiers Engaged On This Field." GARY BAYS

At left: This rock reportedly marks the gravesite of a soldier killed at Droop Mountain, although this has never been verified. This rock, as well as a number of others, can be found on private property just over the park boundary near the modern private cemetery, and just a few feet away from the Confederate monument. GARY BAYS

Adjutant Oliver P. Boughner of the 10th West Virginia said those of the 10th West Virginia who fell were "quickly buried on the field of their glory." Major Robert A. Bailey, 22nd Virginia Infantry, was taken to the Mountain House on the battlefield and passed away either November 7 or 11 [sources differ]. As earlier noted, he was later buried at the Old Stone Presbyterian Church Cemetery at Lewisburg. "T.M." correspondent for the *Wheeling Intelligencer* wrote, "Major Bailey...was mortally wounded, and two [of the 22nd Virginia's] captains made prisoners. There was also a major said belonging to the 14th killed, and a Captain and a Lieutenant wounded. We brought in eighty-five prisoners."[118]

Orderly Sergeant John D. Baxter, 10th West Virginia, was moved to the [Beard?] tenant house [which later became the Spice P.O.] along with some other

Federal wounded. Baxter would linger throughout the night and expire the following day. He was buried near the house, although his body was later exhumed and moved to the national cemetery at Grafton, West Virginia.

A 1928 view of the old Spice Post Office on Droop Mountain which was used as a hospital. Sgt. John D. Baxter, Co. F, 10th West Virginia Infantry, reportedly died from his wounds here on November 7, 1863. DROOP MOUNTAIN COMMISSION BOOKLET

The Cost in Blood

Exact casualty figures for the two armies engaged at Droop Mountain remain difficult to ascertain. The Federal army did submit a detailed list by regiment, yet modern research has shown Averell's total figure as being too low. On November 7, the day after the battle, Averell stated his loss at "about 100 officers and men." Later, on November 17, he upgraded his figure to a total loss of killed, wounded, and missing at 119. This final report indicated Averell lost 30 men killed in action, 88 wounded (some of whom were obviously mortally wounded although Averell failed to mention this fact), and one captured. A recent study of available records, however, results in a total loss of 140. This study, incorporating Federal surgeon reports and other sources, shows a total of 45 Federals killed (15 of that number were soldiers mortally wounded), 93 wounded, and two captured. By applying Averell's statements, his two infantry regiments under Col. Moor which struck the Confederate left, as would be expected, suffered the worst. This included the 10th West Virginia Infantry, with two killed and 35 wounded [10 mortally] for a total of 37; and the 28th Ohio Infantry, which lost three killed and some 28 wounded [five mortally], making approximately 31. When combined, the two regiments suffered about 10 killed and 57 wounded [10 mortally] for a total Federal loss on the rebel left of 68, out of approximately 1,175 engaged.

The three mounted West Virginia regiments, fighting dismounted, which attacked the Confederate center and right, tallied some 61 losses, with the 2nd West Virginia at the highest figure with 29, including some nine killed and 20 wounded [four mortally]. Next in losses was the 3rd West Virginia at 24, comprising six killed, five to 15 wounded [one mortally] and one captured. The 8th West Virginia lost three killed, eight wounded, for a total of 11. Additionally, the 1st West Virginia Light Artillery, Battery B (Keepers'), sustained seven losses (two killed and five wounded), and the 3rd Ohio Independent Company Cavalry, during the Confederate retreat, took three losses (two wounded and one captured). No casualties were reported by either the 14th Pennsylvania Cavalry, Major Gibson's Independent Cavalry Battalion, or the 1st West Virginia Light Artillery, Battery G (Ewing's Battery), although a premature report in the *Pittsburgh Gazette* said the 14th Pennsylvania Cavalry "...escaped with slight loss. Half a dozen wounded will cover the regimental loss. No commissioned officers were wounded. The names of the wounded were unknown to our informant, as the regiment, after the battle, pushed rapidly on in pursuit of the enemy—leaving the wounded behind in charge of a detail."[119]

Prominent among the Federals killed were Lt. Joseph W. Daniels of Keepers' Battery; Lt. Arthur J. Weaver, Company K, 2nd West Virginia Mounted Infantry; and Capt. Jacob G. Coburn, Company C, 3rd West Virginia Mounted Infantry, who was mortally wounded.

Total Federal losses in the U.S. army at Droop Mountain may have even been higher than the figures presented herein, as indicated by the adjutant of the 10th West Virginia, who wrote from Beverly on November 13, "...there were a few slightly wounded who are not reported as they are already doing duty." Although this statement was directed at his regiment, it quite possibly applied to the other Federal regiments engaged as well.[120]

It was not unusual after the battle for many of the solidiers to give boastful and ridiculous statistics involving the battle. One such example was Sgt./Lt. David T. Peterson, Company B, 10th West Virginia, who said, "The total loss in our Brigade in killed and wounded are as follows 26 killed and 76 wounded including 2 Lieuts the Reble loss was 60 killed and I don't know how many wounded but there were 35 of their wounded fell into our hands many others got off the field the rebs lost a large number of their officers they lost one Lt. Colnel 2 majors 2 or 3 Capts about 3 Lts. This list are all killed but one and he is mortally wounded he was a Major. We have 81 prisoners including 2 Capts. of the 22nd Reble Va."[121]

Confederate casualties at Droop Mountain are much more difficult to determine than the Federal loss as the southern army was in a state of confusion in the immediate aftermath of the battle, and as far as is known, no formal detailed report of casualties was ever submitted by Gen. Echols. At Droop Mountain Gen. John Echols reported his losses "...275, amounted to 15 percent of his force, and his two largest units, the 22nd Virginia Infantry and the 23rd Virginia Battalion, each suffered losses of 20 per cent of their numbers." When Gen. Sam Jones first heard of the defeat he greatly over-estimated Echols' loss, as did a number of others. But throughout the ensuing days many of Echols' command, lost in the wilds of West Virginia, managed to find their way back to the army, and it was soon discovered the rebel loss was not near as great as first anticipated. Additionally, Edgar's 26th Battalion Virginia Infantry, which was never engaged and was cut off from Echols when the army retreated, eventually located Echols through the brilliant management of Lt. Col. Edgar. Despite such optimism, though, the Confederate army at Droop Mountain unquestionably suffered heavily. Unfortunately, Confederate service records are very incomplete [for example, a list of casualties for the 19th Virginia Cavalry apparently could not be found and may not exist] making an exact figure impossible.

As noted, Gen. Echols said he lost a total of 275 men, while Col. Jackson claimed his loss was about 150, yet Jackson does not indicate if his figure is part of or in addition to that submitted by Echols, although the former is probably correct. From Echols' Brigade, the 22nd Virginia Infantry, which entered the battle 550 strong, as had so often been the case in the past, suffered the worst, and listed a total of 113 losses, although available records show 12 killed, 45 wounded [six mortally], and 43 captured, tallying only 100. Lt. William Bahlmann, Company K, 22nd Virginia said the "22nd lost 15 officers out of 24, or 61%, including two captured and 13 wounded, two mortally." Companies A and K of the regiment had no officers left at all. James H. Mays, Company F, 22nd Virginia, said the 22nd only numbered 300 after the fight.[122]

The 23rd Battalion Virginia Infantry filed a total loss of 61 out of 350 engaged, yet figures list five killed, nine to 11 wounded, and 29 captured, making some 43 to 46 altogether. If these figures are correct then one out of every six members of the 23rd Battalion was a casualty. Edgar's 26th Battalion Virginia Infantry, which was not engaged, reported no losses, although two men were apparently captured in the retreat.

Among Jackson's Brigade, records for the 19th Virginia Cavalry, as noted, are very incomplete, and show only two killed, one wounded, and six captured, totaling nine, a figure obviously much too low for a regiment engaged in the hottest part of the fight. The 20th Virginia Cavalry listed a total loss of 26, with seven killed, five wounded, and 14 captured.

From Jenkins' [Ferguson's] Brigade, the 14th Virginia Cavalry lost 11 to 25 men, including two to three killed, six to eight wounded, and two to 14 captured, while none of the Confederate batteries took any casualties.

By applying modern figures with available records one comes up with a total Confederate loss of about 191, far below the figure given by Echols. While it would not be unrealistic to assume some 84 men eventually found their way back to Echols, the figure of 275 presented by the southern commander is probably fairly accurate, since statistics for the 19th Virginia Cavalry remain a mystery. [See Appendix A for detailed casualty lists].

Laurels for All

As would be expected, both sides boasted of their gains, downplayed their losses,and gave praise to the valiant participants. Gen. Echols was impressed with the "exhibition of cool courage and noble daring" during the fight of Lt. C. Irving Harvie of Jenkins' Brigade, and Capt. Labon R. Exline of Jackson's command. Also, from his own staff, Echols noted the "activity and energy and courage upon the field, in the rallying and encouraging the troops, and in conveying orders" of Capt. R.H. Catlett and Capt. W.R. Preston, Lt. J.W. Branham, Lt. Wood Bouldin, Jr., Lt. E.C. Gordon and Lt. H.C. Caldwell. He also praised Major George McKendree, brigade quartermaster, for his "valuable services on the field."

Colonel George S. Patton praised Capt. William [Waller] Redd/Reed Preston, 14th Virginia Cavalry, Assistant Adjutant General, of Jenkins' Brigade, and Lt. Noyes Rand, Adjutant of the 22nd Virginia, acting as aides, who gallantly carried his orders out intelligently although much exposed. He also noted Sgt. Maj. Monroe Quarrier who he said was "entitled to credit for courage and efficiency."

Colonel William L. Jackson reported, "The officers and men of Captain Lurty's battery exhibited a high order of courage and skill, and both at Mill Point and at Droop Mountain gave indications of what that new company will yet become." He added, "The brilliant fight at Droop Mountain and the subsequent movements of our forces will, I am satisfied, compel the enemy to abandon his designs whatever they are or were, notwithstanding his force, numbering near 10,000, including the force from the Kanawha Valley that was to cut off our retreat."

Lieutenant Colonel William P. Thompson, 19th Virginia Cavalry, wrote, "In my own command the loss was heavy...the battle was skillfully managed and gallantly fought; but the enemy numbered four to our one, and it was but a question of time when our force should retire on the flank, which I had the honor to command." Thompson concluded, "Very few prisoners were taken from us...The enemy confess to a much heavier loss in killed and wounded than we sustained."

General William W. Averell, the Federal commander, reported the "...victory was decisive and the enemy's retreat became a total rout..." He said his loss was "about 100 officers and men," and following the affair, said his "Troops [were] in excellent spirits, with plenty of ammunition." Averell claimed, "The conduct of the officers and men at my command, with a few exceptions, was excellent." Averell made particular notice of Col. Moor, "...whose admirable conduct cannot be too highly commended." A northern correspondent said he personally witnessed Col. Moor's actions during the battle and said, "His coolness and courage were equal to the task, and as officers of several regiments engaged have assured me, to him a large portion of the merit of that fight belongs. His soldierly bearing, his nice sense of honor after the fight, and his devotion to the interests of the wounded, all attracted attention, and drew forth praises from the officers and men upon the field. Each of the regiments engaged were earnest in their expressions of praise of his valor and courage." Averell also believed the Confederate loss at "about 250, one cannon, one stand of colors."

Colonel John H. Oley, 8th West Virginia, wrote, "to the best of my knowledge, all of my officers and men behaved well, and did their duty in the battle and on [the] expedition, and good discipline was maintained."

Colonel James M. Schoonmaker, 14th Pennsylvania Cavalry, claimed the men under his command were "...in excellent spirits over their more than successful trip..."

Despite all the boasting of both armies and their respective commanding officers, the battle of Droop Mountain was over and now Averell and Duffié's biggest challenge was to catch the retreating rebel army at Lewisburg in their pincer movement.

CHAPTER 6

Return of the Victor

November 7 - The Day After

Saturday, November 7, broke "warm and spring-like." Gen. Alfred N. Duffié, having gained knowledge on the night of November 6 of the Droop Mountain fight, pressed forward in his effort to reach Lewisburg. At about 2:00 A.M. a company of rebel cavalry [probably from the 16th Virginia Cavalry] struck his advance. The attack was repulsed, and by 3:00 A.M. Duffié's men were 7 miles west of Lewisburg, ironically occupying the strong rebel fortifications on Muddy Creek Mountain that had been erected in early 1862 by the 22nd Virginia Infantry. The enemy however, was not present, and his infantry took possession of the works. Duffié hurried his army forward in the continued hope of reaching Lewisburg in time to seal off Echols' escape. James Ireland, 12th Ohio, accompanying Duffié's infantry contingent, recalled: "A very windy night; quite cold. Get but little sleep. Camp aroused by 2 o'clock A.M. Get breakfast; start for Lewisburg by 3 o'clock. Our out pickets were attacked fiercely during the night. Move on rapidly. Pass through a beautiful country & healthy people..."[1]

Unknown to Duffié and Averell, Echols wasted no time in getting away. At Frankford, a small town established in 1801 by Frank Ludington and about eight miles north of Lewisburg, Echols halted the remnants of his infantry, artillery and trains for two hours rest and refreshment. Lt. Micajah Woods, Jackson's Battery, recalled, "At Frankford we remained for about two hours, assembling the remnants of our command...the 22nd Virginia, while at Frankford, could not collect 40 men...the column of fugitives was halted & some organization effected."[2] Pvt. Joseph Alleine Brown, Company H, 22nd Virginia Infantry, severely wounded in the Droop Mountain battle, recorded, "...the ambulance of wounded soldiers...reach[ed]...the small town of Frankfort [sic] and the Manse of my cousin the Rev. John C. Brown, the pastor of the Presbyterian Church of Frankfort [sic]. I was dumped out of the ambulance and taken into the house and placed in a comfortable and restful bed. In removing my boot I became aware of the severity of my wound, when a large clot of blood fell out, that had run from the wound in my body and near my heart. After being in bed for a few minutes I saw another wounded member of my Company—Kanawha Riflemen—was also in the room, Arch Young, who was more desperately wounded than I...We were destined to remain here more than two months, receiving the kindest treatment from these people."[3]

While at Frankford Major James H. Nounnan of the 16th Virginia Cavalry arrived and informed Echols of the presence of Gen. Duffié at Meadow Bluff, 15 miles west of Lewisburg, although Duffié was already much closer. Lt. Micajah Woods, Jackson's Battery, remembered, "Intelligence came through courier after courier that Scammon [Duffié] from Kanawha was within 12 miles of Lewisburg & rapidly advancing to intercept us. Another run was made to pass Lewisburg & cross the Greenbrier before his arrival."[4]

State historical highway marker alongside Route 219 at Frankford. General Echols rested his defeated southerners here for two hours during the retreat from Droop Mountain. The town was established in 1801 by Frank Ludington. BRIAN ABBOTT

Presbyterian Church established in 1853 at Frankford. Pvt. Joseph A. Brown, Company H, 22nd Virginia Infantry, was seriously wounded at Droop and left here by his comrades during the retreat. Rev. Samuel Henry Brown, who died in 1857, is buried directly beside the church. BRIAN ABBOTT

Echols passed through Lewisburg and crossed the nearby Greenbrier River between 3:00 and 4:00 A.M., an event vividly recalled by Lt. Randolph Blain, Jackson's Battery, who wrote: "...by 3 [A.M.] we had the artillery passed through town & crossed the river. I stopped in town to tell Aunt good-bye."[5] While Echols passed through Lewisburg the 16th Virginia Cavalry, "having come in from Meadow Bluff, remained there and brought up the rear," and continued to serve as rear guard while the Confederate army moved south toward Union in Monroe County. Sgt. William J. Dixon, Chapman's Battery, by an undisputed road and without support, managed to rejoin the command near Greenbrier Bridge with the artillery piece that had been detached on the old road with Lt. Col. George M. Edgar's 26th Battalion Virginia Infantry.[6] Confederate artillerist Lt. Micajah Woods added, "Our escape must only be attributed to our rapid running. The whole distance from the battlefield to Greenbrier bridge—32 miles—was made in full trot and often at a full run. Hundreds of infantry & dismounted men took to the bushes and large numbers who are infantry will come up."[7] The situation was perhaps best summed up by a postwar historian who wrote that the Confederates covered 28 miles over mountain roads in 24 hours to reach Droop Mountain, yet took only 11 hours to get back from the battlefield, marching a total of 56 miles in 42 hours, and fighting a 7 hour battle.[8]

Among the many soldiers, as well as citizens, lost or separated from the southern army during the retreat was Pvt. George Henry Clay Alderson, Company A, 14th Virginia Cavalry, who wrote of his episode as he awoke about two miles above Frankford: "...in early morning Joe Bob McClung [Pvt. Joseph Allen McClung, Company K, 14th Virginia Cavalry, also known as "Joe Bob" and "Whistling Bob"] and I concluded we would go to my home on Muddy Creek. We went west to Sinking Creek, down Sinking Creek, and crossed the old turnpike near Sammon's."[9] Samuel D. Edmonds, Company B, 22nd Virginia Infantry, after spending the night on top of a mountain where he built a fire to keep warm, awoke, and attempted to locate the army. He wrote, "the path we traveled was froze & raised up about 2 in. Poor Jim Starcher [Pvt. James Starcher, Company B, 22nd Virginia Infantry] had lost his shoes & just sock left & his feet was bleeding on the way. So we came to house & got breakfast & an old lady gave Jim Starcher a pare [pair] of rag [rug] slippers. We traveled on. We were about eight or 10 days getting our brigade. Citizens kept us off of our course, tell us if we kept on in that course that we meet the Yankees sure. Then we go another course."[10] Although not a soldier, and if one accepts the story that he hid in a log cabin on Droop Mountain to avoid military service [rather than the tobacco chewing story], Cornelius Stulting supposedly hid in a Droop cave and "...all day and all night the cannon roared back and forth over the mountain and the family sat in fear [at the Stulting home near Hillsboro], scarcely able to pray, even, lest Cornelius be caught in his hiding place. But before dawn he staggered in, his hands and clothing torn and his bare legs badly scratched. He had hidden in a cave through the day and in the cover of darkness he had run down the steep cliff-like side of the mountain. He was alive and unhurt but his little field, ploughed ready for seed, was ruined by cannon shells."[11]

General Averell was also in motion early on the morning of November 7. The previous night, while encamped at Renick's Valley, he had issued the following day's order of march to the regimental commanders, and informed the men that at daylight each would receive "enough ammunition to supply forty rounds to each cartridge box." Eager to bag Echols, Averell placed the 14th Pennsylvania Cavalry in the advance, which moved out in a "steady but brisk gait." Accompanying the Pennsylvania troopers was one of Ewing's pieces as well as Lt. Denicke of the 68th New York Infantry, signal corps detachment. Denicke was placed with the advance in order to relay any matters of importance to the rear, as the purpose of the Federal advance party was to determine if Echols had made a stand north of Lewisburg or retreated through the town, moving southward.

At 8:00 A.M., the 14th Pennsylvania Cavalry, one section of Ewing's Battery, the two infantry regiments and Keepers' Battery were placed in motion. An hour later, at 9:00 A.M., the 8th West Virginia Mounted Infantry, followed by the remainder of Ewing's Battery, the 2nd and 3rd West Virginia mounted infantry regiments, Gibson's Independent Cavalry Battalion, the ambulances, and the trains, also set forth. Major Gibson detached one squadron of his battalion to serve as rear guard and march behind the wagons.[12] Correspondent "Irwin" noted the column moved slow in order to save the horses, and that the entire route was strewn with abandoned Confederate rubbish.[13]

As the Federal column proceeded down the mountain Averell noted he "could see the smoke of several campfires on mountains eastward indicating a wide dispersion of the Confederates." As earlier noted, the column also located the cannon abandoned by Chapman's Battery while descending the mountain, and near Falling Springs (Renick) Averell reported his "troops in excellent condition, with plenty of ammunition." Col. Schoonmaker reported he moved his horsemen rapidly southward through Frankford, and Sgt. Thomas R. Barnes, 10th West Virginia Infantry, wrote in his diary, "...marched passing through Frankford."[14]

While Duffié and Averell moved toward Lewisburg, Gen. John D. Imboden had his men in the Shenandoah Valley region up at dawn. His command, consisting of elements of the 18th Virginia Cavalry, the 62nd Virginia Mounted Infantry, and Capt. John H McClanahan's Company Virginia Horse Artillery, moved in the direction of Warm Springs.

At 7:00 A.M. the last of Echols' men [excluding his rear guard], yet in great confusion, passed through Lewisburg en route to Union. None too soon, either, as Gen. Duffié managed to enter Lewisburg at 9:00 A.M., just as elements of the 16th Virginia Cavalry, Echols' rear guard, were departing [Averell claimed Duffié arrived at 10:00 A.M. and Capt. Denicke of the signal corps said the time of arrival was 11:00 A.M. Not only were both men incorrect on Duffié's arrival time, but Denicke also incorrectly stated Duffié had four regiments of cavalry, while Frank Reader, 2nd West Virginia Mounted Infantry, wrongly believed Duffié had four infantry regiments and four artillery pieces]. James Ireland, 12th Ohio, said Duffié's infantry, "Arrived at Lewisburg by 11 o'clock A.M. having marched 17 miles. Our cavalry charge[d] in town; find no enemy there. Town in our possession without a fight. They had gone out to meet Genl. Averell on the Frankford Road; who engaged them. Defeated them badly, putting them to flight in all directions. Killed, wounded, and captured 1,000 of them. Taking a portion of their artillery & train."[15] One newspaper claimed Duffié captured 36 of the rebel rear guard.[16] Pvt. George H. C. Alderson, 14th Virginia Cavalry, one of the many "lost" Confederates, spotted Duffié's column, writing "...learned that the Federals had passed on toward Lewisburg. We saw their wagons. Coming down Sinking Mountain we passed off to Joe Bob's father's and got our breakfast. When I got home (at the Feamster place) my face was as black as any Negroes."[17]

At Lewisburg, Duffié sent Major John J. Hoffman, 2nd West Virginia Cavalry, forward to closely press the retreating enemy, as Duffié advanced his entire column south of Lewisburg toward the Greenbrier River. He soon afterward managed to overtake Echols' rear guard and capture "110 head of cattle, 2 caissons, and a few prisoners." Echols had burned the bridge spanning the Greenbrier River and placed blockades in the road, which kept Duffié's forces from capturing him. Realizing further pursuit was useless, Duffié had his men fall back to Lewisburg, where he managed to capture the Confederate camps, destroy "large quantities of quartermaster, commissary, and ordnance stores," burning them as he lacked adequate transportation. He also captured and burned the knapsacks of the 22nd Virginia Infantry, which had been abandoned in their flight, and tents of one southern regiment. A citizen would write in a letter, "The 22nd and Edgar's battalion & D[errick's] lost all their clothing and all they had burnt. James' trunk and all his company papers and a good many other things but most of his clothes were at home. He was sick so was not there. I heard [Capt. John P.] Donaldson [Company H, 22nd Virginia Infantry] was wounded. Browne's beau—[2nd Lt. Woodson A.] Tyree [Company C, 22nd Virginia Infantry] is wounded in the shoulder—he is at John Alexander's."[18] James Z. McChesney, 14th Virginia Cavalry, claimed "...all of Echol's infantry had to burn their knapsacks and nearly everything they had on their retreat,"[19] which disputed Duffié's claim to the damage. Another view was recorded by William Ludwig, 34th Ohio Mounted Infantry, who said, "Our brigade saw no fighting for the Rebs run leaving knapsacks, ammunition, and burning some of their artillery carriages. We also captured the colors of the 22nd Virginia Rebble [sic] Regt."[20] Even Lt. Micajah Woods, Jackson's Battery, deplored the loss to his comrades, stating, "The troops on leaving Lewisburg to fight left all their camps standing, with all their baggage—which was immense. All was burnt & the loss sustained by the men & officers in clothing and material is very heavy. I am proud to say that our Battery has come off without the loss of a pound of baggage of any description. Even our tents [are] saved. And not a man [of the battery] was out of his place at any time during the fight or retreat."[21]

However, Duffié's command not only destroyed military items, but also brought havoc to the citizenry, in direct violation of their orders. Confederate artillerist Micajah Woods wrote, "Soon after the enemy occupied Lewisburg they burnt the Southern Methodist Church and several large houses on Main Street which had been denoted to government purposes. I learn that two or three friends' residences were also burned but whose they were I have not heard." He also remarked that as all the men of the area departed with the army "the women are left helpless to the tyranny of the foe."[22] James Ireland of the 12th Ohio confirmed the destruction, writing, "The town of Lewisburg ransacked by the soldiers for forage. Some soldiers act very disgracefully, taking things which do them no good."[23] A report in the November 12, 1863, issue of the *Gallipolis Journal* claimed, "...our boys had a brush with the rebs at Lewisburg, Va., and that the latter were nicely cleaned out. One or two houses were destroyed in our attempt to shell the scoundrels out. The rebels were driven from the place, and some prisoners taken. Without the loss of a man on our side..."

While Gen. Duffié was occupied at Lewisburg, Gen. John D. Imboden had passed through Warm Springs at 1:00 P.M. on the Huntersville road. At that point he gained his first knowledge of the Mill Point and Droop Mountain fights, but not that Echols had been defeated at Droop. Determined to engage Averell's rear he advanced rapidly, but, at the foot of Back Creek Mountain a courier arrived from Covington, carrying a dispatch from a Capt. Skeen. The message told of the rout of Echols at Droop Mountain and his subsequent retreat through Lewisburg into Monroe County. It also contained information of the Federal occupation of Lewisburg, vastly over-estimated at 15,000 men. Realizing it was now useless to advance by way of Huntersville to assist Echols, Imboden changed his course and moved down Jackson's River toward Covington. He also issued an order to the Rockbridge Home Guard of Lexington, under Col. James Woods Massie, and the Cadets of Gen. Francis H. Smith, at Millboro, to move to Clifton Forge.

In the meantime, Averell's advance, under Col. Schoonmaker, had arrived eight miles north of Lewisburg, where he was able to view large fires within the town. Suspecting it was Echols evacuating the town and destroying his stores, Schoonmaker covered the next six miles at a trot. But at about 2:00 P.M., when within two miles of Lewisburg, he discovered the smoke was emanating from the campfires of Duffié's men. Col. Schoonmaker immediately placed his 14th Pennsylvania Cavalry in camp within the suburbs of Lewisburg and sent two squadrons of the command into the village to "secure the town and remove all troops from it." He felt Lewisburg was in a state of chaos, with Duffié's men plundering and destroying much property. Shortly afterward, with the assistance of the regimental provost marshal, he was able to restore a semblance of order. A report in the *Ironton Register* claimed the 2nd West Virginia Mounted Infantry "...were in the advance, and were among the first to greet the 2nd [West] Virginia Cavalry at Lewisburg."[24]

During the advance of Averell's Brigade to Lewisburg, over what he described as "an excellent road" [the Seneca Trail], Major Thomas Gibson's Independent Cavalry Battalion had been busy as Sgt. Joseph S. Hess, Company A, 1st West Virginia Cavalry, managed to capture six armed rebels with the assistance of only one man, "overcoming them by his daring." Major Gibson also sent out a foraging party under 2nd Lt. Herman Scharenberg [Company C, 16th Illinois Cavalry], acting adjutant, and they brought back "one rebel prisoner, 50 head of beef cattle, and 125 sheep." As the brigade passed through Frankford, the Federal soldier who had killed the member of the 14th Virginia Cavalry called "Miles" during the Droop Mountain battle, and took possession of his pencil, notebook, and a letter written from his sister, struck up a conversation with a young lady. Upon presentation of the letter the lady screamed with grief, as she realized the document had belonged to her brother, and she had written it.[25]

Correspondent "Irwin" said that as Averell's men neared Lewisburg, they came to the camp of the 22nd Virginia Infantry, "screened from view in a grove in a sink-hole."[26] Averell's column finally arrived at Lewisburg between 4:00 and 5:00 P.M. [times vary according to source—correspondent "Irwin" and Echols both say 4:00 P.M., while Averell says 5:00 P.M.], where the commanding officer learned for the first time that Duffié had arrived too late to cut off the retreating Echols. But Duffié had managed to destroy the enemy camps and secure a stand of colors. Averell also mentioned Duffié captured a few rebel stragglers and some abandoned material. While at Lewisburg Averell also learned, falsely, that Gen. Robert E. Lee had promised to reinforce Echols "at or near Dublin." With that knowledge in hand he decided to move quickly against Dublin and the Virginia & Tennessee Railroad.

In actuality, during the day Gen. Sam Jones, commanding the Department of Western Virginia and East Tennessee at Dublin, sent a report to Gen. Samuel Cooper at Richmond, which described Echols' terrible defeat and stated Echols was being closely pursued, primarily by mounted Federals. Jones felt Echols would retreat via Salt Pond Mountain if possible, but would be unable to outrun the Federal cavalry. He quickly dashed off a telegraph to General Lee requesting reinforcements, stating, "...will need all I can get. No time to lose."

Although Jones and Echols did not know it at the time, no reinforcements would be forthcoming, and the only men within distance of giving any possible support were those of Imboden and a brigade in the vicinity of Princeton in Mercer County, which included the 36th Virginia Infantry and Bryan's Battery. Sgt. Milton Wiley Humphreys, a member of Capt. Thomas A. Bryan's Company Virginia Artillery (also known as the Monroe Artillery), wrote from his camp near Princeton: "On November 7th I received orders to move immediately to Princeton with all the horses and wagons. Gen. Echols had been defeated by Averell at Droop Mountain in Pocahontas Co. After some hours spent in catching horses and picking up, we moved to Princeton, and the Battery moved immediately after the Brigade for the Narrows of New River. I was left behind in charge of a caisson and the luggage. I started then and came up with the Battery at daylight on the 8th at the Narrows. The Battery encamped on the south side of the river until the 10th."[27]

To the east of Echols, Imboden received a telegram from Gen. Jones during the night, but as it was written in cipher, and as Imboden did not have the key to decode it, he was unable to read the message. On the positive side of affairs for Imboden, however, at 7:00 P.M. some 800 men, comprising the Rockbridge Home Guard and the Cadets, were already moving on a 12 mile march to consolidate with him.

In the evening, as November 7 drew to a close, Averell, having covered some 20 miles during the days march, and Duffié bivouacked at and "just beyond Lewisburg" while Echols' battered army halted for the night at Salt Sulphur Springs, just south of Union in Monroe County. The Reverend S. R. Houston, a resident of Union, wrote in his diary that Echols' men "much disorganized and demoralized, passed through town today and encamped near Salt Sulphur. Farmers driving off their stock."[28] He also felt the rebel loss at Droop Mountain was 21 killed and 130 wounded. Randolph Blain, Jackson's Battery, said the southerners spotted "a large smoke...in the direction of Lewisburg & since we have heard rumors of the property burnt." Blain also noted, "...the whole command is now scattered, there being some 200 infantry. Our two Reg. 16th & 14th [Virginia cavalry] follow as rear guard."[29]

November 8

During the windy night of November 7 Gen. Averell had issued the order of march for November 8 to his brigade. At 8:00 A.M., one of the infantry regiments would move out, followed by Battery B, 1st West Virginia Light Artillery, and the other infantry regiment. Then, at 9:00 A.M., the 8th West Virginia would lead off, trailed closely by the 2nd West Virginia, the 3rd West Virginia, Ewing's Battery, Gibson's Battalion, the 14th Pennsylvania Cavalry, and the ambulances and trains.[30] Averell also sent out Lt. Denicke of the 68th New York Infantry, signal corps detachment, accompanied by Lt. John R. Meigs, of the Engineer Corps, to study the feasibility of crossing a ford, about three miles east of Lewisburg, on the road to White Sulphur Springs. Denicke was to keep in touch with Averell by proper signals. Denicke also sent Lt. Merritt of the signal corps to accompany a company of cavalry to Edgar's Ford [Ronceverte] on the Union road, about 5½ miles south of Lewisburg. Merritt was instructed to keep in touch with headquarters, but near the ford they were "bushwhacked by a number of the enemy in ambush."

Averell, encamped at Lewisburg, had received further false information that Echols had concentrated his forces at Union in order to make a stand. Taking this under consideration on the morning of Sunday, November 8, he decided to move his brigade to White Sulphur Springs and ordered Gen. Duffié to move south and engage Echols. The two commands set out to the ford across Greenbrier River, two to three miles east of Lewisburg. En route Averell's men located two rebel camps designed for winter quarters, which were aflame. One of the camps was situated on a hill beyond Lewisburg "and the other hid away in the ravine alongside the road."[31] At the Greenbrier River the Federals found the enemy had destroyed 500 barrels of flour from the mills, which were floating in the river. James Ireland, 12th Ohio, with Duffié, wrote: "We fell in to move we knew not where. March toward White Sulphur. Cross Greenbrier River in a flatboat. The disgraceful plunder of the Caldwell or North house–burning of barn, etc."[32] Duffié, promptly obeying orders, was immediately beset with problems. At Greenbrier River, four miles south of town, he had to leave his infantry, the 12th and 91st Ohio regiments. Many of the men of those two commands were new recruits who had spent much of the previous winter in camp and were therefore unaccustomed to long, enduring marches. As a result, they were exhausted and footsore, unable to march over 10 miles a day. Additionally, they were without supplies, as their wagons had previously broken down. Averell later relayed Duffié also told him the command had only one day's rations. A full report of the situation was quickly sent back to Averell.

Duffié placed Col. Freeman E. Franklin of the 34th Ohio in charge of the two mounted regiments and sent them toward Union, with Col. Franklin placing his 34th Ohio Mounted Infantry in the advance. The possibility of a clash mounted, as evidenced by Rev. S. R. Houston of Union, who wrote in his diary: "[Col.] Jackson's 600 cavalry in battle array in our very midst, having heard enemy advancing in force. Excitement of the people now intense."[33] At Second Creek, about 8½ miles north of Union, rebel pickets opened upon the 34th Ohio. However, no serious damage befell Duffié's men, who managed to capture a few rebels in the encounter. James Ireland, 12th Ohio, gave a somewhat different account in his diary, claiming the infantry remained with Duffié on the march. He wrote, "Here [at Caldwell] we took the Union & Red Sulphur road [Monroe Draft]—find the enemy have made a disorderly retreat leaving any amount of forage on the route—flour, etc. which is destroyed. Marched some 8 miles & countermarched back across the river again & camp on this side of Lewisburg."[34]

The rebel prisoners told Duffié that Echols was in full retreat toward the Narrows, where, according to citizens of Union, Echols would be reinforced by Generals John Stuart Williams and Imboden, as well as the 36th Virginia Infantry, which had reportedly left Princeton during Echols' retreat through Lewisburg. Very little of this information would prove to be true; although Imboden was a potential threat and some troops, including the 36th Virginia Infantry and Bryan's Battery, had indeed moved to the Narrows of New River.

On that very day, Gen. Sam Jones had written Confederate Secretary of War James Alexander Seddon reporting Echols' Brigade had been nearly destroyed at Droop Mountain. Jones, in somewhat of a panic, also told Seddon he had only two infantry regiments, two batteries, and a cavalry battalion between Averell and the Virginia and Tennessee Railroad. Requesting reinforcements, he estimated the Federal force at Lewisburg the previous day as 7,000 strong. Confederate States President Jefferson Davis responded to his request, writing that no reinforcements would be possible unless local defense and militia were to be had, and "it only remains to concentrate on the best position and make entrenchments if they will avail. I suppose General Robert Ransom Jr. is advised, and if possible, will cooperate."

The David Creigh Incident

It is believed that during this time on November 8 one of the strangest incidents of the campaign transpired. Numerous versions of the story have been told, but only the most commonly accepted version is presented here. Supposedly, a man dressed as a Federal soldier was roaming through the country alone south of Lewisburg and went first to a Mr. Dunn's house, one-fourth of a mile from the David Creigh residence, and robbed the women of their watches, and stole other items from the house. Then he went to the David Creigh home, known as "Montescena," located about two miles south of Lewisburg. The house, built in 1834, was of Greek revival design, large and imposing, brick, with white columns, massive chimneys, lawn and trees [the house is yet standing along Rt. 37, just west of Fairlea, and a short distance from the Greenbrier East High School].

Mr. Creigh, an area citizen, was informed of a Federal soldier plundering his house and using abusive language on his wife and sick daughter. The soldier was upstairs breaking open drawers and trunks when Creigh, carrying a small pistol in his pocket, arrived. He found his daughter's belongings scattered on the floor and some being flaunted in his wife's face. The soldier then attempted to break open a trunk belonging to a teacher hired by the family. Creigh asked the man to stop and told him it belonged to the teacher. The soldier rose up, drew his pistol, cocked and pointed it at Creigh, then said, "Go out of this room. What are you

Creigh marker located in the Old Stone Presbyterian Church Cemetery at Lewisburg. David S. Creigh, a citizen of Lewisburg, was accused of killing a Federal soldier during Averell's occupation of the town and hanged by the military near Brownsburg, Virginia, June 9, 1864; as a result he became a martyr to the south. BRIAN ABBOTT

doing here? Bring me the keys." Creigh and the soldier then fired their pistols at each other and a bullet grazed Creigh's face and lodged in the wall. A scuffle ensued and both fell down the stairs to the front door, landing with the soldier on top of Creigh. As both arose together Creigh made an effort to get the Yankee pistol, but it accidentally discharged and hit the soldier, producing a "profuse flow of blood." The soldier struggled to the portico, where he fired another shot at Creigh, the bullet missing him and lodging in the upper part of the front door. A Negro woman servant standing nearby with an axe screamed, "Master, he will get up," and begged Creigh to strike him with the axe. Creigh did as she requested and killed the soldier, then placed the body in a hidden burial place.

When Gen. George Crook came through the area the following year Creigh was arrested by Averell, and later by Gen. David Hunter, given a mock trial and then carried away with the army and executed [against the objections of Crook and Averell] at Brownsburg, Virginia on June 9. Although it has long been argued that the man Creigh killed was not actually a soldier, and that may well be true, official records do show a Pvt. Henry Rumble, Company G, 1st West Virginia Cavalry, was missing in action on November 8 [at Droop Mountain], whereas another source says he was killed by a citizen of Lewisburg on that date. On the other hand, Federal accounts of the incident vastly differ from those given by people of pro-Confederate sentiment, as evidenced in the diary of Pvt. Benjamin F. Zeller, Company L, 8th Ohio Volunteer Cavalry, who apparently witnessed Creigh's execution at Brownsburg, Virginia in 1864. Writing on June 11, Zeller said, "At sun-rise a rebel who had been arrested at Green Briar river by the 2nd Cavalry for the murder of a Union soldier out of the 34th Ohio, was tried and found guilty, and suffered the extreme penalty of the law by hanging until he was dead, upon a small tree in a wheat field near the camp. His history relating to this case was as follows: He was a man somewhat advanced in years, a man of wealth, owning twenty-six hundred acres of land, and other property. The murdered soldier went to his well, which is some distance from the road, to get a drink, and fill his canteen with water; while stooping over the well drawing water, the old Rebel slipped up behind him, hit him over the head with an axe, then drew the revolver from the soldiers belt, and shot him twice; the rebel daughter standing near the horrible scene, with a heart filled with much the same spirit that prompted the damsel of olden times to demand the head of John the Baptist in a 'Charger,' ran into the house, procured a shovel full of hot embers, and coming quickly up, threw them into the face of the dying man. After he was dead, his body was chopped up by this old fiend, and thrown into a well to prevent detection." If Pvt. Zeller is correct in his statement that the murdered soldier belonged to the 34th Ohio Mounted Infantry, then a name which is a possiblity is Pvt. Sobeska Cozard, Company K, who died November 21, 1863. While the actual names and true specifics of the David Creigh Incident may never be known until a more detailed investigation of the incident is undertaken, it remains representative of the horrors of war and how it affected the citizenry.[35]

Return to Safety

Averell had pored over the information sent to him from Duffié and concluded further pursuit of Echols was fruitless considering the condition of Duffié's foot soldiers, various obstacles in the road, and the extra burden placed on Averell's own two infantry regiments, which were bogged down carrying the wounded and prisoners. After carefully assessing the situation Averell sent word to Duffié at Second Creek to retire back to Meadow Bluff, west of Lewisburg, and issued orders to Col. Augustus Moor to take his infantry regiments, the 28th Ohio and 10th West Virginia, as well as Keepers' Battery, and return to Beverly via Hillsboro. Moor was to take all the transportable wounded, dismounted men, horses, and some 82 prisoners from the Droop Mountain fight, and was also told to capture all small arms, cattle, horses, and to destroy the rebel camps, all of which Moor immediately accomplished as far as practicable. Sgt. Thomas R. Barnes, accompanying Moor with the 10th West Virginia, recorded in his diary: "Marched to Frankford, camped overnight."[36] Averell retained the cavalry, mounted infantry, and Ewing's Battery, which continued to move east toward White Sulphur Springs "along [the] eastern border of West Virginia," and eventually to New Creek [Keyser]. Confederate artillerist Micajah Woods noted rebel cavalry pursued both Averell and Duffié, writing: "Col. Ferguson cmdg. our Brigade...followed them with detachments of his command on back roads with vigor, pushing Averell some 30 miles and the other force nearly to Gauley Bridge."[37] If the rebel troopers were ever any nuisance to the Federals, mention was never made of it by the Yankees.

About four miles west of the resort of White Sulphur Springs, Averell encountered two men from Ewing's Battery; one on crutches due to an amputated leg, and both of whom had been captured in the battle of White Sulphur Springs in August. At 10:00 A.M. the Federal column reached the resort, where Averell recovered a number of other soldiers who had been wounded and captured in the late August fight. Frank Reader, 2nd West Virginia Mounted Infantry, recalled, "The brave fellows were rejoiced to see us, and the meeting with their comrades was a happy and an effecting one."[38] One particular exception was an unidentified soldier who had been nursed back to health by a local red-haired lady. She had apparently fallen in love with the soldier and grimly told Averell, "You can't have that soldier. He is mine. I captured him, and nursed him, and made him well, and he is going to stay with me. He is mine." Despite the vehement protests of the beautiful young lady, the soldier was eventually made to depart with Averell, and as far as is known, never did return to White Sulphur Springs.[39] Col. Schoonmaker reported that by noon the command had left the resort and reached the battlefield of August 26-27, where the former combat site was examined and eight additional Federals who had been wounded and captured in that affair were recovered. Before departing, Averell's men also visited the graves of their comrades who had died in the fight.

During the day Gen. John D. Imboden had not been idle, having advanced his men to Covington, where he found some 100 to 200 panic-stricken stragglers and refugees from Col. William L. Jackson's command, having found their way to Covington from Droop Mountain; during the night, additional stragglers would arrive. Imboden supplied them with subsistence funds, and, if they had lost their weapons, equipped them with some Virginia state arms he had located at Covington. Imboden also formed the group into two companies and ordered them to participate in his anticipated fight with Averell. As a precaution he placed a company of his own men on picket at Callahan's.

Sometime during the evening of November 8, Averell's advance, comprised of the 8th West Virginia Mounted Infantry, neared Callahan's. Col. Oley was ordered to send two cavalry squadrons "to and beyond Callahan's to ascertain the position" of some 150 rebels reported in that vicinity. For this purpose Oley chose Major Hedgeman Slack, a 30 year old soldier who had been promoted to major of the regiment March 1, 1863. Slack sent the appropriate scouts forward to investigate. Although Slack's men failed to report any confrontation, Imboden's pickets said that shortly after dark the Federals attacked them, with the loss of one horse shot. They also claimed they were able to repulse the Yankees and hold their own position. Slack's men made no mention of any such encounter and said the rebels fell back upon the sound of their approach. Slack therefore deemed pursuit useless and went into camp at Callahan's, as did Col. Schoonmaker's men, as well as most of Averell's command. Corp. George W. Ordner, 2nd West Virginia Mounted Infantry wrote from Callahan's at 11:00 P.M.: "Camped at this place for the night,"[40] having covered some 55 miles during the days march.

At Petersburg in Grant County, West Virginia, Averell's immediate destination, Pvt. Amos A. Vandervort, Company B, 14th West Virginia Infantry, wrote in his diary November 8: "The 1st W. Va. and the 23rd Illinois reg'ts left here today. Some say for Seneca gap, 20 miles distant, others that they are going to form a junction with Gen. Averill [sic] to surround McNeal [sic]."[41]

Before the day was over, Imboden's situation began to improve, as the Rockbridge Home Guard and the Cadets, consisting of about 800 men and two six-pounder guns, had arrived at Clifton Forge, about 13 miles from Imboden. Additionally, the remnants of Echols' Brigade had continued to move south to "the foot of middle of Salt Pond Mountain" for the night. Micajah Woods of the brigade noted the arrival of inclement weather, writing: "...it is snowing–we are halted but a few minutes and my fingers are so cold that I can scarcely use them." The situation was further detailed as Woods said "the 22nd Regt. now numbers for duty about 100 men,"[42] and fellow Confederate artillerist Randolph Blain, who wrote: "Halted on the top of Salt Pond Mountain for the night where we were enveloped in a snow an inch deep..."[43] Apparently Jackson's Brigade remained in the vicinity of Union, as Lt. Col. William P. Thompson, 19th Virginia Cavalry, wrote his report of the Droop Mountain battle from there on November 8.

November 9 - Clash at Covington

As noted, on the morning of Monday, November 9, snow began to fall in scattered areas of southern West Virginia and southwest Virginia and would continue to do so throughout the day. During the early morning hours, Major Hedgeman Slack, 8th West Virginia Mounted Infantry, was ordered to move his two companies forward on the Covington road and "ascertain what is in that direction." Slack promptly obeyed and soon encountered rebel scouts, who fell back upon his approach. At dawn, Slack's men, numbering about 400, ran into Imboden's pickets and slowly drove them two or three miles. Slack claimed the enemy was constantly reinforced during this period. The rebel pickets finally joined with Imboden, who had developed a defensive line on the crest of a mountain pass about one-and-a-half miles west of Covington. When Slack's troopers came into view, about 1,200 yards distant, Imboden had McClanahan's Battery fire a few rounds, which quickly scattered the Federals. Regretfully for Imboden, this action also had a detrimental effect upon his own men. When his cannon first opened fire, "a large number of Jackson's stragglers ran off to the woods and mountains." A portion of Jackson's men held steady, but the bulk "fled most shamefully when the Federals" were closer than two miles from where Imboden had left them to await orders.

Despite the poor performance of Jackson's stragglers, who were probably reliving the nightmare of Droop Mountain; as Imboden's artillery created confusion by firing a few shots among Slack's men, two companies of rebel cavalry charged the blue-coats. The final result of this skirmish at Covington remains unclear. Major Slack says he had no casualties and fell back to Callahan's as he deemed further pursuit useless, although how one who is being chased can abandon pursuit is a remarkable accomplishment in itself. Averell would later indicate Imboden was reportedly wounded in the affair and Slack captured a rebel lieutenant and 20 of his men. Correspondent "C" of the *Ironton Register* gave a boastful version of events, writing, "Our command had started in the morning, before we discovered that Imboden was so near. Two squadrons were sent out the Covington road and on learning that Imboden had about 1,500 men, we chose a position, and invited an attack. But the great 'Bushwhacker' declined and fled southward."[44] Frank M. Imboden, 18th Virginia Cavalry, probably put it best when he wrote in his diary: "Skirmish this morning with two companies and section of artillery, drove Yanks back."[45]

At Callahan's, Major Slack told Col. Oley that he believed he had engaged Imboden's rear guard, and Imboden was in the process of moving a portion of his command toward the Virginia and Tennessee Railroad in order to support Echols. Oley then ordered Slack to rejoin the main column on the Warm Springs road, an order he promptly obeyed. Oley then proceeded to inform Averell of Slack's report. Based upon this new information Averell concluded, "At Callahan's, on the morning of the 9th, I learned that General Imboden, with about 900 to 1,500 men, was at Covington on his way to reinforce Echols at Union." Averell correctly assumed Imboden was not strong enough to attack him and deemed it "inexpedient" to capture the rebel general.

While Averell's column moved north from Callahan's on the Warm Springs road, the 2nd West Virginia Mounted Infantry was placed in the advance. 2nd Lt. William Schmulze [Schmolze], and 12 men of Company I, 2nd West Virginia, were in the advance and were soon fired upon by a party of rebel bushwhackers. The Federals incurred no damage and captured 15 rebels with their horses and accouterments. Corp. George W. Ordner of the 2nd West Virginia noted this activity, writing in his diary: "Guerrillas fired on us this morning at Callahan's. Nobody injured."[46] The Federals did, however, suffer some injury during the early morning march as an improperly packed percussion shell in one of Capt. Ewing's caissons exploded; injuring three men, creating intense smoke, destroying the caisson, and scattering its contents over a wide area. At 1:00 P.M. Averell reached the Jackson's River road and waited for Major Slack to rejoin him. At this location, known as Morris Hill, Averell left the Warm Springs road and continued north up the Back Creek valley.

Having observed Averell's command, Imboden's men informed the rebel officer that Averell had 3,000 men, seven cannon, and was moving off on the Warm Springs road. Imboden realized there was an obscure country road, which, if utilized by Averell, could possibly place the Federals at his rear. He felt confident Averell would attempt this and fell back to a hill one mile east of Covington. He also sent word to the Rockbridge Home Guard and Cadets to hurry to his assistance, as he intended to give Averell battle in that "strong position." A portion of Imboden's cavalry, under Col. George Imboden, was thrown forward about four or five miles in order to keep watch on Averell, who actually had no intention of striking Gen. Imboden. Averell's only concern, as it had been ever since departing Lewisburg, was to reach the Federal lines at Petersburg and New Creek [Keyser]. At this point he considered Imboden nothing more than a nuisance and was already moving north.

Indeed, the force then stationed at Petersburg in Grant County, under Col. Joseph Thoburn of the 1st West Virginia Infantry, consisted of the 23rd Illinois Infantry; 14th West Virginia Infantry, detachments of the 2nd Maryland Potomac Home Guard and 1st West Virginia Infantry; seven companies of the Ringgold Cavalry (22nd Pennsylvania Cavalry); and the batteries of Rourke and Carlin, comprising 13 guns. Once Averell made it to this location he would be only 42 miles south of New Creek [Keyser].

While Averell and Imboden played cat-and-mouse games, Gen. Duffié had marched his column to Meadow Bluff from his camp at Millville [Bunger's Mill?], about four miles west of Lewisburg. Complications set in immediately as Duffié was unable to receive subsistence supplies due to the poor conditions of the roads, a situation created by destroyed bridges and other assorted obstacles. He also noted "several families of refugees and about 100 Negroes followed the command out from Lewisburg." Making matters worse was a sudden heavy snowstorm that made it impractical to remain at Meadow Bluff. Duffié had no alternative but to depart Meadow Bluff after dinner and begin the return march to Gauley Bridge as quickly as possible. As he crossed Sewell Mountain a five-inch snow fell, making it extremely difficult to move his artillery and train over the already blockaded roads. Sgt. William Ludwig of Company G, 34th Ohio Mounted Infantry recalled, "The weather was very cold and when we crossed the Suel [Sewell] Mountain there was four inches of snow and still snowing."[47] Chaplain Anthony H. Windsor, who did not join the 91st Ohio Infantry until April of 1864, wrote in the regimental history: "On the return trip a violent snowstorm set in, which occasioned much suffering to the regiment before it arrived at Fayetteville."[48] J. E. D. Ward, 12th Ohio Infantry, referring to both this expedition and the Salem Raid in December of 1863, said, "Very cold weather prevailed which occasioned numerous swelled heads, chapped hands and lips, pinched noses, contracted muscles, enlarged joints, rheumatism, and other ills and aches."[49] Pvt. James Ireland, Company A, 12th Ohio, wrote in his diary: "It turned quite cold & windy here. Every now & then some snow squalls...after dinner march over Little Sewell Mt. & camp between it & Big Sewell. This afternoon we have a heavy snowstorm...It snowed some 3 or 4 inches..."[50] Portions of Duffié's command had to dismount and push wagons and artillery pieces up the steep slopes where bridges had been destroyed. A number of wagons broke down, but all were repaired and would complete the return trip.

As Averell, Imboden, and Duffié maneuvered about, Col. Augustus Moor had arrived back at Hillsboro, where he picked up 55 Federal and one Confederate wounded from the Droop Mountain battle, placing them in ambulances and wagons filled with straw. A number of Federals and Confederates were too badly injured to travel, so at least 14 of the Federals were left behind at the Mountain House [three died by the time Moor reached Beverly] and Moor predicted only two would recover. To care for these men Moor left "what rations, hospital stores, and medicine he could spare." Additionally, Assistant Surgeon Jonathan R. Blair, of the 10th West Virginia Infantry, remained at Hillsboro with the two most badly wounded Yanks and nine injured rebels, while the 10th West Virginia marched to Mill Point and camped for the night.

Imboden continued his watchful pursuit of Averell. In the early afternoon he ascertained Averell had left the Warm Springs road and moved toward Huntersville. Suspecting this might be a ruse by Averell in which he would sweep around by Warm Springs to Millborough; Imboden concluded to move as quickly as possible via Clifton Forge to Goshen. Sending scouts to watch Averell's movements and to report to him at Goshen, Imboden immediately began a march that, in a little over 24 hours, would carry him over 40 miles to Goshen, arriving there on November 10. Frank Imboden, 18th Virginia Cavalry, wrote in his diary: "Marched at dark and camped at Clifton Forge. Met Corps Cadets and Rockbridge Home Guards, 600 venerable looking old men."[51]

As November 9th came to an end, Averell's main column went into camp at Gatewood's on Back Creek, in Bath County, having covered some 27 miles during the day [Frank Reader, 2nd West Virginia, wrote the column marched 19 miles]. At Gatewood's, plenty of corn and wheat for the horses was located; and, although it was "a cold, wintry night, there was plenty of rails for fuel for fire." Both Averell and "Irwin" said, "after night burned a rebel camp and potash factory."[52] Col. Moor was yet in the Hillsboro vicinity making arrangements for the wounded and prisoners, while Gen. Duffié was on the James River and Kanawha Turnpike between Big and Little Sewell mountains, moving west towards Charleston. James Ireland, 12th Ohio, recalled: "An uncommon hard night on the soldiers who have no way of sleeping. As disagreeable a night as I have ever passed. Marched 26 miles [during the day]."[53] Gen. Echols was at the foot of Salt Pond Mountain, as indicated by Randolph Blain, who wrote, "...we are now at the foot of Salt Pond Mountain near Giles Court House."[54] Also confirming this location was James Z. McChesney, 14th Virginia Cavalry, who remembered, "I was cut off by the Yankees [at Droop Mountain]...and went through Monroe by way of the Sweet Springs and Allegheny Co., Craig, and Giles to Newport, about 7 miles from Salt Pond, where I found Genl. Echols with the infantry... the 14th Regiment was at Union, Monroe Co., on picket...I [would eventually find] the 14th Regt. at Frankford. It had just returned from a scout of the battlefield of Droop Mountain."[55] Gen. Imboden was en route to Goshen, and Gen. Sam Jones was at the Narrows, where he wrote Secretary of War Seddon that the Federals had camped at Burnt Bridge in Greenbrier County on the night of November 8 and the remnants of Echols' Brigade, in no condition to fight, were at Salt Pond Mountain. Jones remarked to Seddon, "I have asked for reinforcements and I'm sure you will send them if you can." Obviously, Jones had yet to receive the information that no reinforcements were forthcoming.

November 10 - Bushwhacking

On Tuesday morning, November 10, Lt. Micajah Woods of Jackson's Battery wrote a letter from Echols' camp "10 miles from Pearisburg, Giles Co., Va." which read in part: "We reached this camp yesterday evening after an arduous march over the mountains in terribly cold weather. The enemy have not advanced further than Union in this direction. Reinforcements are arriving on the Railroad and we will doubtless check the enemy in any advance, which I have no doubt that they will soon make towards the Railroad. Hundreds of fugitives are coming in & are being again organized. Our entire loss [at Droop Mountain] killed, wounded, and missing will not be over 350 or 400."[56]

Duffié's infantry detail was up by daylight on this "cold snowy morning" and anxious to reach camp at Fayetteville, some 30 miles distant. Departing camp between the two Sewell mountains, it was "slowish traveling cross Big Sewell Mt." And the march was not without incident as Sgt. James Z. McChesney of the 14th Virginia Cavalry recalled, "...the 16th [Virginia Cavalry] Regt. charged the U.S. picket post at Sewell Mt. and captured two of them, one of them proved to be a deserter from Derrick's Bttn. [23rd Battalion Virginia Infantry], Echols' Brig., who had left our own army and joined the Yankee's some months before. He was later tried and shot near Giles C.H."[57] Samuel D. Edmonds of the 22nd Virginia Infantry later said the name of the deserter was "Boler" and "...our men captured him on Picket with Yankee uniform on..."[58] Since no soldier by this name appears on the roster of the 23rd Battalion, and the Edmonds memoir is laced with bad spelling and grammar, the man in question was Pvt. Josiah M. Boling of Company B, who deserted from the regiment in the retreat from Charleston in 1862 and was captured by the Federal army in early 1863. He later joined the 2nd West Virginia Cavalry as Pvt. Joseph M. Bolen of Company F.

Also during the early morning hours, while yet camped at Gatewood's, a dog ran a buck into one of Averell's picket posts; the pickets snared the buck. Soon afterward, Averell began his destruction of the various area salt works, first having Col. Oley and a company of the 8th West Virginia Mounted Infantry destroy an extensive salt work near Gatewood's in the Back Creek Valley, which had been in active operation the previous day. The Federal command also destroyed another Confederate winter encampment and Averell ordered Major Thomas Gibson to take his Independent Cavalry Battalion, move by way of Huntersville to either Marlin's Bottom [Marlinton] or Edray, and make contact with Col. Moor. Having accomplished this Gibson was to go from Edray to the vicinity of Hightown or Monterey and rejoin Averell on the march to the Federal camp at Petersburg in Grant County.

Averell proceeded to move his main column in the direction of Monterey. Col. Schoonmaker reported that during this march bushwhackers managed to shoot one of his best wagon horses. Correspondent "Irwin" wrote, "...our train [was] fired into by a bushwhacker...he broke his leg and was caught."[59] Frank Reader, 2nd West Virginia, said that during the day's march there was "considerable bushwhacking" and one of his regiment was killed and another wounded.[60] Despite such hazards the march was basically without interruption, and some of the soldiers took notice of some fine, large trout basking in the sunshine in a stream.

At 10:00 A.M. Col. Moor departed the vicinity of Hillsboro and Mill Point and moved in the direction of Beverly. On Elk Mountain he was ambushed by some 60 bushwhackers under the leadership of a man by the name of McCoy. Several of Moor's cattle were wounded and his prisoners were fired upon. Moor detailed one company from each of the two infantry regiments to ascend the hill to his front and rear; once accomplished, they drove McCoy's men down the opposite slope. The incident created a one hour delay in Moor's march, after which he continued toward Beverly "without further molestation, accident, or even straggling." Sgt. Thomas R. Barnes, 10th West Virginia, reported the column "camped on Big Elk River" for the night.[61] Later in the day Major Gibson arrived at Marlin's Bottom as ordered and found Col. Moor had already passed through about two hours previously. Gibson opted to leave his men under the command of his senior captain and personally advanced with a squad of his battalion to Edray, where he found Moor's rear guard and discovered the colonel was yet three miles ahead. Locating Lt. Col. Moses B. Hall, 10th West Virginia Infantry, Gibson gave a verbal account of Averell's movements and success for him to relay to Col. Moor.

General Imboden arrived at Goshen, or as Frank Imboden said, "Command sent to Goshen Pass,"[62] where his scouts informed him Averell was moving very rapidly up the Back Creek toward Monterey. This information prompted Imboden to dismiss the Rockbridge Home Guard and the Cadets, as they could be of no further service to him. Gen. Sam Jones, who had learned no reinforcements per se were forthcoming, arrived at Salt Pond Mountain and was greeted with good news, which he immediately relayed to Secretary of War Seddon. He found Echols' Brigade was not nearly as battered and decimated as first thought; many of the absentees and stragglers from the Droop Mountain battle had continued to find their way back to Echols; and Lt. Col. George M. Edgar's 26th Battalion Virginia Infantry, which had not even been in the fight, was then approaching Salt Pond Mountain via Sweet Springs in Monroe County. Edgar was expected to arrive that very day. Lt. Micajah Woods noted, "Edgar's Battn. took a different route by way of Sweet Springs—was overtaken & forced to make a very rapid march to save itself. Several hundred fugitives are with it"[63] Jones promptly supplied Echols "with the necessary arms and clothing" and stated, "I think Echols will have them in fighting condition in a few days."

Lieutenant Colonel Alexander Scott, 2nd West Virginia Mounted Infantry, relayed that Averell crossed the Staunton and Parkersburg Turnpike near Hightown in Highland County, and near sundown came to the intersection of the road leading from Monterey to Franklin. At that location, then known as Green Hill [Greenhill], the Federals went into camp. Corp. George W. Ordner, 2nd West Virginia, said his regiment "camped on the farm of Captain White, who commands a company of Guerillas." He also noted the command had marched 21 miles on this "very cold" day.[64] Col. Schoonmaker, 14th Pennsylvania Cavalry, said his men set up camp at 5:00 P.M., some eight or nine miles south of Monterey. Correspondent "Irwin" noted that during the White Sulphur Springs march in August the command had been bushwhacked in this same area and had a "score to settle." While camped here a rebel lieutenant was captured, the men found a number of apples "hidden in the ground," abundant forage was located for the horses, and mountain-muffins were consumed for supper. He also said, "...with a soft bed of hay after supper, before our 'big' fires we had a luxurious night's rest"[65]

Duffié's infantry column of the 12th and 91st Ohio volunteers, took a cutoff on Big Sewell Mountain and arrived at Bowyers Ferry about 3:30 P.M. After crossing the New River the Ohio boys finally returned to their camp at Fayetteville shortly after dark. James Ireland, 12th Ohio, noted: "'Boys never more anxious...having but little sleep during the whole trip. Many of them have worn out their shoes & are barefooted. Co. F goes by Gauley Bridge"[66]

Word of the Droop Mountain fight had also reached the Federal camp at Petersburg in Grant County, as Pvt. Amos A. Vandervort of the 14th West Virginia wrote in his diary: "The news is that Gen. Averill [sic] has had a fight with the rebels under Jackson and completely routed them—they fled in all directions without arms"[67]

November 11

At 7:00 A.M., Wednesday, November 11, "a clear cold day," Averell's brigade resumed the march to Hightown from Capt. White's farm. At the point where the road diverged to Monterey, the Federals destroyed another Confederate winter encampment and the 14th Pennsylvania Cavalry was sent around "by that route to meet us [Averell's brigade] at the point where the Crab Bottom road strikes the South Branch." The remainder of Averell's brigade continued up the valley to Hightown, where the command halted at noon. At this location Correspondent "Irwin" noted: "Here is another of the splendid views to be met with in the mountains, and as each season has its own peculiar beauties and charms, yet for grandeur, the winter scenery of the mountains cannot be surpassed, when earth's huge billows are capped with snow, and a wilderness of mountains is spread out as far as the eye can reach."[68] Col. Schoonmaker's 14th Pennsylvania Cavalry passed through Monterey at about 1:00 P.M.; during this march he "rounded up enough sheep to feed the entire command for the remainder of the expedition."

At Petersburg, Col. Joseph Thoburn of the 1st West Virginia Infantry, commanding the 2nd Brigade, had led a relief column in search of Averell, as indicated by a letter written by Pvt. Aungier Dobbs, Company A, Ringgold Cavalry, which stated: "General Averell is on the move toward Staunton with a large force of Cavalry and the most of this command is out on a ten-day Scout to support Averell."[69]

While at Hightown Averell received word of a Confederate force purportedly camped at Crab Bottom, about nine miles beyond Monterey. The Federals moved out in the hope of surprising the enemy, but upon arrival at Crab Bottom found the Ringgold Cavalry and Federal infantry under the command of Col. Joseph Thoburn of the 1st West Virginia Infantry. "Irwin" noted that Thoburn had also heard reports of the presence of rebels in "the Gap" [Buffalo Gap] and when approached at Crab Bottom by Averell's men Thoburn's troops "were wide awake and drawn up in line ready to receive us."[70] Averell reported Thoburn, with an infantry brigade and two artillery pieces, then joined him at Hightown. Averell did not feel he needed assistance and ordered Thoburn to return to Petersburg. Thoburn's caution and movements had not been without merit, as Imboden had arrived at Buffalo Gap and ascertained that Col. James A. Mulligan [actually Thoburn] had brought a fresh force of 800 men from Petersburg and definitely planned a juncture with Averell at Monterey that evening. Imboden suspected the purpose of such a combined force was to launch a raid on Staunton.

Back at Dublin, Gen. Jones wrote Adjutant and Inspector Gen. Cooper that the Federals had left Lewisburg on November 10 [Averell actually departed November 8] and the mounted portion had left with Averell toward Warm Springs, while his infantry returned through Pocahontas County and Gen. Duffié moved back toward the Kanawha Valley—all rather belated information. Rebel artillerist Micajah Woods, in camp near Newport in Giles County with Gen. Echols, wrote: "No idea of the captured as hundreds who are missing will come up & are arriving hourly. The 22nd infantry which at Frankford couldn't collect 40 men, now numbers about 400..."[71] The 14th and 16th Virginia Cavalry regiments remained in the vicinity of Union.

During the day Major Gibson's battalion continued to move east via Dunmore toward Hightown in order to rejoin Averell; while near sunset Col. Schoonmaker's 14th Pennsylvania Cavalry linked up with Averell and went into camp. "Irwin" said the Federal command camped on the south side of the Franklin road, and records indicate some sort of action took place nearby as Pvt. Benjamin Starr, Company K, 3rd West Virginia Mounted Infantry, was wounded in action "near Franklin." Corp. George W. Ordner of the 2nd West Virginia said his regiment marched through Hightown and New Hampshire [New Hampton] and camped at Forks Church. Col. Moor's command, according to Sgt. Thomas R. Barnes of the 10th West Virginia, camped at Conrad's; while sometime during the evening Gen. Duffié's men arrived safely within the Federal lines at Gauley Bridge.[72]

That night, although his men were thoroughly exhausted, Imboden took precautions against an anticipated attack on Staunton by Averell and Thoburn by sending 150 men 16 miles away to the top of Shenandoah Mountain, apparently to watch for any Federal approach. Averell, of course, was not the least bit interested in striking Staunton.

November 12

During the early morning hours of Thursday, November 12, a "fine warm day," Imboden followed his advance of 150 men to Shenandoah Mountain with an additional 750 men and four artillery pieces from McClanahan's Battery. To supplement this force he also called out the Augusta Home Guard. Such defensive preparations soon proved unnecessary as Imboden had gained information that Averell and Mulligan [Thoburn] had joined forces, bringing the combined command to 4,000, and was moving toward Hardy County rather than Staunton.

Indeed, Averell was on the move; his advance broke up a group of guerrillas who were preparing to ambush Thoburn at Crab Bottom. The Federals also destroyed 400 gallons of apple brandy at one distillery and one barrel at another. An advance guard from the 8th West Virginia Mounted Infantry destroyed the same salt works at Franklin in Pendleton County that Averell had eliminated in August, as the rebels had already begun to repair them. A contraband also told the advance guard of another salt-petre work situated in a nearby ravine and the Federals destroyed that, too. In addition, they drove off a party of guerrillas.

Major Thomas Gibson's Independent Cavalry Battalion rejoined the brigade at Franklin, also; and Col. Schoonmaker said his 14th Pennsylvania troopers passed through Franklin in a "brisk march." The 2nd West Virginia Mounted Infantry, apparently Averell's advance, reached Petersburg during the morning.

General Imboden soon realized Staunton was in no danger from Averell and sent three couriers to Capt. John Hanson McNeill, of McNeill's Rangers, in Hardy County. Imboden had sent McNeill into Hardy County two weeks previously with 200 men. Imboden's couriers were to inform McNeill "to obstruct roads about the north of Franklin, and take position in the cliffs and bushwhack the enemy as he passed." To add to this harassment of the Federals, Imboden also sent out 40 fresh troops to various points in the mountains. Imboden felt he could do no more, as his men were "much jaded, having marched nearly 200 miles in six days through the mountains and over bad roads." He did feel he had already inflicted respectable damage on Averell and was convinced he had saved all the furnaces, some six or eight in full blast, on the western side of Rockbridge and Botetourt counties. He also claimed he had reliable information that the Federals he had repulsed at Covington "were furnished with several day's rations" and had been ordered to move down Jackson's River, "burn the depots and bridges, destroy the furnaces, and retreat by Millborough and Warm Springs" to Averell's main body. Despite this claim, there is no indication of such a plan in Averell's various reports of the campaign.

As Imboden called off his pursuit of Averell, Col. Moor, having marched 25 miles from Conrad's, arrived safely at Beverly at 4:00 P.M. with, "colors flying and drums beating in the most perfect order, having marched 222 miles in a little over eleven days," besides the fight at Droop Mountain, (which had cost them nine hours marching time, not to mention the hour delay with McCoy's bushwhackers). Lt. Col. John J. Polsley, 8th West Virginia Mounted Infantry, commanding the post at Beverly, wrote: "Colonel Moor arrived here...with the wounded and prisoners. Our loss in the battle at Droop Mountain...was twenty-two killed and about fifty wounded. Lieutenant Hager was the only officer of the Regiment wounded. The enemy lost about sixty-killed, three hundred wounded, one hundred prisoners and a great many missing. We captured three pieces of artillery [actually only one], one of which, a brass Howitzer, Colonel Moor brought with him. The enemy were completely defeated, routed, scattered, and demoralized."[73] Sgt. Thomas Rufus Barnes, 10th West Virginia Infantry, wrote a letter stating: "We subsisted on the country for meat, and foraged for the horses. We brought most of our wounded in. Dr. Blair was left with those that could not come. There was no less than 400 cattle...besides sheep, chickens and turkeys, and 200 beef brought in with us...we had three days snow while on the march."[74]

Averell's Brigade, according to "Irwin," had passed through Franklin and camped on a large bottom on the river where an abundance of corn and hay for the horses was located. While George Ordner of the 2nd West Virginia said his comrades "camped two miles below" Forks Church. Frank Imboden of the rebel 18th Virginia Cavalry wrote in his diary: "Reached Buffalo Gap at dark to find command had gone to meet Averell on his return,"[75] although Imboden had already given up the chase.

At Camp Narrows on the New River Pvt. George T. Shaver, Company F, 36th

Virginia Infantry wrote: "There has been a fight near Lewisburg a few days ago [Droop Mountain]. It is reported our men was much cut to pieces but I don't know how it is. There is but little said about the fight. It is reported the Yankees had left Greenbrier..."[76]

At the camp of Gen. Echols along Stoney Creek in Giles County, the soldiers were forced to watch the execution of deserter Pvt. Josiah M. Boling of the 23rd Battalion Virginia Infantry who was captured in the attack on Federal pickets on Sewell Mountain on November 10. Samuel D. Edmonds, Company B, 22nd Virginia Infantry wrote, "they had him corte marsheld [sic] and he was to be shot that evening. So they marched us in a level place and formed us in a hallow square. Then the garde [sic] set in at ende [sic] and stopped him in middle of square to a stal [sic]. The chaplin [sic] and him nealed [sic] down and prayed–The officer of the garde [sic] went to Boler and tied a white hankerchief [sic] rounde [sic] his head–Then stepped back and ordered the guard to make ready [,] tuk [sic] aim and fier [sic]. Down went Boler [Boling]. Then they marched the Brigade write [sic] past Boler [Boling]. It was a sight to see what a quantity of blood run from him. Twelve loaded ½ blank. There was a love letter founde [sic] in his pocket stating that him and a young lady down on the Kanawha River was to get married in two weeks from the date he was shot."[77]

Confederate artillerist Micajah Woods, Jackson's Horse Artillery, Jenkins' Cavalry Brigade, wrote from Dublin Depot, Virginia: "The temperature has been bitter cold for several days. Some snow has fallen. The troops have suffered severely."[78]

November 13

General Alfred N. Duffié's column reached camp at Charleston at 10:00 A.M., "lucky" Friday, November 13. Duffié said he was "gone 11 days...marched an aggregate distance of 250 miles...captured 34 prisoners and about 50 horses; one wagon; 140 cattle; 102 stand small arms; a large quantity of ammunition; tents for a regiment; knapsacks for a regiment (with clothing, etc. in them); two artillery caissons; 10 or 12 wagons and some quartermasters stores." Duffié's bodyguard captured one enemy battleflag during the retreat. He also praised Col. Carr B. White and Capt. A. H. Ricter, AIG on his staff, for their services. Duffié made mention that many of the rebels were now scattered throughout the woods, or at their homes, and predicted many would soon desert to the Federal lines. He gave his losses in the campaign as two men of the 2nd West Virginia Cavalry, that were captured by the enemy, who had shot their horses out from under them during the Confederate retreat. Also, one man of the 34th Ohio Mounted Infantry was wounded in the expedition. Duffié also claimed such good care was taken of the horses that not more than a half dozen of them were left behind as broken down. Sgt. William Ludwig of Company G of the 34th Ohio claimed the Federals captured two enemy artillery pieces at Droop [actually only one] and "fifteen cases of shells."[79] Chaplain Anthony H. Windsor of the 91st Ohio would claim the expedition was "chiefly to be remembered on account of the hard marching undergone, the inclemency of the weather, and the rapid flight of the rebels at our approach."[80]

Averell's column apparently arrived at Petersburg sometime during the day, as Col. Schoonmaker noted his troopers entered town at 5:00 P.M. George W. Ordner, 2nd West Virginia Mounted Infantry, said he arrived at Petersburg in the evening, having marched 28 to 30 miles during the day. The 8th West Virginia Mounted Infantry, or at least a detachment, possibly serving as rear guard, apparently did not arrive until the following day. Indeed, Col. Oley noted that on November 13 he sent two squadrons, under Capt. Jacob M. Rife, Company F or M, by way of the Seneca Trail, via Circleville, from Franklin; in search of the enemy. Finding no opposition, they would reach Petersburg on November 14. Correspondent "Irwin" made mention of this in relating that a detachment of the 8th West Virginia was sent through the North Fork, while the rest of the brigade started for Petersburg. On the march to Petersburg the brigade passed through Mill Creek Valley, a good, loyal neighborhood containing the homes of Capt. Henry A. Ault's "Swamp Rangers." Averell's Brigade arrived at Petersburg and rested for two days. Pvt. Amos A. Vandervort, Company B, 14th West Virginia Infantry, noted in his diary: "A considerable force of Averill's [sic] command came in here today. Among them is the 3rd W. Va. reg. They told the mournful news that Capt. Cobin [Capt Jacob G. Coburn, Company C, 3rd West Virginia Mounted Infantry] was mortally wounded in the late fight at Droop Mountain. How sad the thought, yet sadly true that thousands have fallen in defence of their country, 'every passing breeze' brings the mournful news of some fallen friend."[81]

General Imboden was at Staunton, where he wrote a detailed report of his activities in the campaign, although Frank Imboden of the 18th Virginia Cavalry wrote in his diary, "General returned from mountains and camped near Augusta Springs."[82]

November 14 and 15

Colonel John H. Oley's 8th West Virginia Mounted Infantry detachment arrived at Petersburg on Saturday, November 14, where along with the remainder of Averell's brigade, he reported they would rest for two days. One exception was the 2nd West Virginia Mounted Infantry, which received orders to march at 7:00 A.M. the following day. At Petersburg Col. Schoonmaker inspected the horses of the 14th Pennsylvania Cavalry in the morning and found that "with the exception of the greased heel, which had broken out in a number of cases," the mounts were in almost as good shape as when the expedition began.

Unfortunately, news of the Federal losses was slow to reach their loved ones. Sirene Bunton, of Upshur County, whose two brothers, Burnham (Birney), who died in 1862, and Walter, served in Company E of the 3rd West Virginia Mounted Infantry, wrote in her diary on November 14: "There has been two or three battles lately and I suppose the Third [3rd West Virginia Mounted Infantry] and Tenth [10th West Virginia Infantry] were in them. We cannot hear who was wounded. Our troops were victorious."[83]

But with Duffié having returned from the expedition, news circulated quickly throughout the Kanawha Valley. The *Gallipolis Journal* of November 19 mentioned, that on "Saturday last" about 100 rebels "...exhibited their ragged carcasses on the river bank at or near Scary [Creek], fifty of whom were ferried over the river. They robbed a small store at Scary and seemed to be only intent on stealing and plundering the people in the valley. Their repulse at Lewisburg [Droop Mountain] has probably saved the valley from a raid this season, which is now becoming too far advanced to admit of any very extensive operations."

While Averell remained in the safe confines of Petersburg the Confederate command of Imboden rested for two days at Staunton. Gen. Sam Jones, writing from Dublin, developed an optimistic attitude based on new information, writing that the Droop Mountain battle was not nearly as disastrous as first thought, and felt the Federals had suffered more in killed and wounded. He reported Averell's troops had returned to Beverly and Monterey, and had left their wounded at Hillsboro. He said the Kanawha force [Duffié], "returned to their former stations in haste" and claimed all the Federal troops fell back for lack of subsistence and information of large Confederate reinforcements. Jones stressed the Yankees "complained of their losses and fruitlessness of [the] expedition." While stating he felt the total strength of the Federals in Greenbrier County had been about 8,000, he noted his scouts had since been to Huntersville and Elk Mountain and the entire Confederate army in southern West Virginia had moved back and reoccupied all points they had held prior to the Droop Mountain battle. Indeed, this point was confirmed by Confederate artillerist Micajah Woods, who wrote his battery had returned to Lewisburg in the evening "after an arduous and long march from Giles."[84]

Micajah Woods also described the impact on the citizenry left by the Federal occupation, writing, "It grieves me to tell you of the misfortunes of the people of this region and especially of the damage suffered by our friends and relatives. The stay of the enemy, not long but every minute a day in horror, will long be remembered by many families once comfortable now destitute and helpless in all this country. Fences are burnt all along their lines of march, thousands of bushels of corn used and destroyed...drove off their horses, cattle and sheep...take all, even milk cows...committed many depredations...Uncle David lost three horses only leaving one. Cousin Jane Price lost one Negro boy and a horse. Indeed every citizen has lost something valuable."[85] This situation was confirmed by some Federal combatants as well, exemplified by Pvt. Amos A. Vandervort of the 14th West Virginia Infantry at Petersburg, who wrote, "Averill's [sic] force has been buying lots of goods, and taking corn from the citizens. I suppose the citizens wish they had never espoused the rebel cause, they are now paying for their folly"[86]

Although Imboden had ended his pursuit of Averell, the men he had sent to Hardy County earlier would soon strike an indirect blow against Averell, who had become over confident within the safety of Petersburg. The Federal commander at New Creek (Keyser), Col. James A. Mulligan of the 23rd Illinois Infantry, was as equally lax in his vigilance as was Averell, and late on the morning of Sunday, November 15, "a cold dreary" day, he sent to Averell's relief a supply train. Comprised of some 80 to 90 wagons, 200 horses, and four sutler wagons, the train was loaded with commissary and quartermaster stores, and was protected by some 75 to 100 foot soldiers and cavalrymen. The train itself was approximately one to two miles in length, and camped that evening near Burlington. The following day this train would be attacked by the force Imboden had earlier sent to Hardy County. Pvt. Charles F. Miller of McNeill's Rangers recalled his fellow rebel raiders made an all night march and halted about one mile from Burlington, on the New Creek and Petersburg Pike. Imboden had also sent off Capt. Hill to Barbour County in an attempt to capture a supply train from Grafton to Beverly, as he had received a report that only about 500 Federals were then at Beverly. Obviously, he must not have been aware of the return of Col. Moor's column to Beverly.

The 14th Pennsylvania Cavalry and the 2nd West Virginia Mounted Infantry were supposed to depart Petersburg for New Creek on November 15 but a severe rain and snow storm struck in the morning creating a delay of one day. Indeed, at 10:00 A.M. marching orders for the two regiments were countermanded. Pvt. Amos A. Vandervort, 14th West Virginia Infantry, noted in his diary, "Averill's [sic] force is still here. They are using up the corn belonging to the citizens, I don't pity the owners of corn for they are all rebels. Averill's [sic] force have orders to leave in the morning."[87]

While the Federals continued to rest at Petersburg, the Confederate army gradually regrouped, as evidenced by a letter which stated: "Echols...lost most all [at Droop] but they were scattered and have come in and is still coming. They have neither blankets nor clothes–lost most of their guns. Tell the girls to beg for them–the girls is begging all over the county with great success."[88]

November 16 - Attack on the Federal Supply Train

On Monday morning, November 16, the "weather was very cold and at daylight snow was falling." Pvt. Amos Vandervort, 14th West Virginia Infantry, stationed at Petersburg, wrote in his diary the day was "...cloudy and cold. Snow is visible on the mountains." At 7:00 A.M. Col. James M. Schoonmaker moved his 14th Pennsylvania Cavalry, as well as the 2nd West Virginia Mounted Infantry, out of Petersburg en route to New Creek, and would cover some 20 miles [during the day] on the Burlington road, to a point 11 miles west of Romney on the Northwestern Turnpike.

Events of a more serious nature began to transpire at about daybreak, or 7:00 A.M. according to the officer in charge, as the relief supply train sent out from New Creek for Averell resumed the southern trek along the crooked, narrow mountain road to Petersburg. In command of the supply train was Capt. Clinton Jeffers, Company B, 14th West Virginia Infantry, who claimed he had a force of 90 soldiers under his charge. For precautionary purposes he placed 40 men in the advance, under 1st Lt. George H. Hardman, Company C, 14th West Virginia Infantry; Jeffers remained at the center of the train with 10 men, four of whom he threw forward between himself and the advance guard as a signal party; and at the rear he assigned 40 men under 2nd Lt. David C Edwards, Company H, 2nd Maryland Potomac Home Guard.

Despite recent attacks along this route the Federal infantry, cavalry, and teamsters accompanying the train were completely unaware that an approximate 300 man force of McNeill's Rangers, as well as a portion of the 62nd Virginia Mounted Infantry, led by Capt. John Bean Mooman [White's 41st Battalion Virginia Cavalry], had been observing the train from a safe distance nearby. Indeed, Pvt. Charles F. Miller of McNeill's Rangers said his force numbered 105 men in this episode, including 45 dismounted (which included Miller and several from other commands), under the leadership of a Lt. Cunningham of the 7th Virginia Cavalry, "who had lost an arm" in service. The Rangers also had 60 mounted men under 1st Lt. Jesse Cunningham McNeill, son of Capt. John Hanson McNeill.

Private Miller remembered that due to the inclement weather it was decided to delay their attack on "Mulligan's Irishmen" until the train was well under way. The rebel force, sent to Hardy County days earlier by Imboden, had been waiting on just such as opportunity to catch the Federals unprepared and arranged an appropriate ambush. About three and a half miles south of Burlington, the 45 dismounted rebels took position about 20 feet from the pike and Lt. Cunningham took command of the right wing of this force. Pvt. Miller said, "...we were all standing in full view on open ground; but being on the side of the hill above the road," McNeill's men would not be spotted by the Yankee's until Lt. Cunningham ordered them to halt. Cunningham told his men not to fire unless ordered, as he was convinced the Federals would view his line and surrender.

Sometime between 8:00 and 9:00 A.M. [Capt. Jeffers gives the latter time] the center of the supply train passed by an abandoned house situated about midway between Burlington and Williamsport, at a location alternately known as Pierce's Gap, Pierce's farm, and Pierce's Lane, and so named for a Mr. Pierce, a "notorious rebel" and former occupant of the abandoned house, who served as a guide in this expedition with McNeill [although Pvt. Miller refers to him as Price].

With the Federals marching in close order, the head of the column "made a sharp turn in the woods beyond" and was approached by Lt. Cunningham of the rebel force who yelled, "Halt there boys! Where are you going?" Rather than surrender the Federals passed to the front and opened fire, while the Rangers "fired point-blank and charged.." In the first volley Lt. Hardman and one enlisted man were killed instantly, and some other Federals were wounded. This situation caused the guard to scatter in confusion, and several bluecoats were taken prisoner. Fortunately, 2nd Sgt. Silas W. Hare, Company I, 14th West Virginia Infantry, rallied the men and had them fall back from the road into the woods, skirmishing as they went, although Sgt. Hare was wounded in the process. Pvt. Charles Miller of McNeill's Rangers recalled, "on the west side [of the pike] there was a heavy timber. The enemy took shelter behind large trees, and for a moment it looked as if we had more than we had bargained for. However, they soon broke and ran for the top of the mountain and gave us trouble in unhitching the teams."

The opening shots had been the pre-arranged signal for the remainder of the 62nd Virginia detachment, hidden inside the abandoned house near the middle of the train, to burst out and open a severe fire upon Capt. Jeffers and the defenders of the center of the train. Capt. Jeffers said that when the advance of the train was fired upon he immediately sent forward his remaining six men at the center, and started himself to the rear to bring up the 40 men under Lt. Edwards. He said he went only a few steps when the rebels poured out of the old house "and immediately commenced a fierce attack upon the center of the train, while at the same time a body of rebel cavalry made an attack upon the rear guard."

The Federal rear guard, warned by the sound of gunfire, managed to throw up a defensive line just in time to stop the assault on the rear by McNeill's Rangers. Capt. John H. McNeill had hoped to get his men in open hand-to-hand combat with the enemy. As he advanced to size up the situation he waved his hand and yelled, "Come on, boys!" Alongside Capt. McNeill was Pvt. David M. Parsons and 2nd Sgt. Joseph L. Vandiver, who encountered a wooden fence situated between them and the Burlington-Williamsport road. This situation presented an immediate obstacle to Capt. McNeill's objective, having lost the element of surprise, and the Federal rear guard under Lt. Edwards fell back and took position in the edge of the woods where they kept up a rapid fire on the rebels.

According to Pvt. Miller of McNeill's Rangers, as the rebels reached a gate to the fence on the side of the pike which opened inside a field, Sgt. Vandiver was shot through the thigh and his horse killed. Reportedly, as his horse fell it propped open the gate. Capt. McNeill rode up to the fence near the gate and fired both barrels of his shotgun, a weapon he used only at close range. Although caught in a severe fire, "Neither he nor his horse was hit. Captain McNeill enjoyed a scrap like this as much as he did a fox chase. I do not think that he thought for a moment of being killed." Capt. McNeill reportedly dismounted and personally opened the gate, his men immediately rushing through the opening and dispersing the Federals, whose rifled muskets could not compete against McNeill's short-range cavalry weapons. Many of the Federals fled to the nearby woods, where their rifles proved more effective in sniping at the rebels.

Mr. Pierce [Price], although not a Ranger, participated in the attack on the rear and "advanced to charge at the gate, not knowing at the time that the fence was thrown down above that point. He carried a shotgun and engaged in the fight." Pvt. Charles F. Miller, McNeill's Rangers, said the rebel infantry "was on the pike and behind the fence over which they charged into the ranks of the enemy, who stood their ground well for a time, then ran. They were soon made prisoners, mounted on horses, and brought out."

Captain Jeffers said that as firing became general along the entire line he found himself cut off from his own men and was pursued by a party of rebels who shot his horse out from under him as he made his escape. Jeffers then started for the advance and found they had fallen back and taken position behind a fence, "from which they kept up a spirited and telling fire upon the rebels, who were now busily engaged running horses off from the train and attempting to fire the wagons." This was confirmed by Pvt. Miller of McNeill's Rangers, who said that when the Yankee advance had been driven away and the firing had begun in the rear, Lt. Jesse McNeill had ordered all men to the rear, excluding three (of whom Miller was one) to unhitch the teams. Miller said they unhitched the teams "as rapidly as possible under the fire of the men on the hilltop, who were using long-range rifles." Miller mounted a saddle horse and rode to Lt. McNeill, who had come to a halt where he could view Capt. McNeill's assault on the rear. Lt. McNeill ordered Pvt. Miller to cross an open piece of ground and ask Capt. McNeill for orders, to which Capt. McNeill replied, "Get out of here at once," which the Rangers proceeded to do.

Responding to the sound of the ambush when it began, a Federal reinforcement stationed at Burlington, led by Capt. Alfred Lindsey Hoult, Company K,

14th West Virginia Infantry, arrived and drove McNeill back. Reportedly, it was during this exchange that 2nd Lt. David C. Edwards, Company H, 2nd Maryland Potomac Home Guard was wounded. But Capt. Hoult's men arrived too late to actually participate in the fighting although they did help to extinguish the wagon fires. Capt. Jeffers added that the rebels only managed to burn five wagons and injure two due to the severe rifle fire of the Yankee force on the hilltop. He said the rebels began to leave with their plunder, "which consisted only of horses."

Officially, in the attack on the Federal supply train the rebels reportedly burned five wagons, captured 245 horses and 43 teamsters. Pvt. Miller of McNeill's Rangers said all wagons "[were] set on fire" and they captured 245 horses, "all with new harness." But they failed in their attempt to destroy the wagons and goods, "owing to their haste and being constantly fired upon by infantrymen sheltered behind trees." The *Staunton Vindicator* claimed the raiders captured 80 wagons and brought out 245 good wagon horses. Two days later Frank Imboden of the 18th Virginia Cavalry would write in his diary that he had received information the raiders took 80 wagons, 245 horses, and 30 prisoners. On the other hand, among Federals involved, Capt. Jeffers said he lost "150 horses" to the raiders; Col. Joseph Thoburn at Petersburg reported McNeill took 200 horses, four prisoners, and 20 wagoners and negroes; while Pvt. William Davis Slease, Company C, 14th Pennsylvania Cavalry, said when his regiment reached the scene they found the train guards had been killed and Col. Schoonmaker's troopers recaptured the wagons and escorted them to safety.

Casualties in the encounter are sketchy as well, although Capt. Jeffers reported the total Federal loss at 19, which included two killed, six wounded, and one missing from the 14th West Virginia Infantry, as well as six wounded and four missing in the 2nd Maryland Potomac Home Guard. Among the Confederates the 62nd Virginia Mounted Infantry had five men wounded, four of them only slight. The fifth, Pvt. James F. Keister, Company I (2nd), died from his wounds the following January. This may be the man Capt. Jeffers made reference to when he said one rebel, "supposedly mortally wounded," fell into Federal hands. McNeill's Rangers had one casualty, 2nd Sgt. Joseph L. Vandiver, who took a bullet in the thigh during the attack on the rear guard.

Despite whatever differences in loss of property and life, at about 11:30 A.M. Gen. Averell, en route to New Creek, received word of the attack via two soldiers, one a member of the Ringgold Cavalry, who had been sent by "a lieutenant, who heard that the train was captured and driven off" by 1,500 rebels, although upon careful reflection, the messenger changed the figure to 500 Confederates. Since Lt. Col. Frank Thompson's 3rd West Virginia Mounted Infantry was at the head of the column, Averell ordered Thompson and Major Thomas Gibson's Independent Cavalry Battalion to vigorously pursue the raiders in the direction of Moorefield. As the two commands had only enough rations to get to New Creek they were to subsist on the country and the Ringgold trooper, who claimed knowledge of the country, was to serve as a guide.

Lieutenant Colonel Thompson's mounted command kept "at a brisk walk, and a trot, when possible." Shortly after leaving the New Creek pike they encountered and removed a timber blockade across the road.

This movement by Averell's detachment of some 1,000 to 1,200 mounted troops under Thompson and Gibson was recalled by Josiah Davis of the 3rd West Virginia, who wrote, "about 12 miles from Petersburg, heard of wagon train capture. The 3rd West Virginia turned toward Moorefield hoping to head off and recover train—went to junction of Strasburg and Winchester roads and camped, sending a detachment to the junction of Romney road with Winchester road, but were too late, the enemy having passed."

By 1:30 P.M. Col. James A. Mulligan, 23rd Illinois Infantry, commanding the 2nd Division at New Creek, had learned of the attack on the supply wagon train and took immediate action by having Col. J. N. Campbell, 54th Pennsylvania Infantry, commanding the 1st Brigade, send out the 54th Pennsylvania, under Lt. Col. J. P. Linton; two companies of the 15th West Virginia Infantry, led by Major Wells, and one section of Capt. Moore's Battery, to pursue the raiders by getting behind them. Mulligan also sent mounted messengers to Averell and Col. Thoburn at Petersburg to intercept McNeill in the South Branch Valley, but received word ten minutes after writing the dispatch that Averell already knew of the attack and had sent mounted troops in pursuit.

Therefore, Col. Mulligan advised Averell that if he was near Williamsport, "the enemy might be intercepted by you were you to move east across Patterson's Creek Mountain, entering the Moorefield Valley by the Williamsport and Moorefield road."

At 2:00 P.M., only ten minutes before Col. Campbell received his orders, a messenger from Gen. Averell arrived at Petersburg and informed Col. Joseph Thoburn, 1st West Virginia Infantry, commanding the 2nd Brigade of the attack and Averell's detachment which was sent in pursuit. Thoburn responded by ordering out all available cavalry (about 100 men) to start down the South Branch to co-operate with Averell. This was remembered by Samuel Clarke Farrar of the Ringgold Cavalry, who wrote: "A messenger from Gen. Averell informed Col. Thoburn about 2:00 P.M. of the capture, and that he was sending one of his regiments [3rd West Virginia Mounted Infantry] and a battalion of cavalry [Gibson's] towards Moorefield to intercept and cut off the enemy's retreat. [In conjunction] Col. Thoburn ordered Capt. [Andrew J.] Greenfield [Company B, Ringgold Cavalry] to take all the available cavalry of our battalion (about 100 men) and proceed down the South Branch and co-operate with the force sent by Averell." Also recalling this scene was Pvt. Amos Vandervort, 14th West Virginia Infantry, who wrote in his diary, "Our train coming from New Creek was attacked by about three hundred rebels. Lieut. Hardman of Co. C and ten others were said to be killed by the first volley. It is sad to record such occurrences. It is well said that in these perilous times, 'the sighs of distress are borne upon every breeze,' would that peace did reign."

Lieutenant Colonel Thompson's 3rd West Virginia arrived at Moorefield between 3:00 and 3:15 P.M., left Major Gibson's Battalion in town to guard the roads in the direction of New Creek, and proceeded towards Wardensville, as Thompson felt the rebels would employ that route for their escape.

Captain Greenfield and his cavalry battalion arrived at Moorefield between 3:35 and 4:00 P.M., where he found Major Gibson, who told him that Lt. Col. Frank Thompson's 3rd West Virginia Mounted Infantry had advanced in the direction of Wardensville, and that Thompson had instructed him to remain at

Moorefield and guard the roads in the direction of New Creek.

Night overtook Thompson's men near the intersection of the North River pike and he concluded to go into camp, claiming his horses had traveled over 33 miles with no feed since morning; the horses had also just came off a 15 day's raid and many could not keep up with the command; and as Confederates might be in the vicinity it would be impractical to leave the horses behind. Therefore, he halted the command to feed the horses one mile down the North River pike; left one squadron to guard the Wardensville pike; and sent one squadron one mile down the pike below camp.

Captain Greenfield of the Ringgold Cavalry had been told by Col. Thoburn to catch up with Lt. Col. Thompson and instruct him "not to rest until they heard from McNeill, and cut off his retreat, or until all the intersecting roads from the direction of the South Branch were occupied." This would have cut off McNeill's retreat. Capt. Greenfield recalled, "I proceeded on and overtook [Lt.] Col. Thompson [about one half-hour after Thompson went into camp] at the intersection of the North River Pike, he having gone into camp for the night. I informed him that my instructions were to co-operate with him. He said that his command was worn out as they had just came off a 15 day's raid and could go no further. I then proceeded on with my command to the Grassy Lick Road, ten miles below the Wardensville Pike and about 18 miles from Moorefield, where I thought the rebels might cross and I might intercept them. I here received what I considered reliable information that they had crossed the road at 3:00 P.M. with horses only. I reported my information at once by messenger to [Lt.] Col. Thompson, and suggested that we proceed and try to overtake them by daylight next morning. He replied that as they had six hours' start and his horses were worn out, it would be useless to try to overtake them."

Captain Greenfield concluded to end the pursuit, writing: "As Averell had reported to Thoburn, their force to be 400 to 500 strong I did not feel safe in following and attacking them with my small command of 100 men. I have since learned their force did not exceed 200 men..."

The reports of Lt. Col. Thompson and Col. Thoburn confirmed most of Greenfield's account, verifying that Greenfield ended the pursuit at about 9:00 P.M., went into camp, and returned to Petersburg via the Moorefield and New Creek road the following day. Thoburn did mention that a wagoner who escaped from McNeill at Brock's Gap told him Capt. McNeill did not cross the pike until after midnight and passed within 600 yards of Lt. Col. Thompson's camp, and McNeill's force did not exceed 100 men. However, Thompson felt the rebels had crossed the pike at 3:00 P.M., as earlier noted, "they having crossed at a place where there were no roads."

The entire Federal pursuit was called off, the rebel raiders having struck the supply train, vanished into the mountain passes, returned to the Shenandoah Valley, and dispersed of their haul. Afterwards, the 62nd Virginia joined with Imboden and McNeill's Rangers returned to their camp in Rockingham County. Pvt. Charles F. Miller of McNeill's Rangers remembered that following the attack the rebels "rode all day and into the night, and camped in the hills."

The repercussions of the attack on the supply train were severe. Capt. Greenfield of the Ringgold Cavalry correctly stated, "the guard was entirely insufficient in numbers for such a train." Col. Mulligan angrily charged the attack was due to "...a want of precaution, a want of skill, and a want of fighting, [and] I have ordered charges to be preferred and forwarded against the commanding officer of the escort." Gen. Benjamin F. Kelley was infuriated and convened a formal inquiry into the affair, and the service record of Capt. Clinton Jeffers, who was in charge of the train guard, shows he was court-martialed and dismissed March 1, 1864. Most importantly, Gen. Averell and Colonels Thoburn and Mulligan had gained a valuable lesson in never underestimating the enemy.[89]

Imboden was convinced Averell would winter his cavalry at Moorefield, where forage and supplies were abundant. He also suspected Averell had 3,000 men and Mulligan [Thoburn] at least 1,000, a combined force much too large for his small command to cope with. However, Imboden did vow to "harass" Averell throughout the winter, then began a march to the lower end of Rockingham or Shenandoah to investigate what further actions he could take prior to the arrival of winter. His efforts against Averell were all but ended.

Sometime around 3:00 P.M., at about the same time Lt. Col. Thompson had arrived at Moorefield, the 14th Pennsylvania Cavalry encamped 20 miles short of New Creek, while Averell's Brigade, having marched about 20 miles distance, camped in the evening on the farm of a Mrs. Williams, who had a son in the Confederate service with McNeill's Rangers [possibly Pvt. Vincent Osceola Williams] and two bitterly secessionist daughters. While encamped on the Williams farm plenty of corn and hay for the horses was procured.[90]

Final Safety

Averell's brigade started the march from the Williams farm to New Creek at 7:00 A.M. on the "clear and cool" morning of November 17, and arrived at their destination about 4:00 P.M. in the afternoon, with the exception of the 3rd West Virginia Mounted Infantry and Gibson's Battalion, which didn't arrive until the following day. Correspondent "Irwin" reported "...our ears were gladdened by the music of the steam-whistles on the locomotives of the B & O railroad."[91]

Averell had in tow about 150 captured horses, 27 prisoners [exclusive of those returned with Moor and Duffié], and several hundred cattle that had been captured during the march. Col. Schoonmaker said his 14th Pennsylvania Cavalry had begun the march during the morning and arrived at New Creek at about 4:00 P.M., where they "selected appropriate campgrounds." He remarked that his men were "in excellent spirits" and the "horses will be better than when we started after a few days rest and care."

On the march from Petersburg to New Creek Major Gibson had been ordered to report with his battalion to Lt. Col. Frank Thompson, 3rd West Virginia Mounted Infantry, who was on his way to Moorefield in pursuit of McNeill's Rangers. Gibson had complied with this order and arrived at New Creek on November 18. He reported his Independent Cavalry Battalion "from October 28 to November 18 [had] marched 330 miles, took 15 prisoners, 20 horses, 50 head of beef cattle, and 125 sheep." The 3rd West Virginia Mounted Infantry of Lt. Col. Frank Thompson traveled via the Grassy Creek road and Romney, and arrived at

New Creek "on the evening of the 18th," while the remnants of the relief supply train which had been attacked en route to Averell on the 16th arrived at Petersburg.

Averell's other regimental officers reported their accomplishments in the campaign as well. Col. Oley, 8th West Virginia Mounted Infantry, said his men arrived at New Creek on November 17 "in better condition than when we left Beverly, seventeen days before." Frank Reader, 2nd West Virginia Mounted Infantry, said "since leaving Beverly, 17 days, we marched 296 miles, a part of the time suffering intensely from the cold, constantly subjected to the sudden attacks of bushwhackers, and having fought one of the most gallant and triumphant little battles of the war."[92]

Correspondent "Irwin," believed to have been with the mounted West Virginia regiments, said the command was "in the saddle 17 days, covered 300 miles, although there was not a single complaint or case of sickness." He also said the horses, on the average, were in better condition than when the expedition began. He boasted they captured 200 horses, 300 cattle, 500 sheep, brought out a number of contraband, including some waitors from the White Sulphur Springs hotel, and created a dread of "Averell and his Yankees."[93]

Despite the success of Averell's command, the families of the soldiers still had no knowledge of the accomplishments as late as November 22, when Sirene Bunten wrote from Upshur County: "The Third [3rd West Virginia Mounted Infantry] has been in another battle. We have not heard whether any were killed or wounded, wish we could."[94]

On November 19 Gen. Echols wrote his report of the battle from Lewisburg, while the Federals at Petersburg remained on alert, as one soldier wrote, "It is rumored that we will be attacked here soon by Imboden [,] Echols and others."[95] Of course, Echols and Imboden had no intention of striking Petersburg, and the next day, November 20, the 23rd Battalion Virginia Infantry arrived in Lewisburg following a long march from Giles County. Just as Gen. Sam Jones had stated, the Confederates had reoccupied all positions held in West Virginia prior to the Droop battle, and if there was any real fear of Averell it certainly didn't show the following month when Averell launched his third, and most successful, raid against the Virginia and Tennessee Railroad. Echols and a host of other Confederates chased after him in an effort to give battle, but Averell managed to elude all pursuit.

Epilogue

Viewing the Droop Mountain campaign in retrospect, it must be understood that it was quite common during the Civil War for victors to exaggerate their gains and for the defeated to minimalize their losses, sometimes so well the vanquished appeared to be the conqueror. But with the passing of time and proper investigation, the basic truths eventually find their way to the surface. To be more explicit, the advantage of hindsight has enabled modern historians and journalists to better assess the accomplishments, or lack thereof, of Gen. Averell's November 1863 raid.

Consequently, there can be but little doubt that Averell won a combat victory at Droop Mountain, as he badly bloodied and disorganized Echols' army. President Abraham Lincoln made reference to the victory in an effort "to boost the sagging morale of the Federals in Tennessee, who had recently been defeated at Rogersville by General Sam Jones." This was confirmed by Gen. Ambrose Burnside, who said the Droop Mountain victory "greatly encouraged his troops." In addition, a number of modern historians have claimed Averell's victory was "...to some degree responsible for the creation of the state of West Virginia," as "...it grew out of the interrelated political and military strategies that gave West Virginia its present borders, and it broke a stalemate that had existed between the two armies since the first year of the war." Supposedly following Averell's victory, the Confederates were never again strong enough to mount an offensive in West Virginia, and were therefore, cleared of the area.

Despite all this, which is questionable in part, Averell has been criticized for not achieving his main objective—destruction of the Virginia and Tennessee rail line. Averell gave as his primary reasons for aborting the raid as the footsore condition of Duffié's infantry, a lack of food and forage in the area, the extra burden of transporting the wounded and prisoners from Droop Mountain, and notice the Confederates had received reinforcements and were ready to confront him.

While these factors held true to some degree, Averell has been justifiably attacked by historian Stephen Starr in his excellent work on the Union cavalry, writing, "...Averell was pursuing an enemy he had driven from a strong defensive position 'in total rout' less than forty-eight hours before; he had at his disposal the four regiments he had taken with him on this raid, plus a battalion of cavalry, Duffié's two cavalry regiments, and the two infantry regiments, presumably not footsore or without rations of his own command, a respectable force of upwards of four thousand men. If Averell had had the drive of a Sheridan, or a Forrest, or a Custer, he could have left his prisoners and captures at Lewisburg, to be guarded by Duffié's footsore infantry, and gone after Echols and on to the [New River] bridge with the rest of the combined command."

Making matters worse was the dissension between Averell and Duffié, each attempting to blame the other for the failure to bag Echols. Duffié even went so far as to claim, "Had General Averell, instead of attacking the enemy in force and making a general engagement [at Droop Mountain] engaged him lightly, detaining him until my command reached Lewisburg, it is my opinion that we might have captured almost the entire rebel force." Obviously, there was no loss of love between Averell and Duffié. But perhaps Stephen Starr was correct when he assumed that possibly Averell ended the raid due to "his intense desire...to return to the Army of the Potomac, from which he felt unjustly exiled. He had won a fine tactical victory at Droop Mountain and may have wished to avoid any possibility of tarnishing his achievement."

As for Gen. Echols, while Gen. Benjamin F. Kelley, "the department commander in West Virginia, informed Washington that all organized bodies of Confederates had been driven from the new state," the rebels, as earlier noted, "returned within 10 days to their former posts." Indeed, Echols had survived the Federal offensive, and in so doing, prevented Averell from his primary objective. And as also noted, the southern army had managed to elude Averell and Duffié's pincer movement fairly intact. Afterwards, he made an unsuccessful effort to corner Averell during the "Salem Raid" the following month, and was more than willing to give battle.

And as to whether or not Averell cleared southern West Virginia of any further Confederate activity, one must remember that in early 1864 the theater of war moved to the Shenandoah Valley of Virginia, where Echols and other rebel commands from West Virginia won a brilliant victory at New Market. Obviously, this was not the picture of a defeated army.

Echols was unquestionably denied at Droop Mountain, but as to whether or not he lost the battle by neglecting the obscure road on his left flank, which has never been substantiated, is irrelevant, as the fact is he did not have enough manpower to defend his entire line. Every time he moved some men from the center or right to the left, he weakened those points. Echols may have had the advantage of terrain, but Averell had superiority of numbers. Additionally, Gen. John D. Imboden, "who had hurried up to Covington to block Averell's path toward the Virginia Central," was infuriated with the performance of the "panic-stricken refugees" from Echol's command in the fight at Covington, and unsuccessfully attempted to bring charges against Col. William L. Jackson.

In summary, Averell did achieve a decisive military victory at Droop Mountain but failed in his primary objective. Whether or not his expedition eliminated any further Confederate threat in West Virginia, thereby helping to solidify the creation of the new state of West Virginia, remains open to conjecture. On the other hand, Echols, although severely beaten, held his command together and eventually was sent to another theater of war where his men performed admirably.[95]

CHAPTER 7

Faded Warriors

Most of the primary participants in the Droop Mountain battle and campaign went on to successful careers after the war, although a few would not survive the bloody civil conflict. Third time proved a charm for Gen. William Woods Averell, as he finally succeeded in striking the Virginia and Tennessee Railroad during his "Salem Raid" in December of 1863. Performed during bitter cold weather and over treacherous terrain, Averell described the successful expedition as having "marched, climbed, slid, and swum 355 miles." In the effort his command broke the line at Salem, Virginia, burned the station, destroyed five bridges and a mile of telegraph wire before being driven back to Covington by the Confederates, where Averell lost a number of prisoners and horses before escaping over the mountains. He developed a reputation as a capable division commander during the first half of 1864 in the Shenandoah Valley, but during the fall found himself the victim of the wrath of his commander, Major Gen. Philip H. Sheridan, Army of the Shenandoah. Sheridan felt Averell had been lax in his pursuit of the routed Confederates at Fisher's Hill, Virginia on September 22, 1864, and relieved him from duty. Averell "sat out the rest of the war, recuperating from severe dysentery, malaria, and a seriously damaged ego."

Averell held no further military command, although there was an unsuccessful effort to have him appointed to the Department of the Gulf. Yet by war's end he had received the brevets of major general of volunteers and brigadier general of regulars. He resigned from the military May 18, 1865, returned to Bath, and became breifly involved in the Averell Coal and Oil Company, with coal interests in the area of the mouth of Poca River near Charleston, West Virginia. From 1866-69 Averell served as U.S. consul general at Montreal to British North America (French Canada). In late 1869 he became involved in the steel industry and in 1870 turned his interests to asphalt, being elected president of the Grahamite Pavement Company. In this capacity he was noted for the pavement of a number of streets in New York City, but by 1873 became disgruntled with his business partner and formed the Grahamite Asphalt Pavement Company. The Panic of 1873 hit him hard financially but during the winter of 1873-74 Averell formed the New York and Trinidad Asphalt Company, and during the next few years was involved in pavement projects on New York's Fifth Avenue, Philadelphia, and Washington, D.C. In 1880 Averell was named president of the American Asphalt Pavement Company and in 1883 he "developed and perfected an asphalt conduit for an underground electrical system" for which he received a U.S. patent. Gen. Averell married in 1888, at the age of 53, and the same year was reinstated in the U.S. Army and placed upon the retired list. In 1891 Averell began writing his memoirs, entitled *Ten Years In The Saddle* but, unfortunately, would die before the work was completed [it would finally see print in 1978] but, as noted, he managed to invent a number of devices which made him financially independent. With his return to service in 1888 Averell served as inspector general of the Soldier's Home in Bath, New York until 1898 and died at Bath February 3, 1900.[1]

General William W. Averell, Federal commander at Droop Mountain as he looked about 1890. AUTHOR'S COLLECTION

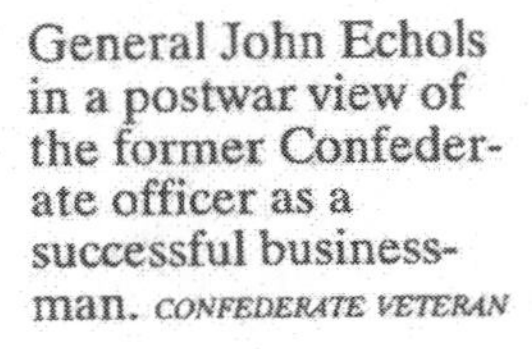

General John Echols in a postwar view of the former Confederate officer as a successful businessman. *CONFEDERATE VETERAN*

Averell's opponent at Droop Mountain, Gen. John Echols, spent late 1863 unsuccessfully chasing Averell during his "Salem Raid." In 1864, he fought at New Market in the Shenandoah Valley and at Cold Harbor near Richmond. In the fall he was assigned, for the second time, to command of the Department of Southwest Virginia and in the spring of 1865 was promoted to major general and given command of the Confederate army in the Valley and Southwest Virginia. Echols was on the march to reinforce Gen. Robert E. Lee at the Appomattox surrender, after which he moved the remnants of his command to North Carolina. He then escorted Confederate President Jefferson Davis to Augusta, Georgia, and returned to Greensboro, North Carolina, where along with his staff, he was paroled, with the army of Gen. Joseph E. Johnston.

Echols, who had served on the Board of Visitors of the Virginia Military Institute from 1858-61, was declared an honorary graduate of that institution July 2, 1870 [he had resigned from the school in 1841 for unspecified reasons]. He also served on the Board of Visitors of Washington and Lee University. At the close of the war he resided in Kentucky and Staunton, Virginia, where he developed a prominent law practice, became president of the National Valley Bank of Staunton, and president of the Chesapeake, Ohio, & Southwestern Railroad Company. His business and railroad interests kept him active in both Kentucky and Virginia, and it was said Echols "rarely made an enemy and never lost a friend." He died from Bright's disease at Staunton May 24, 1896, and is buried there.[2]

From Averell's brigade which fought at Droop Mountain, Lt. Col. Alexander Scott of the 2nd West Virginia Mounted Infantry (5th West Virginia Cavalry) contracted a severe cold during Averell's famed "Salem Raid" in December of 1863 and never fully recovered. He was mustered out with the regiment and returned to private life at Pittsburgh, Pennsylvania. Scott died May 29, 1870, at the age of 49, from the effects of the cold he caught during the war.[3]

Lieutenant Colonel Frank (Francis) W. Thompson of the 3rd West Virginia Mounted Infantry (6th West Virginia Cavalry) married January 14, 1864, and was promoted to colonelcy of the 6th West Virginia Cavalry April 21, or May 15, 1864. He commanded Averell's Brigade in the battle of Moorefield, West Virginia and was mustered out August 18, 1864. Following the war he went into private life and served as a city councilman at Morgantown, West Virginia for many years and was mayor of Morgantown in 1868. He also served as Justice of the Peace and was so popular he was told he could "hold the office as long as he cared to do so." In 1873 he "established the Eureka Steam Flouring Mill along the Walnut Street wharf [in Morgantown]. By using a steam engine he surpassed production capacities of local water-powered mills." Thompson was yet engaged in the milling business in 1883. Following a long illness, Thompson died July 14, 1900, and is buried at Oak Grove Cemetery in Morgantown.[4]

Colonel John Hunt Oley of the 8th West Virginia Mounted Infantry (7th West Virginia Cavalry) served the last year of the war with his regiment in the Kanawha Valley of West Virginia, where they patrolled against and paroled ex-Confederates following the surrender. On October 14, 1865, Oley was appointed brevet brigadier general to rank from March 13, 1865, "for gallant and meritorious service in West Virginia and the Shenandoah Valley." After the war Oley resided in Charleston until moving to Huntington in 1871, where he became an associate of railroad magnate Collis P. Huntington. He became resident agent for Huntington's Central Land Company and was made treasurer and local land sales agent for the city of Huntington, West Virginia during his first year in town. The following year he became the town's city recorder, a position he held until his death. In 1872 he aided in the organization of the Trinity Episcopal Church and served 16 years as its treasurer. Oley hired an organist and chair for the church out of his own personal funds. He was a member of Huntington's first school board and from 1874-88 "was periodically in charge of the city's schools." Such Huntington schools as Oley Elementary and Oley Junior High were named after him. He never married and died of heart disease March 11, 1888 after "complaining of a pain in his chest." John H. Oley was laid to rest, at the age of 58, in Spring Hill Cemetery in Huntington, with over 3,000 people attending his funeral.[5]

Among Averell's troopers at Droop Mountain, Col. James Martinus Schoonmaker of the 14th Pennsylvania Cavalry commanded a brigade when barely 21 years of age. His guitar became a symbol for his regiment "even after it was ruined by water in crossing a stream and was tied together by its strings and carried as an oriflamme throughout the rest of the war." On July 1, 1864, he was promoted and given command of the 1st Brigade, Averell's Cavalry Division, Army of the Shenandoah and won the Congressional Medal of Honor while leading a charge against the Star Fort at Winchester, Virginia September 19, 1864. He "made a strong plea against the order to burn the town, and was honored there as a friendly hero on a visit after the end of the war." At the close of hostilities he was sent to Indian country and mustered out at Ft. Leavenworth, Kansas on August 31, 1865. Immediately afterward Schoonmaker engaged in a "short period of commercial employment at Louisville, Kentucky" and then returned to Pittsburgh where he prospered in the mining and shipping of coal on the Ohio River, which led him into the coke business in the Connellsville area. He later sold his interests to the H. C. Fricke Coke Company.

Postwar view of the leader of the 14th Pennsylvania Cavalry Colonel James Martinus Schoonmaker. DROOP MOUNTAIN COMMISSION BOOKLET

While in the coke business Schoonmaker became involved in numerous railroad projects, including the Pittsburgh & Lake Erie Railroad. On January 29, 1877, he was elected a member of the Board of Directors of the Pittsburgh & Lake Erie Railroad Company, and remained in that position for fifty years. By January 3, 1906, he was elected Vice President in executive charge of the company and on February 15, 1918, was elected Chairman of the Board of Directors. While involved in the railroad business Schoonmaker also engaged in the banking field, was a Vice President and a director of the Union Trust Company of Pittsburgh, a director of the Mellon National Bank and of the Union Savings Bank, a director of the Union Storage Company, a philanthropist, which included services as President of the Pennsylvania Association for the Blind, and at the time of his death, was President of the Western Pennsylvania School for the Blind.

Schoonmaker was also very active in veterans affairs and at the first Memorial Day ceremony for the Grand Army of the Republic (GAR) he was made one of it's leaders and remained so until his death. In 1913, he actively promoted erection of the Soldiers and Sailors Memorial Hall in Pittsburgh and, at the 50th Anniversary of Gettysburg, he was selected as the outstanding Civil War veteran of Pennsylvania and made chairman of the event. He was involved with the Union Veterans Legion and Army and Navy Legion of Valor, and was made an honorary graduate of the Virginia Military Institute in recognition of his military courtesy in the Shenandoah Valley during the war. In 1914, during World War I, he helped organize the Pittsburgh branch of the National Security League and was a member of the Council of National Defense. Col. Schoonmaker passed away October 11, 1927, following an operation for appendicitis. He was remembered for, "his distinguished services as a soldier and his eminent work as one of the builders of western Pennsylvania [which] assure him of a prominent place of honor in the annals of his native state, Pennsylvania, and especially in those of his native Pittsburgh."[6]

Rounding out Averell's mounted troops was Major Thomas Gibson (Jr.) who led Gibson's Independent Cavalry Battalion at Droop Mountain. On June 18, 1864 he married at Pittsburgh, Pennsylvania and on February 19, 1865 suffered a spinal injury at Ashby's Gap, Virginia. He was honorably discharged from the 14th Pennsylvania Cavalry at Harrisburg, Pennsylvania, December 13, 1865, to date from August 24, 1865. From about 1880 until 1892 Gibson resided in Minneapolis, Minnesota, and in 1895, while a resident of Soldiers'Home at Minneapolis, he filed for an invalid pension complaining of heart and lung trouble, rheumatism, catarrh, defective eyesight, total deafness in the right ear and defective hearing of the left ear, and his spinal injury sustained during the war. He had at least 2 daughters and died January 20, 1910.[7]

Captain Julius Jaehne, who led his Company C, 16th Illinois Cavalry in the pursuit of Echols' retreating rebels, was dishonorably dismissed from the service July 11, 1864, "for neglect of duty, in not having the rolls and records of his Company in readiness for its muster out of service, thereby working prejudices to the interests of the enlisted men of his Command, and violating the orders of the War Department."[8]

Averell's infantry at Droop Mountain included Col. Thomas M. Harris and his 10th West Virginia Infantry, and in June of 1864, the regiment was transferred to the Shenandoah Valley, where, during the summer, Harris commanded a brigade under Gen. George Crook. Harris was later placed under Gen. Sheridan, "where he distinguished himself in command of a division of the Army of West Virginia at Winchester and Cedar Creek." Noted for gallantry at Cedar Creek, Harris was brevetted Brigadier General. On December 10, 1864, his division was sent to Petersburg, where Harris was commissioned a brigadier general effective March 29, 1865. Three days later Harris broke the Confederate lines around Petersburg and took Fort Whitworth, an act that earned him a brevet major general for "bravery and gallant conduct."

By a forced march, Harris and his division were thrown between Gen. Lee's army and Lynchburg, which compelled Gen. John Gordon to abandon his plan to slip out of the subsequent surrender. "In recognition of his services on the field" he was proffered the Lieutenant Colonelcy of the 37th Regulars, but he declined due to his advanced age.

After the war Harris served on the commission that tried the Lincoln conspirators. He was mustered out of the army in 1866, returned to his native town and resumed medicine, and served one term in the West Virginia legislature in 1867. From 1869-70 he was Adjutant General of West Virginia and from 1871-77 [or 1875 according to another source] served as Pension Agent at Wheeling, having been commissioned by President Ulysses S. Grant. He continued his practice of medicine, retired to private life in 1885 and authored several essays and religious tracts, as well as a highly prejudiced account of the trial of the Lincoln conspirators in 1892 titled *History of the Great Conspiracy*. Harris remained active in local civic and religious affairs. He died at Harrisville, West Virginia at noon on September 30, 1906, and is buried there. "He died rich in the love and esteem of all who knew him."[9]

Colonel Augustus Moor of the 28th Ohio Infantry, Averell's other infantry leader at Droop Mountain, known as the "portly German colonel," went on to fight at New Market and Piedmont in the 1864 Shenandoah Valley campaigns. He reportedly did not achieve the rank of brigadier general until after he retired due to the opposition of political enemies in Ohio. In any event, he was made brevet brigadier general to rank from March 15, 1865, for gallantry at the battles of Droop Mountain and Piedmont. Moor became the first president of the Deutsche Verein at Cincinnati in 1869 and died in 1883.[10]

From Averell's artillery at Droop Mountain, Capt. John V. Keepers later complained of partial deafness in both ears, which he attributed to his own guns in the fight at Droop. He felt it was temporary and did not seek treatment. On or about July 17, 1864, he injured both hands at Ashby's Gap, Virginia, and treated them himself as there were no medical facilities available in the field. Keepers was honorably discharged October 1864, upon expiration of term of service.

From 1864 to 1876 Keepers was a resident of Newcastle, Pennsylvania. He divorced September 1876, and remarried November 9, 1876, and then resided at Youngstown, Ohio from 1876 to 1879. From 1879 to 1887 he again resided at Newcastle, Pennsylvania and in 1893 said he had no income, as being a millwright by trade he was unable to work due to his deafness and crippled hands, a result of his war injuries. Keepers died at Allegheny City, Pennsylvania, February 5, 1905, at the age of 84.[11]

Captain Chatham Thomas Ewing, Battery G, 1st West Virginia Light Artillery was soon afterwards arrested for spreading "rumors" and was married June 29, 1864 and had four daughters. He was honorably discharged "about January 24, 1865" and returned to Pittsburgh to practice law, where he complained his "eyes were diseased" and he had headaches and constipation, caused by his White Sulphur Springs wound. He moved his family west due to failing health, the result of wounds and the hardships of war. He first located at Des Moines, Iowa but in 1871 moved to Thayer, the healthiest town in Kansas, where he became printer of the Thayer *Head-Light*. Ewing had to take many prescription medicines for his health relating to his "gunshot wound in the abdomen" incurred during the war at White Sulphur Springs and died July 22, 1892, from disease of the bowels, a result of his gunshot wound of the abdomen.[12]

Lieutenant Howard Morton of Ewing's Battery, a "soldier by inheritance" who served gallantly at White Sulphur Springs, Droop Mountain, and New Market, brought charges against Capt. Ewing, but was himself dishonorably discharged in May of 1864 for giving false certificates that men had enlisted as veterans, thereby obtaining false bounties. Despite this, it was said "his skill and courage received the plaudits of his comrades and commanders." When his services to West Virginia ended he returned to Pennsylvania and was commissioned a Major in the 5th Pennsylvania Artillery, where he "served with bravery and skill." In 1895 he was a married resident of Pittsburgh and Vice-President of the Pennsylvania Society of the Sons of the American Revolution.[13]

Tragedy was to strike Averell's Chief Engineer, Lt. John R. Meigs. Following the Droop Mountain battle Meigs participated in the "Salem Raid" and later was involved in the combat at New Market, Piedmont, and Diamond Hill. He was Chief Engineer of the Middle Military Division, and Aide-de-Camp to Major Gen. Phil Sheridan, who considered Meigs a favorite and "liked the bright, young officer as well or better than any other on his staff." In such capacity he served with Sheridan in the Shenandoah Valley Campaign August 17 to October 3, 1864, and was breveted captain September 19, 1864 for "gallant and meritorious services at the battle of Opequon (Winchester), Virginia," and also breveted major September 22, 1864 for "gallant and meritorious services at the battle of Fisher's Hill, Virginia."

The brilliant career of Meigs was cut short on October 3, 1864, when he and two assistants conducted a military survey and ran into three Confederates on scout outside of Dayton, Virginia, located near Harrisonburg. It is unclear as to whether or not Meigs drew his pistol, but he was shot in the head and killed by one of the three rebels. One of the assistants escaped and informed Sheridan, who was enraged by the act and called it murder by Confederate guerrillas. Sheridan sought reprisal by ordering the burning of structures within a five mile radius of Dayton, which became known as the "Burnt District." Sheridan's officers convinced him to revoke the order but not before 17 houses, five barns, and some outbuildings had been burned. The situation had been defused and Lt. John R. Meigs was dead at the age of 22.[14]

General Alfred Napoleon Alexander Duffié, who through no fault of his own, failed to cut off the retreating army of Gen. Echols, served during 1864 in the Shenandoah Valley area until October, when he was relieved of duty stemming from complaints filed against his command while camped around Cumberland, Maryland. A few days later he was captured by Confederate partisans near Bunker Hill and not paroled until February of 1865. After his exchange Duffié was assigned duty in the Division of Missouri and mustered out in August of 1865. In the postwar years he served as U.S. consul at Cadiz, Spain and remained there until his death from tuberculosis November 8, 1880. His remains were interred at Fountain Cemetery, West New Brighton, Staten Island, New York.[15]

Colonel Carr B. White of the 12th Ohio Volunteer Infantry, Duffié's infantry commander, was mustered out July 11, 1864, with brevet rank dated from March 15, 1865, "for gallant and faithful services at the battle of Cloyd's Mountain, Virginia."[16] According to pension papers, Carr B. White died at Georgetown, Ohio, September 30, 1871, of phthisis pulmonalis, an asthmatic tuberculosis of the lungs. This condition was supposedly contracted in the fall of 1861 from exposure in the campaign in the mountains of West Virginia.

With a couple of exceptions, most of the leading Confederate principals at Droop Mountain fared well after the battle. From Echols' Brigade, Col. George S. Patton of the 22nd Virginia Infantry would continue to often take charge of the brigade, so much so that it was often called Patton's Brigade. During 1864 he fought gallantly at New Market, Cold Harbor, Kernstown, and numerous other actions until mortally wounded and captured at Winchester, Virginia September 19, 1864. He sternly refused to have his injured leg amputated and subsequently passed away September 25th. He is buried at Winchester. Although no valid documentation has ever been found, it is claimed Patton was promoted to general but the news did not reach his command until after his death.[17]

Other notables in the 22nd Virginia Infantry included Capt. John K. Thompson, Lt. William F. Bahlmann, Lt. John P. Donaldson, and privates James Henry Mays and Samuel D. Edmonds. Capt. John Koontz Thompson of Company A, who led the valiant last stand on the Confederate left at Droop Mountain, received a severe wound in the side, shot through the lung, at the battle of Cold Harbor. After recovery from his wound he reportedly led the remnants of the 22nd Virginia from the field at Winchester in September. Only 22 years of age when the war ended he was paroled and returned to his Putnam County farm in West Virginia. Unable to decide whether to practice medicine or law he opted to be a farmer. In 1880, Thompson served in the West Virginia legislature and in 1897 was a U.S. Marshall. From 1901-05 he served as U.S. Marshall for the southern judicial district, and later spent two terms as West Virginia Tax Commissioner, refusing a third term. Thompson was on the West Virginia school board and retired to his farm at Poca. Stricken with rheumatism in 1917, he was unable to tend to the farm, sold it, and moved to Point Pleasant, West Virginia in 1919, where he was cared for by his nephew, Dr. Hugh Barbee. Due to his numerous war wounds he was almost bedridden the last ten years of his life and a "bruise by spent ball in face caused most annoyance in his last six years." He never married and died January 3, 1925, at Point Pleasant of senility of age. He is buried in the Lone Oak Cemetery at Point Pleasant.[18]

Lieutenant William Frederick Bahlmann, Company K, 22nd Virginia Infantry, was permanently disabled by his Droop Mountain wound. He reportedly was Provost Marshal at Bristol, Tennessee in the fall of 1864, and then spent the remainder of the war as Enrollment Officer at Lewisburg. In 1875, he taught German and Latin at Central Missouri State Teachers College, then served nine years as Superintendent, Independence, Missouri schools. He returned to Central Missouri State College for 13 years, attended numerous veterans' reunions, and in 1924 gave his address as Charleston, West Virginia. He died March 17, 1930, at the age of 93.[19]

Lieutenant John P. Donaldson, Company H, 22nd Virginia, was captured at Cold Harbor and spent the rest of the war in a prison camp. Afterwards he be-

William F. Bahlmann, Co. K, 22nd Virginia Infantry, C.S.A., wounded and captured at Droop Mountain, shown here in his later years as a teacher and family man. LARRY LEGGE, BARBOURSVILLE, WEST VIRGINIA

came a businessman in Cincinnati, Ohio and was a member of Dry Goods Firms in Philadelphia, Pennsylvania in 1901. He was stricken with an illness and went to a sanitarium in Atlantic City, New Jersey. He passed away July 22, 1901.[20]

Among the enlisted men of the 22nd Virginia were Pvt. James Henry Mays and Pvt. Samuel D. Edmonds. After being wounded in the hand at Cold Harbor, Pvt. James Henry Mays was sent home to recover. He spent the final months of the war with his company performing detached duty in Monroe County. He moved to Charleston, West Virginia thence to Kentuck, Jackson County, West Virginia, where he died of complications from diseases December 4, 1928.[21] Pvt. Samuel D. Edmonds, Company B, fought at New Market with the 22nd Virginia and in 1916 resided in the Old Soldiers Home at Richmond, Virginia, where he died January 4, 1926. His remains are at Hollywood Cemetery in Richmond.[22]

Also from Echol's Brigade, Major William Blessing of the 23rd Battalion Virginia Infantry continued to serve with Echols in the Shenandoah Valley and at Cold Harbor until he became a patient at Harrisonburg Hospital in August of 1864. Lt. Col. Clarence Derrick of the battalion was wounded and captured at Winchester, Virginia, September 19, therefore, when Blessing returned to service, he commanded the battalion until the end of the war. In 1870 Blessing was a married farmer in Smyth County, Virginia with 5 children.[23]

The 26th Battalion Virginia Infantry's Lt. Col. George Mathews Edgar, also remained with Echols' Brigade during 1864, fought well at New Market, and was seriously wounded by a bayonet and temporarily captured at Cold Harbor. In that engagement his battalion was nearly destroyed, but both Edgar and his boys bounced back and participated in the march on Washington, D.C. and the Shenandoah Valley. He was captured at Winchester, Virginia, September 19, 1864 and in late October was released, exchanged, and back in command of his battalion by 1865. When Echols disbanded his army following Gen. Lee's surrender in April, Lt. Col. Edgar and his staff joined Echols at Charlotte, North Carolina to accompany President Jefferson Davis until the surrender of Gen. Johnston. Soon after the close of the war Edgar returned to Virginia and the teaching profession. He became an outstanding instructor at such places as the University of Virginia, Oakland College, Arkansas Industrial University, Florida State Seminary, University of Alabama, and Occidental College in California, just to name a few. His later years were spent with his family in Paris, Kentucky, where he collected various memoirs of his battalion and attended numerous veterans' reunions. Following a short illness he passed away October 13, 1913, at Paris, Kentucky, where he is buried. A detailed review of his career would reveal an academic genius.[24]

The "Gallant" Edgar in a rare postwar photo of the Lieutenant Colonel and two brothers who served in his 26th Battalion Virginia Infantry. The former soldiers are, left to right, Pvt. William Wallace Jones, Lt. Col. George M. Edgar and Sgt. James Madison Jones. Both brothers served in Company B of the battalion. This photo was taken at the Talcott, West Virginia home of William, but the date is unknown. Edgar passed away in 1913, William in 1915 and James in 1917. MRS. ERNEST A. (MARGARET) MARTIN, RUPERT, WEST VIRGINIA

Rounding out Echols' Brigade at Droop Mountain was Capt. George Bierne Chapman of Chapman's Battery. He remained with Echols at New Market, Cold Harbor, and the many battles in the Shenandoah Valley until he, like Col. Patton, was mortally wounded at Winchester, Virginia September 19, 1864. Hit by shrapnel, he was taken to Staunton, thence to Charlottesville, in the hope of recovery, but tetanus set in and he died September 28. His body was laid to rest in Charlottesville at the University of Virginia.[25]

From Jackson's Cavalry Brigade at Droop Mountain, Col. William Lowther Jackson was active in the 1864 Shenandoah Valley campaign and saw action at Gap Mountain, Winchester, Fisher's Hill, and Cedar Creek. On December 19, 1864, he was finally promoted to brigadier general. When the war in Virginia ended, he refused to surrender and went to Brownsville, Texas, where he was paroled July 26, 1865. An unreconstructed rebel, he temporarily emigrated to Mexico before returning to West Virginia, where he found ex-Confederates could not practice law. As a result, he moved to Louisville, Kentucky. He soon afterward gained appointment as a jurist and remained on the Kentucky bench until his death at Louisville, March 24, 1890. He is buried there in Cave Hill Cemetery.[26]

Postwar view of the former Confederate cavalry officer, Colonel William Lowther Jackson. LIBRARY OF CONGRESS

Lieutenant Colonel William P. Thompson of the 19th Virginia Cavalry finished out the war reportedly as colonel of his regiment. Due to the "Test Oath," which prevented former Confederate soldiers from practicing law in West Virginia, he moved to Chicago in order to pursue his profession. Failing health forced him to return to West Virginia where he entered into the oil business with his brother at Parkersburg, under the name of "J. N. Camden & Co." The business became the largest distributor of lubricating oil in the United States. Camden then sold the business to Camden Consolidated Oil Company, which identified with the Standard Oil Company. Thompson became vice-president and later president of the company. In 1881, he moved to Cleveland, Ohio and was first secretary and later vice-president of the Standard Oil Company. Early in 1887 he moved to New York, resigned the vice-presidency of the oil company, and became Chairman of the "Domestic Trade Committee," which supervised domestic trade business of the Standard Oil organization in America. In June of 1889, Thompson resigned from his duties with Standard Oil and was elected President of the National Lead Trust. He also served as a Director of the Ohio River Railroad Company of West Virginia at Clarksburg, the Weston & Midland Railroad, and Director in the Mercantile Trust, New York, of the U.S. Bank of New York, and of the American Pig Iron Warrant Surety Company. He was "among the leading financiers and best businessmen of New York and the country" and was a great fancier of horses and horse-breeding, which he did on his farm at Brooksdale, New Jersey. Ironically, Thompson discovered his wife, Mary Evelina Moffett, in the Levels, within view of the Droop Mountain battlefield. At the time of his death in Brooksdale, February 3, 1896, his fortune was estimated at from five to six million dollars.[27]

The 20th Virginia Cavalry's William Wiley Arnett concluded his wartime career with the regiment. After refusing to take the "Test Oath," he resumed his law practice in Berryville, Clarke County, Virginia in 1868. He was nominated for State Senator from that district but declined and was immediately nominated and elected to the Virginia Legislature from Clarke County. In 1872, Arnett moved to St. Louis, Missouri and established a reputation as a successful criminal and abortion lawyer. Returning to Wheeling in 1875, he became one of the most prominent attorneys of West Virginia. During his career as a distinguished lawyer he made a "whale of a speech" against the Republican party at Hillsboro, near the Droop Mountain battle site. He died February 15, 1902.[28]

Captain Warren Seymour Lurty of Lurty's Company Virginia Horse Artillery (Lurty's Battery) commanded Jackson's Battalion Horse Artillery after Major Jackson was wounded at Nineveh Church, but was captured two hours later at Cedarville (or Front Royal), Virginia, November 12, 1864. He remained a prisoner of war at Ft. Delaware until released June 17, 1865. He was a postwar lawyer in Harrisonburg, Virginia and in 1874, President Ulysses S. Grant appointed Lurty District Attorney for Virginia's West District and in 1890, he was appointed U.S. Marshall for the Oklahoma Territory. Lurty died February 2, 1906, of congestion of the lungs and is buried in Woodbine Cemetery in Harrisonburg, Virginia.[29]

Colonel Milton Jameson Ferguson, who led Jenkins' (Ferguson's) Cavalry Brigade detachment at Droop Mountain, was captured on Laurel Creek in Wayne County, West Virginia, February 15, 1864, along with 39 of his men. He remained in various Federal prison camps until exchanged in late 1864. After the war, Ferguson settled in Louisa, Kentucky, where in 1868 he was elected circuit judge for the Big Sandy Judicial District. Following a six-year term he retired and gave his support to the construction of the Chatterawha Railroad. Ferguson died April 22, 1881, known for the motto "to do justice to all the world, never to forsake his friends and fear no man." He is buried in Wayne County on the old homestead.[30]

Major James H. Nounnan of the 16th Virginia Cavalry led the regiment at various times throughout the remainder of the war. He died October 1, 1900, at the age of 66 at the Lee Camp Old Soldiers Home at Richmond, Virginia.[31]

Out of the 14th Virginia Cavalry, Col. James Addison Cochran led the regiment intermittently until the end of the war, although often hampered by a hernia. After the war this "gallant soldier" resided in Culpeper, Virginia as a Postmaster and newspaper editor, passing away there August 17, 1883.[32] Lt. Col. John Alexander Gibson of the 14th Cavalry, who was present at the Droop Mountain fight, married in January of 1864 and commanded the regiment by April. He was wounded in the arm at Monocacy, Maryland, July 9, 1864, but remained with the 14th Cavalry until wounded again and captured at Cedarville, Virginia, November 12. He was released from a Federal prisoner of war camp in July of 1865 and became a farmer, distiller, Justice of the Peace, Deputy Sheriff, Postmaster, hotelkeeper, and member of the Rockbridge, Virginia County Board of Supervisors at Lexington. Gibson died August 2, 1906, at Timber Ridge and is buried in the Timber Ridge Presbyterian Church Cemetery. He was remembered as "a faithful and gallant soldier."[33]

Major Benjamin Franklin Eakle, who may have actually commanded the 14th Virginia Cavalry at Droop Mountain, remained with the regiment and was in command during May of 1864. Although wounded through the body at Monocacy, Maryland July 9, he was back with the regiment within six weeks, only to be wounded and captured at Cedarville, Virginia, November 12. Following his release from a Federal prisoner of war camp in July of 1865, he became a farmer in Greenbrier County, West Virginia and from 1865-72 served as a hotel manager at Richmond. From 1872-94 he was Chief Clerk and General Manager at White Sulphur Springs, West Virginia, and also worked as a farmer and stockman. Eakle died at White Sulphur Springs, July 22, 1898 and is buried at the Old Stone Presbyterian Church at Lewisburg, West Virginia. His comrades said he "behaved with the most commendable coolness and courage."[34]

The man who commanded the Confederate rear guard during the retreat from Droop Mountain, Capt. Edwin Edmunds Bouldin of the 14th Virginia Cavalry, commanded the regiment at Chambersburg, Pennsylvania, July 18, 1864 and was in charge of Gen. John McCausland's rear guard during the subsequent re-

The grave of Major Benjamin Franklin Eakle, 14th Virginia Cavalry, is located in the Old Stone Presbyterian Church Cemetery at Lewisburg. Eakle commanded at least a portion of the 14th Virginia Cavalry in the battle of Droop Mountain, survived the affair, and died in 1898. His grave is only a few yards away from that of Major Robert Augustus Bailey. BRIAN ABBOTT

treat to Moorefield, West Virginia. On August 7, 1864, while in command of the 14th Cavalry, he was captured in the battle at Moorefield. He was eventually exchanged and rejoined the regiment at Five Forks, Virginia, April 1, 1865 and the subsequent retreat and the final charge at Appomattox, where he escaped to Lynchburg with 30 of his men. In the postwar years he assisted the Federal government in marking the lines at Gettysburg and served as a lawyer at Danville, Virginia. Bouldin passed away at Danville, October 29, 1912.[35]

Completing Ferguson's cavalry detachment that served at Droop Mountain was Capt. Thomas Edwin Jackson of Capt. Jackson's Battery Virginia Horse Artillery. Jackson's guns proved effective at New Market and Cold Harbor, and he participated with a section of his battery on the raid to Chambersburg, Pennsylvania. He lost his guns and many of his men in the fight at Moorefield, West Virginia, August 7, 1864, and was later wounded and captured at Winchester, Virginia, September 19, 1864. He married in Pulaski County, Virginia, January 10, 1866, but died there following a short illness December 28 of the same year.[36]

Jackson's Sr. 1st lieutenant, Randolph Harrison Blain, was wounded by a shell burn of the face at Gardner's Farm, Virginia in late May of 1864, and then sent to private quarters in Richmond, where he was declared unfit for duty in June of 1864. In 1873 he graduated from the University of Louisville, Kentucky with a B.L. and then practiced law in Louisville. Blain authored legal works and served as Judge of Louisville Police Court. He passed away June 9, 1929 and is buried at Cave Hill Cemetery.[37]

Major William L. McLaughlin who was in charge of all the Confederate artillery on the field at Droop Mountain, continued to command an artillery battalion at New Market, Cold Harbor (where he was wounded), Monocacy, Winchester, Fisher's Hill (where his horse was killed), and Cedar Creek. On March 18, 1865, Gen. William N. Pendleton said, "Major McLaughlin is a vigorous and gallant officer who will make, it is believed, an efficient commander at Fort Clifton and it may well be not only to give that command, but to accord him in connection with it an additional grade." He served as a lawyer at Lexington, Virginia from 1870-98 and Rector of Washington and Lee University and Trustee there 1865-98. McLaughlin was a member of the V.M.I. Board of Visitors and member of the Lee-Jackson Camp, United Confederate Veterans, at Lexington, Virginia. He died at Lexington, August 18, 1898, and is buried there in the Stonewall Jackson Cemetery. Comrades remembered him as a gentleman of high character and intelligence .[38]

General John Daniel Imboden, the cavalry officer who commanded the Shenandoah Valley District and chased Averell's army following the Droop Mountain affair, held the Federal army in check, resulting in a Confederate victory at New Market. Following the death of Gen. William E. Jones at Piedmont, Virginia, June 5, 1864, Imboden found himself in command of his own, William L. Jackson's, and John McCausland's brigades. With these men he fought Gen. David Hunter's advance until Gen. Jubal Early reached Lynchburg. Imboden led the cavalry command of Gen. Jubal Early that threatened Washington, D.C. and the subsequent operations against Gen. Philip Sheridan in the Shenandoah Valley. Late in 1864, he contracted typhoid fever and spent the remainder of the war on prison duty. After the war he was prominent in the development of the mineral and mining resources of Virginia. Imboden died at Damascus, Virginia, August 15, 1895.[39]

Although not a bastion of future leadership, at least three Federal officers and one Confederate officer on the field at Droop Mountain went on to achieve some degree of generalship. Out of Gen. Averell's command Harris, Oley, and Moor would aspire to similar rank, as would William L. Jackson in Echols' force, making a total of six present and future generals on the field at Droop Mountain. The question of whether or not Patton achieved that rank is still debated and remains open for discussion.

The final resting place of Captain Alpheus Paris "Captain Dod" McClung, Company K, 14th Virginia Cavalry, in the Old Stone Presbyterian Church Cemetery at Lewisburg. McClung was wounded at Droop Mountain but survived the war and died in 1912. BRIAN ABBOTT

Veterans of the Civil War gathered at Enon, Nicholas County, West Virginia, October 28, 1926. The men are, left to right: George Henry Clay Alderson (Co. A, 14th Virginia Cavalry), William H.H. Neil [Neal] (Co. D, 22nd Virginia Infantry), John James Halstead (Co. C, 22nd Virginia Infantry), Wyatt Meador (57th Virginia Infantry), Jilson H. Neal (Co. D, 22nd Virginia Infantry and Co. K, 14th Virginia Cavalry), and J.C. Laughary (Co. F, 67th Pennsylvania Infantry). Alderson fought at Droop Mountain and both Neals probably did as well. CARNIFEX FERRY STATE PARK MUSEUM

CHAPTER 8

Droop Mountain Battlefield State Park

[Written by Michael Smith, Superintendent, Droop Mountain Battlefield State Park, excerpted from Where People and Nature Meet. *Used by permission of Mr. Smith. Additions and revisions by Terry Lowry]*

Droop Mountain Battlefield, West Virginia's oldest state park, was dedicated on July 4, 1928, as a memorial to the men who took part in what is generally considered to be the largest Civil War battle fought on West Virginia soil. Located in Pocahontas County, five miles south of Hillsboro, the park offers scenic mountain vistas, hiking and picnicking, but is most noted for it's historic significance as the site of a fierce struggle which took place between the North and South on November 6, 1863.

The battle resulted from the movement of the Union army of Gen. William Averell, whose intent was to clear the southeastern section of West Virginia of Confederate troops and then strike at the Virginia and Tennessee Railroad. He was successful in forcing the army of Gen. John Echols into Virginia, but failed to reach the railroad.

In the years after the battle, as the bitterness and hardships of the war subsided, many veterans who had taken part in the Droop Mountain battle often gathered in small groups to share their memories. They began to hope for a memorial so that others might not forget that brave men had here shed their life's blood for the causes in which they believed so strongly.

John D. Sutton, who as a private in the 10th West Virginia Infantry had been in the forefront of the final charge that forced the Confederate retreat, became a leader in the movement for a memorial. As a member of the House of Delegates he persuaded the West Virginia legislature to adopt, on January 25, 1927, House Joint Resolution No. 8, calling for a five-member committee to examine the battlefield, temporarily mark the battle lines, and secure an option on not less than 50 acres of land. In compliance with this, Gov. Howard M. Gore, on April 21, 1927, appointed the Droop Mountain Battlefield Commission. The commission then elected Sutton as chairman at its first meeting, which took place in Charleston. At the meeting Sutton was authorized to "employ a competent engineer to aid and assist in the preparation of a map of the battlefield" designating the important positions.

The commission met at Marlinton on July 18, 1927, and first visited the battlefield on July 19. The commission met in session at the farm residence of Mrs. Rebecca B. McCarty, whose property entailed the major portion of the Confederate positions. The group inspected and surveyed more than 2000 acres, and after viewing the scenery, opted to the legislature "a battlefield state park," which would encompass about 140 acres of the McCarty farm, but would not include 15

The original Droop Mountain Battlefield Commission about 1928. Sutton fought in the Droop Mountain battle with the 10th West Virginia Volunteer Infantry. DROOP MOUNTAIN COMMISSION BOOKLET

On July 4, 1928, a crowd gathered for a battlefield celebration at Droop Mountain Park. The present lookout tower would be about where the clump of trees are found in the center of the photo. The old Spice P.O. can be seen at the right of the photo. DROOP MOUNTAIN BATTLEFIELD STATE PARK

acres around the residence and one-half acre of a cemetery. On December 28, 1927, the commission met at Weston and planned publication of a park booklet, which was released the following year. In January 1928, some 125 acres of land were purchased, and during the following summer 29 signs and several monuments were erected.

On July 4, 1928, the park was officially dedicated in a ceremony which reportedly attracted some 10,000 spectators, including more than 1000 automobiles. Some surviving men of the blue and gray were present, including Matthew John McNeel (Company F, 19th Virginia Cavalry); Noah Davis McCoy (Company F, 19th Virginia Cavalry); Richard Lafayette Diehl (Company C, 22nd Virginia Infantry); all of the Confederate army, and W. Tyler, John D. Sutton (Company F, 10th West Virginia Infantry), and Peter McCarty (Company I, 3rd West Virginia Cavalry), of the Union army. Droop Mountain officially became West Virginia's first state park and was dedicated as a shrine to the men who had fought and died there 66 years before.

The park quickly became a favorite place for picnics, family reunions, and other gatherings, but few facilities were developed until the Civilian Conservation Corps (CCC) established a camp there in 1935. In the meantime, most of the park's forest had been ruined by a combination of the American chestnut blight and a forest fire which had followed the severe droughts of the early 1930's. The fire burned underground in the cranberry bog for several weeks. A small sawmill was later set up to salvage as much as possible, with most of the lumber sawed being used at Watoga and Babcock state parks, which were then beginning developments of their own. As the logs were sawed, bullets fired during the battle were often found embedded within the wood.

On July 4, 1930, the monuments to Sgt. John D. Baxter and Lt. Henry Bender were formally dedicated, and plans were announced for a monument to Major Robert A. Bailey. This monument apparently was never created, nor was a monument planned for November 11, 1930 to 1st Lt. Josiah Woods Price, Company F, 19th Virginia Cavalry, a veteran of the battle who died in 1918. On July 4 of the

4th July Celebration

Pocahontas County will meet at the Battlefield on Droop Mountain July 4 1928, to celebrate the day and to Inaugurate the State Park.

Be there by nine in the morning. Park your cars on the State Grounds. Families will bring dinners for themselves with something over the strangers within our gates.

PATRIOTIC CELEBRATION

Program will be announced later. Speaking, Games, Social Parties, Feasting.

ALL DAY CELEBRATION

A good way to spend the Fourth of July. Battlefield Commission and other Distinguished Visitors expected.

An ad from the June 21, 1928, issue of the *Pocahontas Times* for the grand opening of Droop Mountain State Park.

following year a stone marker in memory of two unknown Confederate soldiers buried nearby, as well as to all the rebel soldiers who fought at Droop Mountain, was presented to John D. Sutton and Droop Mountain Battlefield State Park by Mrs. James E. Cutlip, of Sutton, West Virginia, representing the Capt. Edwin Duncan Camden Chapter (Sutton, West Virginia), of the United Daughters of the Confederacy. Capt. Camden served in the 25th Virginia Infantry, C.S.A. (which was not in the Droop Mountain battle), and the monument is located near the Confederate artillery position (modern cemetery), just across the park boundary on private property.

In the spring of 1935, a small city of tents was erected near the park monuments as the beginning of the CCC's Camp Price, and development of most of the park's present facilities quickly followed. Henry J. Johnson, former CCC enrollee, member of the U.S. armed services and aide to Camp Price's Capt. E. R. Howery, recalled in the November 1982 issue of *Wonderful West Virginia* magazine: "We

This photo of CCC workers was taken at the Bloody Angle and shows the PX, supply room and company office of Camp Price. To the right can be seen the two white markers of Baxter and Bender. DROOP MOUNTAIN BATTLEFIELD STATE PARK

One of the two original log portal entrances to Droop Mountain Battlefield State Park erected in 1935 by the CCC. The portals eventually deteriorated and were removed by 1961. DROOP MOUNTAIN BATTLEFIELD STATE PARK

arrived . . . during the summer of 1935, sleeping in tents until late fall when our barracks, mess hall, dispensary, officers' quarters and company headquarters buildings were completed . . . although the mountaintop area was a veritable wilderness, some vestiges of the battle were evident. The rock breast-works remained in place around the brow of the mountain looking toward Hillsboro. We unearthed hardware from rusted out rifles, as well as bayonets, while excavating for the erection of various park structures." Johnson went on to describe Droop Mountain as ". . . almost solid limestone with a very shallow cap of soil. In our camp area the average depth of soil was 14 inches. This made excavating difficult, and almost all of our structural foundations required dynamiting. This depth factor must certainly have created problems during the battle for burying the dead. During the creation of our camp, one sinkhole was discovered in which more than 50 bodies had been buried." Johnson also confirmed it was "not unusual to discover minie balls in logs used for construction."

The CCC workers built a tremendous log fence and two entrance portals along the highway, while a system of hiking trails and roads, and two large picnic areas were developed. The major structures erected were a six-sided log observation tower with spiral staircase, two picnic shelters, three rustic cabins and numerous service buildings. In addition, a forest improvement program was started, and trees were placed in most of the cleared areas.

On November 6, 1935, the 72nd anniversary of the battle took place at the park, and two surviving veterans of the Civil War, both 91 years of age, were present. The two were Matthew John McNeel (Company F, 19th Virginia Cavalry), who was on detached duty the day of the battle and did not participate, and Pvt. George H. Alderson (Company A, 14th Virginia Cavalry), who actually fought in the affair. Ironically, when Alderson was first introduced to McNeel he did not like him as he thought McNeel had served in the Federal army; once McNeel's Confederate affiliations were made clear the two became close friends. Mr. Alderson "accepted a flag on behalf of his comrades and a veteran of the war for the Union forces." Confederate and U.S. flags were also presented by various historical groups "as the beginnings of a battlefield museum." Thus the park was the center of intense activity that lasted until October 1937, when the CCC gave orders to dismantle Camp Price.

This photo is believed to have been taken at the 1935 celebration. The two surviving Confederate soldiers are believed to be Matthew John McNeel (Company F, 19th Virginia Cavalry) on the left and George H.C. Alderson (Company A, 14th Virginia Cavalry) on the right. Alderson died in 1936 and McNeel in 1938. DROOP MOUNTAIN BATTLEFIELD STATE PARK

An additional 139 acres of land was acquired in 1936, doubling the size of the park. Later land purchases enlarged it further to the present 287 acres, including the McCarty residence, which was destroyed. A proposed swimming pool was never built and the log watchtower was originally planned to be built from stone. Administration of the park was shifted in 1937 from the Forestry Division of the Game, Fish and Forestry Commission to the Division of State Parks of the Conservation Commission.

After Camp Price was gone, the first caretaker assigned to the battlefield is believed to have been Joe McMillion, possibly while the park was still administered under the Forestry Division. He was followed by Robert Bean in 1940 and by Madeline Bean in 1944. Mrs. Bean was probably the first female caretaker in the state park system. A major difficulty during her term was that she had no vehicle and could hardly maintain the three rental cabins without it. The cabins had been available to the public since 1937 and were quite popular, with many guests returning year after year. The park residence, originally the forestry headquarters, was upgraded during the early 1940's with the addition of stone foundations, chestnut siding, and an extra bedroom, while a stone cellar was built nearby.

Throughout the 1940's and 1950's a succession of caretakers and superintendents were in charge of the park, with most staying three years or less. Major activities within the park continued to be family reunions and cabin rentals. Several dozen locust picnic tables were built during this time, one of which was 56 feet long and was, on many Sundays, covered from end to end with great piles of food.

Another park system "first" occurred in 1957 when Gordon Scott, a former employee of Watoga, moved to Droop Mountain and thus became the first black caretaker. Scott's term was short, however, for he died in 1959 after a long illness.

Scott was replaced by William Davis, who soon made several basic changes in the park's operation. The first change was the opening of a sign shop that began constructing signs needed by other parks within the state. Sign-making became one of the park's major activities until 1968, when the shop was moved to a more central location at Carnifex Ferry Battlefield State Park near Summersville.

Another significant change was made in 1961 when the cabins were closed and two were torn down. The third cabin was used for storage until 1974, when it was converted to a Civil War museum. Also, in 1961, a new chestnut rail fence was built along Rt. 219 to replace the original log fence, which had deteriorated beyond repair and the portals of which had fallen down several years earlier. New playground equipment and toilets were added in each of the picnic areas, and the road system, which had been a problem for many years, was much improved. Period atmosphere was added in 1965 with the purchase of a reproduction cannon.

After the sign shop was moved in 1968, activity at the park slowed, with few changes taking place until the mid-1970's, when a new water system was installed to eliminate the old hand pumps in the picnic areas. A Civil War museum was opened in 1974, but was closed in 1983 after many of the artifacts were stolen by vandals. Renewed interest caused it to be reopened in 1985, with several newly discovered items being added each year since.

The success of the tree planting program by the CCC 50 years before became evident in 1985, when it was realized that the scenic view from the lookout tower had disappeared. Many of the trees surrounding the tower were then cut in order to restore the vista. The tower was straightened and repaired several times over the years as the original logs rotted and had to be replaced.

From a historical standpoint, the battlefield has suffered somewhat from a lack of attention since its early beginnings. As an example, the 29 signs erected by John D. Sutton in 1928 were allowed to deteriorate and were never replaced. Many were moved or torn down during the CCC days, and the points they marked and the information they contained are now unknown. However, in recent years the historical background of the park has been re-emphasized and there is great hope that research can restore much of which has been lost. Living history events, collection of data relating to the battle, and plans for interpretive devices are just some of the means being used to direct the park toward its original purpose of keeping alive the memory of those who struggled so long ago.

Thus, the granddaddy of state parks has weathered the changes of almost 60 years, and holds a place as an honored member of the West Virginia state park system. It's mission still is to keep in remembrance the struggles and trials of many brave men, to remind of the simple joys of nature, and to provide a restful place to set aside the many cares of the modern world.

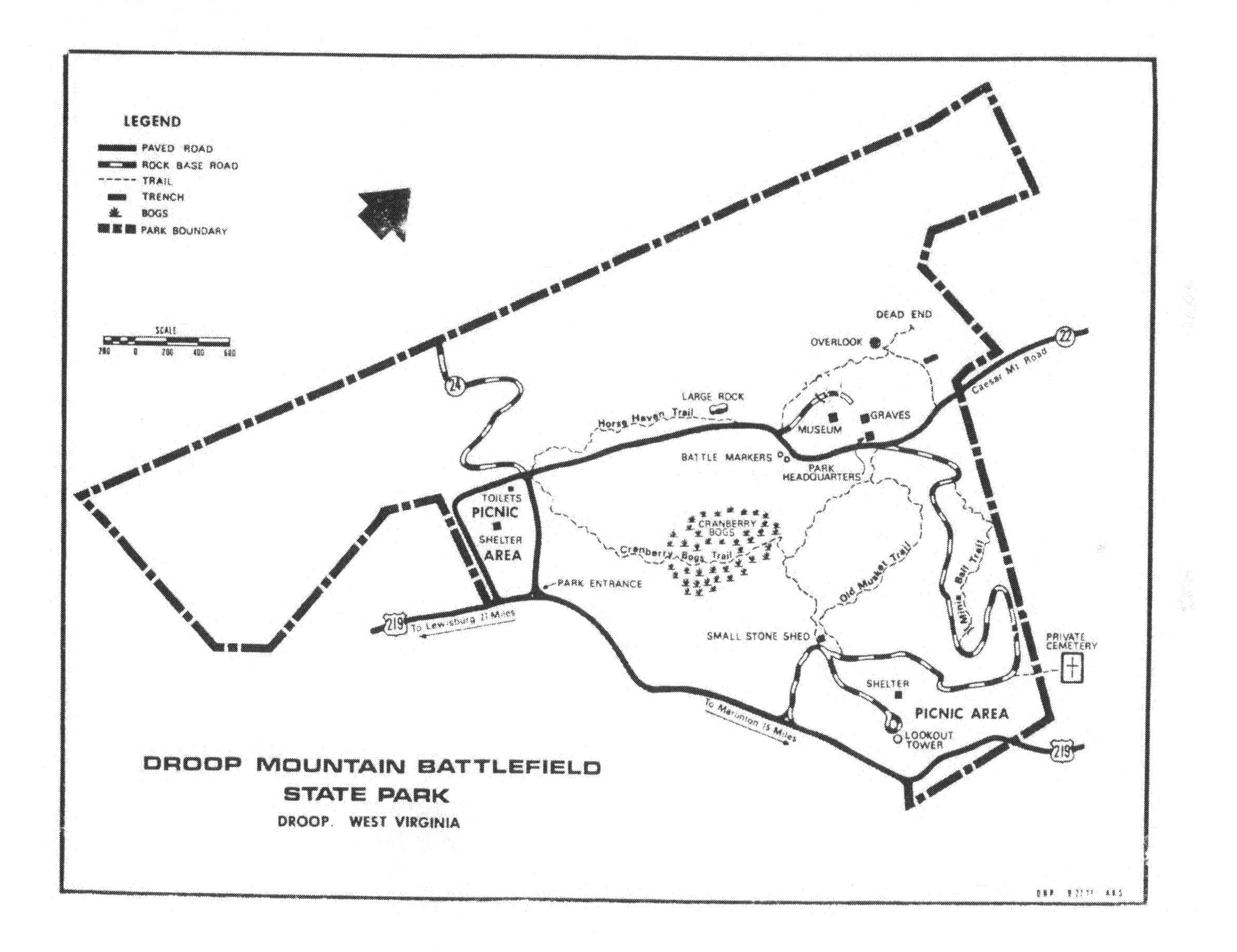
LEGEND
PAVED ROAD
ROCK BASE ROAD
TRAIL
TRENCH
BOGS
PARK BOUNDARY
SCALE
200
0
200
400
600
DEAD END
OVERLOOK
22
Caesar Mt Road
LARGE ROCK
Horse Haven Trail
MUSEUM
GRAVES
BATTLE MARKERS
PARK HEADQUARTERS
24
TOILETS
PICNIC
SHELTER
AREA
CRANBERRY BOGS
Cranberry Bogs Trail
PARK ENTRANCE
Old Musket Trail
Minie Ball Trail
219
To Lewisburg 21 Miles
SMALL STONE SHED
PRIVATE CEMETERY
SHELTER
PICNIC AREA
To Marlinton 15 Miles
LOOKOUT TOWER
219
DROOP MOUNTAIN BATTLEFIELD
STATE PARK
DROOP, WEST VIRGINIA

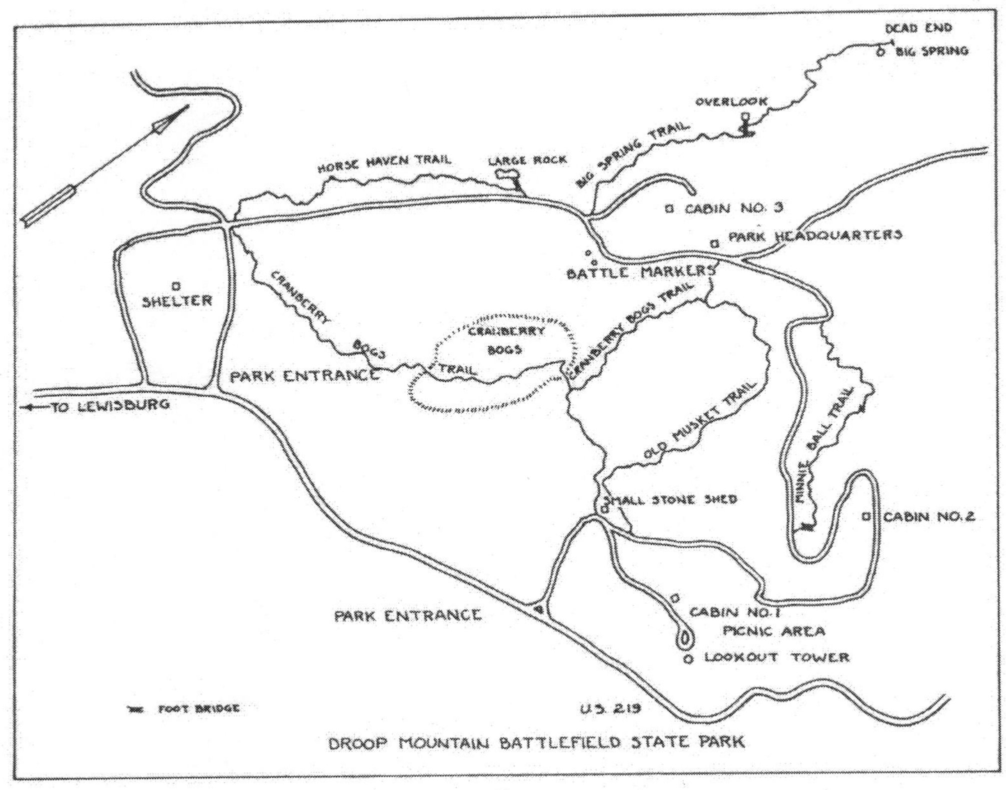

The top map shows markers, trails, and facilities at Droop Mountain Battlefield State Park. The bottom map shows major battle positions in relationship to the park area.

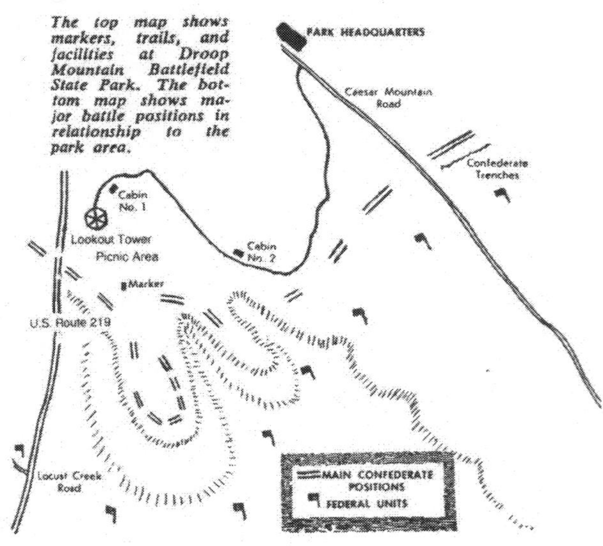

CHAPTER 9

The Ghosts of Droop Mountain

While it is fairly common for ghost stories to arise out of Civil War sites, Droop Mountain probably ranks near, or even at the top, of such areas to spawn wild-eyed stories of ghosts, apparitions, headless soldiers, illusions, and the like. Due primarily to its somewhat isolated, rural location, Droop Mountain battlefield has been the scene of many unexplained happenings since the Civil War battle that took place there in 1863. This is not unusual, considering fog "often rolls over the mountain in waves, there one minute, gone the next," creating an eerie atmosphere conducive to tales of ghosts and the supernatural. As park superintendent Mike Smith so aptly stated, "It's the general feeling of everybody that ghosts of soldiers are nearby. Some people have told about hearing supernatural horse hooves, some so real that they thought they were going to be run over. Other people have said they have been going through the park at night and have been stopped in the road by a Union soldier who won't let them go until dawn."

The earliest known episode of a supernatural occurrence at Droop Mountain took place in 1865 when Betty and Nancy Snedegar, residents of the west side of Droop and daughters of James C. and Rebecca Kellison Snedegar, walked to the east side of the mountain to pick berries. On their return trip they located two guns, apparently lost during the battle two years earlier. As the two girls "started to carry off the guns, rocks were thrown at them but they saw no person. They went on home. As they went to milk, more rocks and clubs were thrown at them. At the house rocks came down the chimney and knocked the lids off the pots. Rocks came through the log walls, but left no holes. There were sheepskin rugs on the floor which started rarin' up . . ." Another account claims the sheepskin rug would stand erect and bawl. The Snedegar sisters then ". . . gathered all the rocks and threw them in a sinkhole several [hundred?] feet deep. The rocks all came flyin' back out." One version of the story claimed "dog irons would come out of the fireplace and race around the room." Reportedly, an uncle came to visit and two rocks hit him in the arm and head, after this he quickly departed. Finally, the guns were returned to their original location and all the problems ceased.

The next known account of a supernatural event took place in 1914 when F. W. Albert spotted a regiment of soldiers on the battlefield, marching two by two, each carrying a gun on his shoulder.

Edgar Walton, an elder resident of Droop Mountain who once lived in the old Sunrise schoolhouse about a quarter of a mile from the park entrance, recalled that in 1920 he and a friend were walking together when they stopped to build a fire and noticed an unusual apparition. Mr. Walton stated, "I never did believe in ghosts and still don't, but we saw something. It was in the form of a man but without a head, and it was drifting along." In another conversation Walton related the same story in more detail and said the incident took place in 1927. He remembered, "I was walking near Spring Creek Mountain with a friend. I wanted to go over the mountain, but he wouldn't come that way. We sat down near an old root pile to rest and broke up a bunch of dry roots to build a fire." Walton continued, "I heard something that sounded like somebody dragging a hand through the leaves. I turned around and there was this soldier standing as close to me as this . . . My friend says, 'Let's get out of here.' The ghost walked to the middle of the gate and it disappeared. It might have had clothes on, but there was no head to it. And it didn't walk. It just drifted along."

Such stories of headless soldiers at Droop Mountain obviously are based upon the actual incident during the battle in which 2nd Lt. Joseph W. Daniels, Battery B, 1st West Virginia Light Artillery, had his head shot off by a Confederate shell. Although a Yankee, most storytellers prefer Confederates as ghosts, therefore, through endless folklore, Daniels has apparently become a rebel.

A great example of the headless soldier story was told by Anna Atkins, who grew up on Droop Mountain. She related that during the battle a mounted soldier encountered an enemy foot soldier about where Gerald Brown would reside in 1981. The trooper grabbed an infantryman by the hair and sliced off his head with his sword. Anna said, "He is supposed to have thrown the head in a pond nearby but no one ever found the head. The body was buried on the side of the road on G. Brown's front property." As a result of this incident, in the years following, "when horses passed there at night, the ghost—a headless man—stood by the horses and held them by the bridle bit until daylight; then they could proceed. One night a driver with a double team was stopped. He urged the team on. Still they stood. Finally he saw the ghost holding the bridle bits of the lead team. Kindly he asked it to release the horses. In desperation he grabbed his blacksnake whip, walked out the wagon tongue, sprang to the back of a horse and came down on the headless ghost several times. The horses bolted and ran, the driver clung to a mane. His buddies, who were part of a wagon team, stopped the wagon down the road. That was the last time the ghost held horses back."

Henry J. Johnson, one of 220 enrollees at CCC Camp Price on Droop Mountain in 1935, recalled that during his tenure, "Tales were told of the sound of marching feet on moonless nights around the perimeter of the camp." Johnson's co-workers, visiting girlfriends in the area, "would scurry through the darkened woods and bring tales of ghostly 'Who goes there?' challenges. The screaming of bobcats at the camp dump and the cries of the large owls that emanated from the darkened forest served to add to the eerie aura surrounding a luckless night time traveler. Only the bravest of CCC workers ventured alone at night into an area alive with memories of those who had fallen and died there."

In 1941 Johnie Keen had been driving a heavy-duty two-ton lumber truck since 3:30 the morning before. He encountered a ghostly apparition "just below the long straight stretch as you reach the foot of Droop on [the] south side in the first curve beyond [the] home known at that time as the Brown's residence." The confrontation took place at about 2:30 A.M. when Keen arrived at the curve and spotted in his headlight beam "about 8 or 12 people riding horses and they were riding fairground or showhorse style." He slammed on his brakes, got out of the vehicle to talk, but the group quickly turned to the left and disappeared. Keen said they "were wearing men's khaki-colored pants, blue jackets . . .[trousers] similar to . . . chauffeurs would wear . . . an army officer. The front horse on the left of the group that approached me was white. Others were bay or darker colored and each were carrying saddle bags as though they were cavalry."

Napoleon Holbrook, who served as park superintendent from 1946-49, recalled a number of episodes that affected him and his family while residing at the park. One morning at 3:00 A.M. he heard a screaming sound that came from the direction of the cranberry bog in the park. Holbrook reflected, "It could have been an animal, but it was real enough sounding that I got up and went to look, but I didn't find anything."

Holbrook experienced a number of incidents, such as the day his son was playing in the road in front of the park office and heard a horse riding toward him. "The noise came so close it scared him and he ran inside." At another time Holbrook's six year old daughter Carole was playing in the pine trees in front of the park residence when she spotted a man in a knee-length Confederate coat lying against a tree as if asleep. Carole stated, "It was very real to me."

Also during Holbrook's residency at the park, Mrs. Holbrook once heard footsteps on the back porch of the park residence. Nap emphasized, "She swears to this day that the footsteps went to a cellar near the back porch, and that the lock jiggled." And in even another episode; one Sunday afternoon, as the Holbrook family was in the park residence, Carole remembered, "We clearly and distinctly heard a shout, 'Halt!' outside . . . it sounded like it came from the front porch. But when we looked there wasn't anybody around."

Two park superintendents, Clyde Crowley, who served from 1950-52, and E. Morris Harsh, from 1952-54, both reported no ghostly experiences during their tenures. However, William Davis, park superintendent from 1959-68, told of the day he was working in the park with caretaker Floyd Clutter when the two heard a scream of distress. Clutter said, "We searched the woods and didn't find anyone . . . Bill had planned to go somewhere, but he stayed at the park because he was certain someone was in trouble." Davis added, "During the time we were at Droop we heard noises we simply considered house noises . . . but I couldn't explain the scream, and it never was explained. It's possible someone could have been on one of the hiking trails, but we searched the area and there didn't appear to have been anyone around."

In 1970 Mrs. John Clutter of Renick, mother of Floyd Clutter, park caretaker from 1968-84, said that while at the park she saw a man standing beside the soldiers' graves in the rear of the park residence. She said, "He stepped behind a tree and didn't reappear . . . I went to look and there wasn't anybody there."

Clarence Murray, a Department of Highways engineer who resided near the park, went squirrel hunting in October of 1972 and heard a sound like horses pulling a wagon, accompanied by jingling sounds. Murray reported, "I have no reason to lie about it . . . I did hear it and it did sound like hoofbeats. It was close, like it was going to run over me." Despite the sound Murray said he saw nothing.

In 1975 Mrs. Jean Murphy (a Clutter relative) of Akron, Ohio, was spending the night at the park when she awoke feeling a presence in the room. Mrs. Murphy told an interviewer, "It wasn't anything I saw . . . it was more like a presence, an evil presence. I kept trying to convince myself I wasn't awake, but I was. I prayed and prayed, and at the first glimmer of dawn, the presence left. I can't sleep in that room anymore."

Caretaker Floyd Clutter made it clear that neither he nor his wife believed in ghosts, but said that on July 16,1977, at about 10 P.M., the Clutters and several relatives had returned from church and were in the park residence kitchen eating. Floyd said, "All of us clearly heard somebody coming up on the porch, we heard the door open and we heard sounds like youngsters talking . . . yet when I went into the living room there wasn't anybody there, and the door was locked." Mrs. Clutter experienced another odd scene during a night of August 1977 as she sat at home alone watching television. She heard a knocking sound on the outside wall near the television and looked out the window to see a hat like the one Mr. Clutter wore floating past the window. She said, "It just floated out into the darkness . . . the porch light was on and I couldn't have imagined it."

The Clutters also mentioned that two summers prior to 1977 they heard a knock at the front door. "It was as natural as you'd want any knock to be," Floyd recalled, "but when I went to the door, there wasn't anybody there." This may be the same event that took place at about 3:30 P.M. on a spring day when Mrs. Clutter heard a loud knock on the front door. "It was like someone was rattling the door to get my attention . . . Floyd had gone to town, and I thought he had come back and didn't have his key. But when I went to the door, there was nobody there." Mrs. Clutter claimed the same occurrence transpired three times that afternoon, but each time there was nothing to be found.

The Clutters have also told of similar encounters, including the sound of car doors slamming shut but no cars being located, and have concluded that at least the ghosts are polite since, "The noises almost always happen in the daytime . . . We rarely hear anything at night," and the ghosts only come around every two or three months.

The headless ghost-soldier story arose again in 1977 when Mrs. Clenston Delaney, daughter of Edgar Walton, along with her husband and sister, spotted a headless, ghost-like figure on the same spot as her father in 1920[27?]. She said it took place one evening while cutting wood near the battlefield, when they "saw an apparition that left them frightened and shaking." The headless ghost, clad in a gray uniform, floated past her making a moaning sound. Mrs. Delaney declared, "It was very odd. I can't explain it. But all three of us saw it."

As recent as 1990, Ron Nelson of Parkersburg, West Virginia, who served as a Civil War living history re-enactor as a Private in Co. F (Night Hawk Rangers), 17th Virginia Cavalry, claimed he saw a dapple gray horse with a grayish blurred vision atop it. "There is no doubt in my mind I saw a gray horse that night on Droop Mountain," Nelson relayed, but "As to whether it was any form of a ghost I cannot say."

Present park superintendent Mike Smith does not believe in ghosts, and said he has yet to see one at Droop, but he also refuses to discount the possibilities. Smith believes the tales make good fodder for tourists and says, "I don't know for sure. I am somewhat skeptical, but I hate to say those tales are not true." Whether or not Smith is correct, it is certain Droop Mountain battlefield will continue to spawn weird tales of ghosts, headless soldiers, apparitions, and things that go bump in the night.

Appendix A

A Report of the Killed, Wounded and Captured at the Battle of Droop Mountain

These names were compiled from various sources, including regimental rosters, newspaper accounts of the battle, personal narratives of the affair, and compiled military service records (CSR). An asterisk (*) following a name indicates the soldier's name was included in a newspaper or personal account of the battle but not found on the regimental roster. The number following the name of the soldier is his approximate age at the time of the battle. The term unknown indicates a specific casualty has been verified but a name has not been located.

FEDERAL

2nd West Virginia Mounted Infantry
(5th West Virginia Cavalry)

Co. A:		
Pvt. Hugh McMannis	23	wounded
Co. B:		
Sgt. William Jenkins	26	wounded-shell wound/left foot
Pvt. John Kerns	25	wounded
Co. C:		
Pvt. Henry Emmerling	26	mortal-left shoulder–died Nov. 20
Co. D:		
Pvt. Andrew Bernard [Parnell]	23	killed
Pvt. Samuel Bowden	34	killed
Pvt. Michael Brubach	22	wounded-flesh/face
Pvt. Edward Doyle	37	killed
Pvt. William L. Hughes	22	killed
Pvt. Charles Ritz	22	killed
Co. E:		
Pvt. Thomas J. Akers	19	killed
Pvt. George Dent	27	wounded-right knee
Pvt. William H. Foulke	--	wounded-slight/head
Pvt. William Garroll	25	mortal-right hand–died Nov. 24 at Grafton
Pvt. S. L. D. Hudson	21	wounded-slight/head
Pvt. Moses Moore	24	killed
Pvt. John Murphy	--	killed
Co. F:		
Sgt. John C. Devlin	27	wounded–discharged for wound
Co. G:		

Ewing's Battery (no casualties reported)

Co. H:		
Pvt. Aden Webb	29	wounded
Sgt. Thomas R. Williams	25	wounded-flesh/right hip
Co. I:		
1st Lt. Charles H. Day	25	wounded-severe gunshot-flesh/ left arm–discharged
Pvt. William Mosby (?)	--	wounded
Co. K:		
Corp. Marcus D. Kenney	25	mortal-right breast–died Nov. 11 at Hillsboro
Corp. Edward C. Malley	29	mortal-right shoulder–died Nov. 9
Pvt. ---- McCarley	--	wounded
Pvt. Thomas McConkey	31	wounded-severe–originally thought to be mortal
Pvt. W. M. McKinney	--	wounded-left lung
Pvt. John Sallyards	47	wounded-shell-flesh/back
1st Lt. Arthur J. Weaver	26(?)	killed

3rd West Virginia Mounted Infantry (6th West Virginia Cavalry)

Co. A:		
Pvt. John H. Courtney	20	wounded-flesh/right leg
Co. B:		
Pvt. William B. Simms	29	killed
Co. C:		
Capt. Jacob G. Coburn	40	mortal-thoracic cavity-right side–died Nov. 24 at Beard's at Hillsboro
Pvt. George W. Heinsman [Harrison]	23(?)	wounded-head-right side
Pvt. William H. Matlick	21	killed
Pvt. John Mooney	30	wounded-left shoulder-thoracic cavity-left side
Pvt. Seth (B. F.) Stafford	24	captured–Nov. 26 at Hillsboro
Pvt. W. H. Wright	--	wounded-flesh wound of back
Co. F:		
Pvt. Lemuel D. Bartlett	20	killed
Pvt. Edward J. Fleming	24	wounded-right hand
Pvt. Abraham McDonald	22	wounded-flesh/right thigh
4th Sgt. Andrew Newland [Verlund]	26	wounded-flesh/chin
1st Sgt. John Webb	26	wounded-severe-thigh/fractured left femur
Co. H:		
Pvt. Josiah Hall	37	wounded-fracture right forearm
Pvt. George Silbaugh	21	wounded-slight/right knee

Co. I:		
Pvt. Simon Tomlinson	31	captured
Co. K:		
Pvt. Enoch F. Basnet	25	killed
Sgt. Franklin Clayton	25	wounded-flesh/left thigh
Pvt. Dudley E. Dent	26	wounded-flesh/left thigh
Pvt. Christopher C. Lipscomb	17	wounded-flesh/left knee
Pvt. E. F. Ranson [?]	--	wounded-head
Pvt. James A. Simonton	24	killed
Pvt. Benjamin Starr	21	wounded-leg—Nov. 11 near Franklin
Pvt. James (V.) Woods	22	wounded-upper third right femur
Corp. Wells Wrick	25	killed

8th West Virginia Mounted Infantry (7th West Virginia Cavalry)

Co. B:		
2nd Lt. Joseph L. Hager	25	wounded-severe/leg/fracture of left ankle-left at Hillsboro and captured—discharged for wound April 21, 1865
Co. C:		
Pvt. Charles W. Angel	23	wounded-right arm
Pvt. Timothy Cook	30	wounded-right thigh
Sgt. Thomas Swinburn	23	wounded-right shoulder and neck
Pvt. Carey Woods	--	killed
Co. G:		
Corp. Richard J. Grinstead	23	killed
Corp. Granville G. Rains	35	wounded-left shoulder
Co. H:		
Pvt. Lorenzo D. Perry	33	wounded-right shoulder—Nov. 5
Co. I:		
Pvt. Jackson Vernatter	18	wounded-head
Co. K:		
Corp. William Lewis	30	killed
Pvt. George H. Siders	27	wounded-dislocation of shoulder

10th West Virginia Infantry

Co. A:		
Sgt. Right Bird Curry	19	killed-gunshot in head
Pvt. James Pickens	21	wounded-gunshot thru left leg and flesh
Pvt. General Jackson Shaw	30	mortal-gunshot
Pvt. Samuel Sweeker[Swecker]	21	mortal-gunshot thru left leg bones fractured-severe

Pvt. George D. Walton	20	wounded-severe gunshot in right knee joint, right side

Co. B:

Served as Provost Guard throughout campaign. David Peterson, 10th West Virginia Infantry, in a Nov. 13, 1863 letter, states: "not a man of Co. B was hurt." Regimental adjutant stated in *Wheeling Intelligencer*: Co. B was not engaged at Droop Mountain.

Co. C:

Pvt. Isaac Burkhammer	21	wounded-gunshot thru left fore arm/left hip/severe/near abdomen
Pvt. Augustus J. S. McDonald	19	wounded-severe left arm/forearm
Pvt. Benjamin Moore	19	mortal-gunshot in left hip, ball retained/severe right shoulder –died Jan. 4, 1864 at Beverly
Corp. George Osburn	23	wounded-severe gunshot thru right arm

Co. D:

Pvt. Charles Bryson	26	killed-gunshot in head
Pvt. Franklin Fisher	21	wounded-flesh-gunshot right thigh, middle third
Pvt. Ezra M. Ours	20	wounded-two flesh wounds-gunshot thru right arm above and below elbow
Pvt. John Queen [Quain]	19	wounded-gunshot thru left shoulder

Co. E:

Corp. Jacob K. Dodd	23	mortal-very severe gunshot left knee, joint retained–died Dec. 24
Pvt. John Forrester	36	mortal-very severe left breast thru lungs/right hand–died Nov. 12
Pvt. Marion Shriver	19	mortal-gunshot
Sgt. Mortimer W. Stalnaker [also listed in Co. D]	--	wounded-slight right hand/ gunshot thru little finger

Co. F:

Pvt. William M. Barnett	19	wounded-severe gunshot thru right leg and knee joint
Ord. 1st Sgt. John D. Baxter	25	mortal-gunshot in bowels–died Nov. 7
2nd Lt. Henry Bender	23	wounded-shot in big toe
Pvt. John Blagg	19	wounded-serious gunshot of right ankle involving joint
Pvt. George C. Gillispie	20	wounded-flesh-gunshot thru left leg
Pvt. Silas M. Morrison	21	wounded-gunshot thru both arms/ upper third both forearms

Pvt. Jacob Riffle	25	wounded-slight gunshot thru left arm/bone (humerous) shattered/lost left arm
Pvt. John Rollyson	27	wounded-slight gunshot thru middle finger, right hand
Pvt. Milton Rollyson	20	wounded-severe gunshot thru left forearm
Pvt. Newlon Squires	23	wounded-slight gunshot top of right shoulder
Pvt. Edward B. Wheeler	27	wounded-severe gunshot thru left shoulder/left chest
Pvt. Addison Wilson	27	wounded-gunshot thru middle of ring and little fingers right hand-first two fingers amputated
Co. G:		
Corp. Michael E. Jeffries	19	wounded-severe gunshot left thigh
Pvt. Coleman Wyant*	--	wounded-gunshot abdomen-flesh-not dangerous-wounded in back
Co. H:		
Corp. Coleman Channel	19	mortal-gunshot
Pvt. Wesley Pullens	19	mortal-gunshot
Pvt. David [Daniel] W. Saunders	30	mortal-gunshot
Pvt. W. Weese [Nimrod]	21	wounded-very severe gunshot in right side, perforating bowels
Co. I:		
On detached duty at Petersburg, West Virginia throughout campaign		
Co. K:		
Pvt. John Calhoun	--	wounded-slight right shoulder
Pvt. John M. Randall	19	wounded-gunshot in left thigh, lower third flesh wound

Regimental adjutant reported from Beverly on Nov. 13: "There were a few slightly wounded who are not reported, as they are already doing duty."

28TH OHIO VOLUNTEER INFANTRY

Co. A:		
Pvt. Otto Briegel	24	wounded-left shoulder
Pvt. Edward Frieler	26	wounded-left thigh
Pvt. Frederick Frieler	26	wounded
Pvt. Henry Schadelman	31	mortal–died Nov. 8 at Hillsboro
Co. B:		
No casualties reported.		
Co. C:		
Pvt. Joseph Autosveiler	--	wounded-right femur fracturing bone

Pvt. Charles Dagan	44	wounded-left thigh
Pvt. Charles Dalhammer	24	mortal-both legs–died Nov. 9 in hospital at Hillsboro
Pvt. Fritzolin Gutzwiller	43	mortal–died Dec. 12 in hospital at Hillsboro
Co. D:		
Corp. Jacob [John?] Deiss	--	wounded-right arm
Pvt. James Heitz	24	killed
Pvt. Christ Meir [Mayer]	33	wounded-glance (?) shot on right foot
Pvt. Frederick Muller	32	wounded-left leg
Pvt. John Weinstein	21	wounded-slight
Co. E:		
1st Sgt. Jacob Frintz	21	killed
Pvt. Hermann Iaeger	--	wounded-right thigh
Pvt. Frederick Schafer	28	mortal–died Nov. 7 at Hillsboro
Pvt. Joseph Traufmann [Haufman?]	--	wounded-slight right hand
Co. F:		
Pvt. Henry Bettsheider	34	killed
Pvt. Sigmund Eicholz	27	wounded-right hand
Corp. Herman Frondhoff	40	wounded-slight right shoulder
Corp. Bernhardt Vogel	39	wounded-mouth and front left side
Pvt. Charles Werner	27	wounded-left shoulder
Pvt. David Wickerschimer	--	mortal–died Nov. 7 at Hillsboro
Pvt. Charles [George] Winges	33	wounded-left shoulder
Pvt. Matthia[s] Zimmerman	38	wounded
Co. G:		
Pvt. Jacob Bohosen [Bohmen]	31	wounded-right arm flesh
Pvt. Frederick Krebs [Kraub]	45	wounded-left shoulder
Pvt. Phillip Wechsler [Wegler]	19	wounded-right leg
Co. H:		
Pvt. Andrew Braumlein [Braeunling]	38	wounded-abdomen flesh
Co. I:		
Corp. Joseph Spinser [Spinner]	23	wounded-left shoulder
Co. K:		
Sgt. John Goettler	--	wounded-right foot flesh

1st West Virginia Light Artillery, Battery B (Keepers' Battery)

Sgt. John Coates	34	wounded-slight right leg/hip
2nd Lt. Joseph W. Daniels	45	killed instantly by shell
Sgt. Richard M. Depew	27	wounded-slight-right arm and right leg
Pvt. John Goff	--	wounded-slight-right arm
Pvt. John Hope	36	wounded-by shell-serious-right leg amputated near knee–discharged

Pvt. James Jackson	22	killed-by shell—died in hospital Nov. 6
Capt. John V. Keepers	42	partial loss of hearing in both ears caused by concussion of the artillery of his own battery—not treated
Pvt. John Kyle Vermillion	20	wounded-left knee [another source says right knee-slight]

3rd Independent Company Ohio Volunteer Cavalry

Pvt. Frederick Donnerline	30	captured
Sgt. Henry Foplie	33	wounded
Capt. Frank Smith	43	wounded-shoulder

1st West Virginia Cavalry

Co. G:

Pvt. Henry Rumble	23	missing Nov. 8—also listed as killed by a Lewisburg citizen

The 14th Pennsylvania Cavalry, the 3rd West Virginia Cavalry, as well as Co. C., 16th Illinois Cavalry, and 1st West Virginia Light Artillery, Battery G (Ewing's) reported no casualties.

Gen. Alfred N. Duffié reported his losses as: two enlisted men of the 2nd West Virginia Cavalry captured in the attack on the Confederate rear guard, and one member of the 34th Ohio Mounted Infantry wounded.

Based upon these figures the Federal army at Droop Mountain lost 45 killed, 93 wounded, and two captured, for a total of 140. There were 15 men mortally wounded, which helps to make up the total of 45 dead. Broken down by regiments these totals would read:

	Killed	Wounded	Captured
2nd West Virginia Mounted Infantry	13	16	0
3rd West Virginia Mounted Infantry	7	14	1
8th West Virginia Mounted Infantry	3	8	0
10th West Virginia Infantry	12	25	0
28th Ohio Volunteer Infantry	8	23	0
1st W. Va. Light Artillery, Batt. B	2	5	0
3rd Independent Co. Ohio Cavalry	0	2	1
Totals:	45	93	2

These figures do not agree with those presented in the *Official Records* by Gen. William Averell. His statement of losses at Droop Mountain reads:

	Killed	Wounded	Captured
2nd West Virginia Mounted Infantry	9	14	0
3rd West Virginia Mounted Infantry	6	5	0
8th West Virginia Mounted Infantry	3	8	0
10th West Virginia Infantry	7	29	0
28th Ohio Volunteer Infantry	3	25	0
1st West Virginia Light Artillery, Battery B	2	5	0
3rd Independent Co. Ohio Cavalry	0	2	1
Totals:	30	88	1

A number of reasons contribute to the discrepancy in numbers. Many soldiers had slight wounds not worthy of reporting to surgeons and therefore have been omitted from both lists. Also, General Averell did not differentiate between the wounded and mortally wounded, possibly because some of the mortally wounded were yet alive at the time he filed his report. Additionally, a few names may be duplicates with variations in spelling.

CONFEDERATE

Confederate casualties at Droop Mountain remain difficult to ascertain due to the extreme confusion which prevailed in the southern ranks in the immediate aftermath of the battle. Many reports were incomplete, and many soldiers, originally listed as missing or having deserted, found their way back to their respective commands soon afterward, having been lost from their commands during the retreat. Casualty records for at least one regiment, the 19th Virginia Cavalry, could not be located prior to publication.

14th Virginia Cavalry

Co. A (2nd):
Pvt. John Aquilla Myles 21 killed
Co. B:
Sgt. Maj. Robert Henry Gaines 23 wounded
Pvt. Thomas C. Harvey -- wounded
Co. C:
2nd Lt. Granville J. Regar -- wounded
Pvt. John Sheridan 16 captured
Co. D:
Pvt. Moses Christopher Beard 39 wounded [possible]
Co. E:
No report

Co. F (2nd):

Capt. William Thomas Smith	23	wounded-also had 3 horses shot from under him

Co. G:

Pvt. Walter C. Barnett	35	captured–Nov. 7 at Lewisburg by Duffié

Co. H:

No report

Co. I:

No report

Co. K:

Pvt. George W. Lewis	27	killed in retreat [22nd Va.?]
Capt. Alpheus Paris McClung	23	wounded

Although this list gives two killed, six wounded, and two captured for the 14TH VIRGINIA CAVALRY at Droop Mountain, other sources have given figures ranging from three killed, to eight wounded, to 14 missing or captured, for a total loss of 25.

19TH VIRGINIA CAVALRY

FIELD AND STAFF:

Acting Adjutant ---- Cranford	--	wounded

Co. A:

Sr. 2nd Lt. James W. Morgan	--	killed

Co. B:

Pvt. Elijah Heater	24	wounded/shoulder and captured
Pvt. John Pierce	21	deserted–Nov. 15
Pvt. Matheson [Madison] Viers	20	deserted–Nov. 4 or 5

Co. C:

Pvt. Henry H. Brookhart	18	wounded-shoulder

Co. D:

No report

Co. E:

Pvt. Jacob S. Hall letter, Dec. 2, 1863, states: "one man wounded in Co. E."

Co. F:

Pvt. William Henry Harrison Galford	21	wounded
Pvt. John Adam McNeil	18	wounded/leg broken/ankle crushed Nov. 3 at Green Bank
Pvt. James H. Morrison [Marison]	20	killed
Pvt. Wilson L. Pugh	17	captured–Nov. 3 at Green Bank by Averell
Substitute for Pvt. Jacob Slaven Woodell	--	captured–Nov. 1, 1863

Co. G:

Pvt. Francis Marion Vicars [Vickers]	21	wounded-gunshot right arm/arm amputated at shoulder joint

Co. H:		
Pvt. Owen V. B. Davis	--	killed [postwar report]
Co. I:		
Pvt. Garland C. Black	33	captured—Nov. 4 or 5 near Marlinton
Pvt. John R. Black	23	deserted—Nov. 4/captured—Nov. 6 near Marlinton on CS records-or Nov. 6 by Averell
Lt. George V. Gay	26	captured—Nov. 8 in Pocahontas Co. by Capt. Rowan
Pvt. William E. Johnson	30	deserted—Nov. ?, 1863
Pvt. James M. Long	27	captured
Pvt. James Marison [Morrison]	--	killed
Pvt. Henry H. Wade	22	captured
Pvt. Thomas Emerson Wood	57	wounded-through body

Casualty figures for the 19TH VIRGINIA CAVALRY are very incomplete and no reliable statements could be located. Even dates of capture given by Confederate and Federal authorities differ. Col. William L. Jackson said he "feared 150 killed and wounded" but did not specify if that figure applied to the 19TH VIRGINIA CAVALRY alone or his entire cavalry command. This list gives only three killed, seven wounded, eight captured, and three deserted, for a total of 21. Considering the 19TH VIRGINIA CAVALRY was engaged in the heaviest fighting on the Confederate left flank this figure is obviously much too low.

20TH VIRGINIA CAVALRY

Co. A:		
Pvt. Philip B. Amos	21	captured
2nd Lt. David B. Burns	20	wounded-hand
Lt. Ulysses Morgan	--	killed [postwar report]
Corp. [or Sgt.] William Lebbuss Straight	23	wounded
Pvt. Aaron Youst	38	deserted—Nov. 4
Co. B:		
Capt. Ezekiel Martin	--	wounded-twice
Pvt. Thomas B. McIntire	--	killed
Pvt. John B. Price	--	killed
Pvt. William G. Straight	--	killed
Co. C:		
Pvt. Andrew S. Gentry	25	captured—Nov. 4 on CS records —Nov. 3 or 6 by Averell
Pvt. Henry Hupman	24	captured—Nov. 10 in Bath Co. by Averell
Pvt. Robert S. McRay	32	captured—Nov. 10 in Bath Co. by Averell
Pvt. D. Blackman Rummell	21	captured

Co. D:

No report

Co. E:

Pvt. John Y. Bassell	17	wounded-severe [also listed in Co. A]
Pvt. Robert C. Ferrell	30	captured–Nov. 4 by Maj. Gibson
Capt. John W. Young	37	wounded-mortal–died Nov. 7 at Lewisburg

Co. F:

No report

Co. G:

Pvt. Franklin Barker	20	deserted Nov. 6–captured Nov. 23 in Pleasants Co. by Home Guard
Pvt. George Barker	18	deserted/captured Nov. 5 or 6 on CS records–Nov. 17 and 23 by Home Guard
Pvt. Wade Hampson Benson	22	captured [deserted?] Nov. 6 in Pocahontas Co.
Pvt. Wade Hampson	22	captured [probably Wade Hampson Benson]
Pvt. Alexander A. Low(e)	21	captured–Nov. 4 on CS records or Nov. 6 by Averell
Pvt. G. A. Roberts	--	deserted–Nov. 6

Co. H:

No report

Co. I:

Pvt. Adam Hinkle	24	wounded [postwar record]
Corp. John Jackson	30	wounded
Pvt. Michael E. McGoldrick [McGoldrich]	--	wounded (3 times) and captured
Sgt. William E. Pae [Poe]	27	captured–Nov. 7 at Lewisburg by Duffié
Pvt. Richard Robinson	21	captured–Nov. 7 at Lewisburg by Duffié

Co. K:

Pvt. Robert Ailstock	18	horse killed in Droop Mt. battle
Pvt. Angus Donald	26	captured–Nov. 7 by Averell
Pvt. Alexander Duncan	23	captured–Nov. 7 at Lewisburg by Duffié
Pvt. James W. Fulwilder	18	killed
Pvt. William E. Gilmore	20	captured–Nov. 6 on CS records or Nov. 7 at Lewisburg by Duffié
1st Lt. N. B. Holland	--	killed
Pvt. Cyrus Morrison Lamb	26	captured–Nov. 6 on CS records or Nov. 9 in Greenbrier Co. by Averell

2nd Lt. Augustus C. Liggott	--	wounded-shot thru head between 4th & 5th pairs of nerves—also captured
Pvt. Angus McNeil	30	captured
Pvt. Alfred Abram Plott	19	captured—Nov. 6 on CS records or Nov. 10 in Greenbrier Co. by Averell
Pvt. David Robinson	35	killed
Pvt. Michael Welsh	24	captured—Nov. 6 on CS records or Nov. 7 at Lewisburg by Duffié

According to this list, it would appear the 20TH VIRGINIA CAVALRY lost eight killed, six wounded (two of whom were captured), 18 captured, and four deserted in the Droop Mountain battle and campaign, for a total loss of 38.

22ND VIRGINIA INFANTRY

FIELD AND STAFF:

Maj. Robert Augustus Bailey	--	mortal and captured—died Nov. 11 at Beard's at Hillsboro

Co. A:

--- Peter Bailey	--	wounded-right elbow
Sgt. William Henry Bailey	--	wounded-leg
Pvt. J. M. Buird*	--	captured
Pvt. Stephen N. Burford**	27	wounded-leg
--- Peter Cartmill*	--	wounded-foot
Pvt. John Walker Childers**	20	wounded
Pvt. Paul Dickerson	28	wounded-gunshot left side and pelvis
Pvt. William S. Dudding**	20	wounded
Pvt. John P. Hamilton**	21	wounded
2nd Corp. William H. Hubbard	--	killed
Pvt. John M. Lewis	23	captured
Pvt. Thornton F. Lovejoy**	37	wounded
2nd Lt. William S. McClanahan	29	mortal—died Nov. 6 or 7
Pvt. Samuel J. McCormac**	23	wounded
Corp. William S. S. Morris	--	killed
Pvt. John J. Short	--	killed
Pvt. Asa W. Smith**	29	wounded
Capt. John Koontz Thompson	20/21	wounded-shot thru neck
Pvt. Thomas B. Turley**	28	wounded

Co. B:

Pvt. John Bennett	18	wounded and captured-minie ball entered front of thigh, 4 inches above knee joint, exit behind one inch higher, passing to inside of bone

Pvt. Michael Brackman**	36	wounded
Pvt. George Hutcheson	21	captured
3rd Lt. Abram Park**	--	wounded
Corp. Samuel A. Reed**	--	wounded
Pvt. Thomas Smith	--	mortal—died Nov. 6 at Lewisburg
Co. C:		
Pvt. James Cavendish	19	MIA/captured—Nov. 8 in Greenbrier Co. by Averell
Pvt. George Criner	--	killed
Pvt. William D. Cunningham	--	wounded
Pvt. John Lawson Duncan	33/35	captured
Pvt. Obediah Hendrick	44	MIA/captured
Pvt. Tolbert P. Hendrick	33	missing
Pvt. George McCutcheon	36	captured—either Nov. 10 in Fayette Co. or Nov. 16 in Kanawha Co. by Duffié
Pvt. Henry H. Skaggs	24	wounded-severe leg—lost leg
Pvt. Joseph B. Skaggs	28	captured
1st Lt. Woodson A. Tyree	26	wounded-severe nape of neck and shoulder
Lt. H. T. Wilkinson**	25	wounded-amputation upper third femur
Musician George L. Wood	20	captured
Co. D:		
Pvt. Joseph T. Hatcher	19	MIA/captured—Nov. 8 in Greenbrier Co. by Averell
Pvt. Jacob L. Hypes	28	captured
Capt. James Monroe McNeil	40	captured
Pvt. George W. Plymale	21	captured
Pvt. Fleming A. Sarver	30	captured
Co. E:		
Pvt. Sampson Boggs	--	captured
Pvt. Hiram A. Booth**	19	wounded
Corp. John T. Booth**	23	wounded
Pvt. Morris Chapman Chandler	--	killed
Capt. George Steptoe Chilton	30	wounded-shot thru right jaw
Pvt. Richard Greenway	--	killed
Pvt. Harvey Henderson	20	captured
Pvt. Moses Hunter	--	killed
Pvt. Shadrach A. Jackson	--	killed
Pvt. Turpin Jones	--	killed
Pvt. William E. Lollis	34	captured
Pvt. Alexander C. Looney	32	wounded-severe [also listed in Co. K]
Pvt. William Rose	--	killed
Pvt. Jessee Shamblin	20	wounded-throat-remained with sick at Lewisburg—captured and paroled by Duffié

Co. F:		
Pvt. Jonathan M. Brown	25	captured
Pvt. Creed Cooper	32	captured
Pvt. George W. Cooper	19	captured
Pvt. Joel Cooper	32	captured
Pvt. John O. Dacon	23	captured
Corp. William A. Gwinn	25	captured
Pvt. Lewis Henderson Lemons**	23	wounded
Pvt. L. D. Meredith**	35	wounded
Pvt. James W. Rusk	25	captured
Pvt. John H. Rusk	27	captured
Co. G:		
Pvt. Albert Allen**	22/23	wounded
Sgt. James O. Bailey**	--	wounded
Sgt. Thomas R. Cooper**	22	wounded
2nd Lt. Floyd S. Gore	24	wounded-arm and leg broken
Pvt. George W. Jackson	23	captured
Pvt. Andrew B. Kincade	31	captured
Pvt. Joseph D. Lester	26	captured
Pvt. Anderson Meadows	37	wounded
Pvt. Harvey G. Morgan**	--	wounded
Co. H:		
Pvt. W. G. Bragg	--	captured
Corp. Joseph Alline Brown	22	wounded-shot in body near heart
Lt. John P. Donaldson	--	wounded-side and shoulder
Pvt. Clarence L. Jackson	17	wounded-severe-hand shattered by minie ball
Capt. Richard S. Q. Laidley	26	wounded
Pvt. Theodosius V. Loving	21	wounded
Pvt. Pete E. Stribling	--	mortal wound and captured-ball entered left side and exited right of spine—died June 17, 1864 at Pearisburg, VA hospital from disease caused by wound
Pvt. Archibald P. Young	21	wounded-scalp
Co. I:		
Pvt. James Henry Allen	--	wounded-shoulder by sabre
Pvt. William Perry Allen	18	killed
Pvt. George Brown	21	captured
Pvt. Andrew Jackson Estep	31	captured
Pvt. Francis Marion Jones	45	captured
Sgt. John Mays	37	captured
Pvt. Ballard J. Smith	32	captured
Pvt. Cary L. Stine	29	captured
Pvt. Carey Toney	20	wounded
Capt. John P. Toney	33	captured
Co. K:		
1st Lt. William F. Bahlmann	27	wounded and captured—minie ball in left knee

Pvt. David M. Beard	22	captured
Sgt. Major Claudius Blake	21	captured
Sgt. George W. Brafford	31	captured
Pvt. Hamilton B. Caldwell	26	killed
Pvt. John Farrall	40	captured
Pvt. Mason V. Helms	28	mortal and captured-shot thru head and body and finger shot off—died Nov. 7
Pvt. Jacob S. Hylton	30	captured
4th Corp. Henry Johnson	25	captured
Pvt. Henry Lewis	19	captured—deserted in retreat Nov. 6
Pvt. James McMillian	26	captured
Pvt. James A. Reynolds	29	deserted
Lt. Martin L. Smiley	--	wounded
Pvt. Martin W. Smith	29	missing and captured
Pvt. Jonathan N. Wilson	20	captured

For the 22ND VIRGINIA INFANTRY a double asterisk (**) after a name indicates the soldier appears as wounded on the Dec. 30, 1863 regimental roll but not on the battle list. Although it is possible some of these are casualties of the White Sulphur Springs battle fought Aug. 26-27, 1863, it is unlikely, as records were well kept following that Confederate victory. Also, discrepancy in the figures supplied by General Echols and those found in the Confederate Compiled Service Records is attributed to the large number of missing soldiers, feared captured, who later found their way back to their regiment.

23RD BATTALION VIRGINIA INFANTRY

Co. A:

Pvt. John Randle	21	captured
Pvt. William F. Shinaut	23	captured-Nov. 7 in Pocahontas Co. by Averell

Co. B:

Pvt. James A. Brammer	22	killed
Pvt. Jeremiah Campbell	--	wounded
Pvt. John W. Dehart	28	captured
Pvt. William H. [or C.] Farmer	21	MIA/captured—Nov. 9 in Pocahontas Co. by Averell
Pvt. John W. Lilly, Sr.	35	captured
Pvt. Charles W. Lucas	--	captured
Pvt. James A. Martin	49	captured
Pvt. John H. McClanahan	21	MIA/captured—Nov. 7 in Pocahontas Co. by Averell
Pvt. William F. [or T.] Phillips	21	captured
Pvt. George Simpkins	43	captured
Pvt. W. A. Simpkins	--	killed

Co. C:		
Pvt. Nathaniel McNeely	29	captured
Pvt. Freeling Sifers	--	deserted—Nov. 7 at Lewisburg
Co. D:		
Pvt. Alvin T. Davenport	27	captured-Nov. 10 at White Sulphur Springs by Averell
Pvt. Henderson Dillion	--	wounded
Pvt. Jacob Kirtner	38	captured
Pvt. Jordan Pack	19	killed
2nd Lt. William M. Witten	--	wounded
Co. E:		
Sgt. William C. Buchanan	--	captured
Pvt. Joseph C. Crabtree	--	wounded [possible]
Pvt. Joseph C. Delung [Delong]	28	killed
Pvt. Alexander Moore	35	captured
Pvt. Ezekiel L. Osborn	52	wounded
Pvt. Thomas O. [or A.] Scipher	32	captured
Pvt. William A. Smith	26	captured—Nov. 7 in Pocahontas Co. by Averell
Pvt. Robert Swiney	--	wounded
Co. F:		
Pvt. Charles S. Copenhaver	25	captured
Pvt. John J. Lampee	18	killed
Pvt. William Lampee	25	captured
Pvt. John C. Parker	38	captured
Pvt. William F. Scott	19	wounded-Nov. 15
3rd Corp. Samuel C. Smith	--	wounded-Nov. 15
Co. G:		
Pvt. John Blackwell	37	captured
Pvt. James M. Boyers	17	captured
Pvt. Samuel Cunningham	--	wounded [possible]
5th Sgt. Samuel Williams	30	wounded
Co. H:		
Pvt. Richard Anderson	18	captured
Pvt. Bethel Coffer	26	captured
Pvt. Wesley Coffer	23	captured
Pvt. William Coffer	23	captured
Pvt. James B. Dunman	--	wounded
Pvt. Nathaniel Davis Hill	38	captured
Pvt. William F. Shelton	--	wounded
Corp. William B. Southern	--	wounded

The official report of the 23RD BATTALION VIRGINIA INFANTRY lists a total of 61 men killed, wounded, and missing at Droop Mountain. Since the most inaccurate statistic is that of the wounded, it can be assumed the 15 men unaccounted for in the total figure were wounded.

26th Battalion Virginia Infantry—not engaged

Co. D:		
Pvt. William L. Brown	19	captured—Nov. 8 at Lewisburg by Duffié
Co. G:		
Pvt. John B. S. Honaker	26	captured—Nov. 6 in Greenbrier Co. by Averell

Captain Warren S. Lurty's Company Virginia Horse Artillery, Captain George B. Chapman's Company Virginia Light Artillery and Captain Thomas E. Jackson's Battery Virginia Horse Artillery reported no casualties in the Droop Mountain battle.

Confederate casualties at Droop Mountain are impossible to accurately calculate due to the large number of missing records and information. Combined with the figures presented by Gen. Echols, Col. Jackson, and the various field commanders, as well as information from service records, the casualty list reads:

	Killed	Wounded	Deserted/ Captured/ MIA
22nd Virginia Infantry	14	53	43
23rd Battalion Virginia Infantry	5	26	29
26th Battalion Virginia Infantry*	0	0	2
14th Virginia Cavalry	3	8	14
19th Virginia Cavalry**	3	7	11
20th Virginia Cavalry	8	6	23
Totals:	33	100	122

*—Not Engaged
**—Records for the 19th Virginia Cavalry not found

Based upon combined information, it would appear the Confederate army at Droop Mountain lost 33 killed, 100 wounded, and 122 captured, deserted or missing in action (MIA), for a total recorded loss of 255. General Echols reported his total loss at about 275, therefore the missing 44 men were probably wounded, as killed and captured records were well kept. Most of the missing figures probably represent the 19th Virginia Cavalry. The final figure for the southern army at Droop Mountain probably was in the vicinity of 33 killed, 121 wounded, and 122 captured.

Combined Federal and Confederate casualties at Droop Mountain would read in the vicinity of:

78 killed, 215 wounded, and 124 captured, deserted or missing, for a total of 417.

Additional Known Confederate Losses in the Droop Mountain Campaign

8th Virginia Cavalry

Co. B:		
Pvt. Seaborn L. Wilson	22	captured—Nov. 7 in Lewisburg by Duffié

18th Virginia Cavalry

Co. F:		
Pvt. Charles H. Lawyer	--	captured—Nov. 9 in Franklin Co. by Averell
Co. G:		
Pvt. Thomas C. Ervin	--	captured—Nov. 10 in Allegheny Co. by Averell
Pvt. Jacob A. Evans	--	captured—Nov. 7 at Lewisburg by Duffié

Casualties In The Attack on the U. S. Supply Train near Burlington – November 16, 1863 –

Confederate

62nd Virginia Mounted Infantry

Co. I:		
Pvt. James F. Keister	--	wounded—mortal—died Jan. 1, 1864
Four other members of the regiment slightly wounded		
McNeill's Rangers:		
2nd Sgt. Joseph L. Vandiver	--	wounded-shot in thigh

Federal

14th West Virginia Infantry

Co. B:		
Pvt. John A. Freeland	18	wounded—mortal—died Nov. 17, 1863
Pvt. William Gardner	20	killed
Pvt. James Graham	35	wounded-pistol ball in right side of back/flesh discharged June 30, 1865 for wound
Co. C:		
1st Lt. George H. Hardman	22/27	killed
Co. D:		
Pvt. James R. Hudgins	20	captured

Co. I:		
2nd Sgt. Silas W. Hare	21/22	wounded
Pvt. James F. Porter	19	wounded-neck
Co. K:		
2nd Sgt. Amos Tyler Morris	22	captured

2nd Maryland Potomac Home Guard

Co. H:		
2nd Lt. David C. Edwards	--	wounded

Capt. Clinton Jeffers, Co. B, 14th West Virginia Infantry, reported a total loss of 19 including two killed, six wounded, and one missing in the 14th West Virginia Infantry, as well as six wounded and four missing in the 2nd Maryland Potomac Home Guard.

Appendix B

Droop Mountain Battlefield Park Superintendents [Caretakers]

Joe McMillion	[Dates unknown]
Robert Bean	1940-43
Mrs. Madelyn R. Bean	1944-46
Napoleon T. Holbrook	1946-49
Henry Lee Harper	1949-50
Clyde Crowley	1950-52
E. Morris Harsh	1952-54
Allan Gordon Brown	1954-55
Ralph Forinash	1955-57
Gordon Scott	1957-59
William Davis	1959-68
Floyd Clutter	1968-84
Michael Smith	1984-

Appendix C

Droop Mountain

Hills blue and silent
  Behind this old battleground;
Hills that once rang with cries of dying men,
  And with the gun's resound.

Once on this cool mountain slope,
  Where grasses green and trees now wave,
Brothers were enemies, friends were foes,
  Who now sleep here in one great, silent grave.

Dusk—failing o'er the battle field,
  Shadows lengthening o'er the hilltops, night—
Sleep on, O gallant men, both blue and gray,
  You gave your all for what you thought was right.

—*Louise McNeill* (17 years of age).
From *Confederate Veteran* November 1928

Nicholas Blues

Written and composed by Capt. James Monroe McNeil, Company D [Nicholas Blues], 22nd Virginia Infantry, Confederate States Army

1. But little do good people know
   What we poor soldiers undergo
   Whilst we are struggling to be free
   From abolition Tyranny

2. Whenever called we have to go
   Through mud and ice, Through rain and snow
   We have to march both nite and day
   On frozen ground we have to lay

3. Our native homes and friends so kind
   Are fare away and all left behind
   No friend to soothe our sorrows now
   Or wipe a suffering soldier's brow

4. Our rations too will compar
With all the other kinds of far
A stern decree has spoke and said
That we must live on beef and bread

5. The battlefield it has no charms
When silence breakes to a clash of arms
The cannons roar - the muskets peal
Proclaims a bloody battlefield

6. The battles rage with fearful roar
Our comrades [fall to?] rise no more
----- are we have to be
to pay the price of liberty

7. The conflict sore [sure] would soon be past
And we would all get home as last
had dixie's sons all been True
To fight against old lincoln's crew

8. Almagamation - They must try
And for the union still they cry
Our brothers now we have to scorn
And hate the day that they were born

9. Two years ago we volunteered
The soldier's fate have always shared
and now we pause to count the cost
And mourn the men that we have lost

10. In Summersville we plighted hands
to fight old Lincoln's thieving bands
All others honors we refuse
We still are called the Nicholas blues

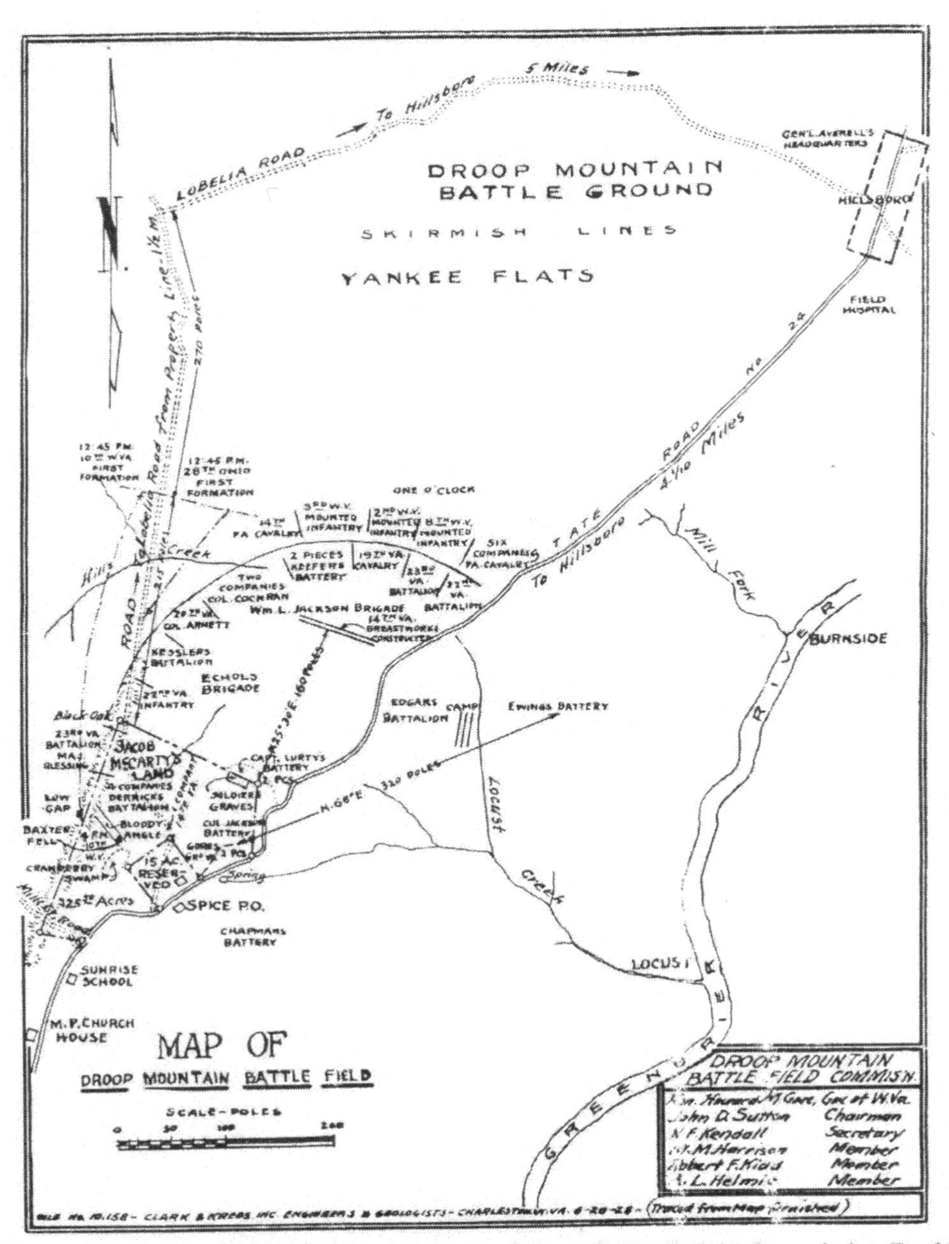

This map, drawn June 20, 1928, for the Droop Mountain Battlefield Commission Booklet, contains a number of flaws and errors. Colonel Jackson's Battery should be Captain Jackson's Battery; the position of Keepers' Battery and the position of Chapman's Battery is questionable; the map places Edgar's Battalion and camp directly between the Confederate and Federal artillery, which would have exposed them to fire from friend and foe (Edgar is not believed to have been anywhere near this position). The field hospital at Hillsboro probably refers to the Beard House, which is on the opposite side of the road. In the semi-circle alignment of troops, the 22nd Virginia Battalion should be the 22nd Virginia Infantry; two pieces of Keepers' Battery have been placed alongside Confederate troops and the name has been misspelled as "Keefer's;" although Major Kesler did command a group of men in the battle, his battalion was reportedly not officially organized or recognized at this time; the 14th Pennsylvania Cavalry is not believed to have had troops on the Confederate left; the 22nd Virginia Infantry on the rebel left should indicate only three companies of that regiment in that position; and in the area shown as Jacob McCarty's land the units designated 23rd Virginia Battalion (Major Blessing) and Derrick's Battalion (four companies) are one and the same organization. DROOP MOUNTAIN COMMISSION BOOKLET

Footnotes

Chapter One

1. Conversations between the author and Tim McKinney, Fayetteville, West Virginia. McKinney is a direct descendant of Joseph V. Rollins, Company A, 22nd Virginia Infantry.
2. Pocahontas County Historical Society, *History of Pocahontas County, West Virginia*, (Dallas, Texas: Taylor Publishing Co., 1981), p 15.
3. Calvin W. Price, *The Pocahontas Times*, Nov. 14, 1935.
4. Roy Bird Cook, "The Battle of Droop Mountain," *The West Virginia Review*, Oct. 1928, p. 14.
5. Calvin W. Price, *The Pocahontas Times*, Nov. 26, 1925.
6. Pocahontas County Historical Society, op. cit.
7. John Alexander Williams, *West Virginia–A Bicentennial History (The States and The Nation Series)*, W.W. Norton & Co., Inc., N.Y., N.Y., p. 56-57.
8. Dallas B. Shaffer, *The Battle at Droop Mountain*, (West Virginia Department of Natural Resources, 1966), p. 3.
9. Theodore F. Lang, *Loyal West Virginia from 1861 to 1865*, (Baltimore: Deutch Publishers, 1895), p. 106.
10. *Ibid*., p. 107-108.
11. *Ibid*., p. 109.
12. *Ibid.*
13. *Ibid.*
14. *Ibid.*
15. Patricia L. Faust (editor), *Historical Times Illustrated Encyclopedia of the Civil War*, (New York: Harper and Row Publishers, 1986), p. 410.

Chapter Two

1. United States War Department, *War of the Rebellion: A Compilation of the Official Records of the Union and Confederate Armies*, (Washington, D.C.: Government Printing Office, 1880-1901), (Hereafter cited as *Official Records*), Series 1, Vol. 29, p. 499-502 (Benjamin F. Kelley to G.W. Cullum, Feb. 18, 1864; G.W. Cullum to Benjamin F. Kelley, Oct. 26, 1863; Thayer Melvin to Gen. William W. Averell, Oct. 26, 1863; Benjamin F. Kelley to Gen. E.P. Scammon, Oct. 30, 1863)
2. *Ibid.*
3. *Ibid*., p. 520 (Thomas Gibson to L. Markbreit, Nov. 19, 1863).
4. *Ibid*., p. 521.
5. *Ibid*., p. 508 (Ernst A. Denicke to William J.L. Nicodemus, Nov. 18, 1863).
6. The bulk of biographical material on Gen. William W. Averell was culled from the following sources: Edward K. Eckert and Nicholas J. Amato (editors), *Ten Years in the Saddle: The Memoirs of William Woods Averell,* (San Rafael, California: Presidio Press, 1978); Ezra Warner, *Generals In Blue*, (Baton Rouge: Louisiana State University Press, 1964) p. 12-13; Patricia L. Faust (editor), *Historical Times Illustrated Encyclopedia of the Civil War*, (New York: Harper and Row Publishers, 1986); William W. Averell Papers 1836-1910, New York State Library Cultural Education Center, Albany, New York.
7. William W. Averell Papers 1836-1910, New York State Library Cultural Education Center, Albany, New York (John B. McIntosh to Averell, Nov. 6, 1863).
8. Diary of Thomas H.B. Lemley, 1st West Virginia Cavalry (copy), West Virginia State Archives, Charleston, West Virginia.
9. *Ibid.*
10. Frank S. Reader, *History of the Fifth West Virginia Cavalry, formerly the Second Virginia Infantry, and of Battery G, First West Virginia Light Artillery*, (New Brighton, Pennsylvania: Daily News, 1890), p. 197.
11. *Official Records*, Vol. 27, Pt. 2, p. 206.
12. *Ibid.*
13. *Official Records*, Vol. 29, Pt. 1, p. 38-39.
14. Log Book of Gen. William W. Averell, William W. Averell Papers 1836-1910.
15. Frank S. Reader, op. cit.
16. Theodore F. Lang, *Loyal West Virginia from 1861 to 1865*, (Baltimore: Deutch Publishers, 1895), p. 210.

17. *Ibid.*
18. *Ibid.*, p. 213.
19. Biographical information on Lt. Col. Alexander Scott was culled from the following sources: Frank S. Reader, *History of the Fifth West Virginia Cavalry, formerly the Second Virginia Infantry, and of Battery G, First West Virginia Light Artillery,* (New Brighton, Pennsylvania: Daily News, 1890), p. 33; Theodore F. Lang, *Loyal West Virginia from 1861 to 1865,* (Baltimore: Deutch Publishers, 1895) p. 207-213; Records of the 5th West Virginia Cavalry; National Archives, Washington, D.C.
20. Theodore F. Lang, op. cit., p. 217-222; George R. Latham Letters 1863 (account of Josiah Davis, Co. F, 6th West Virginia Cavalry), Civil War Manuscripts Collection, West Virginia State Archives, Charleston, West Virginia.
21. Samuel T. Wiley, *History of Monongalia County, West Virginia*, (Kingwood, West Virginia: Preston Publishing Co., 1883), p. 529-531; newspaper obituary for Francis W. Thompson.
22. Theodore F. Lang, op. cit., p. 226-227; Ronald R. Turner, *7th West Virginia Cavalry*, (Manassas, Virginia: 1989); "Lieutenant Colonel John J. Polsley–7th West Virginia Regiment 1861-1865," thesis by Eugene Wise Jones, June 1949, presented to the University of Akron, Akron, Ohio.
23. Biographical information on John H. Oley was culled from the following sources: Rick Baumgartner, *First Families of Huntington*, (Huntington Publishing, April 1977), p. 37-38; obituary for John H. Oley, *The Huntington Advertiser*, Mar. 17, 1888; Roger D. Hunt and Jack R. Brown, *Brevet Brigadier Generals in Blue*, (Olde Soldier Books, 1990), p. 454.
24. William D. Slease, *The 14th Pennsylvania Cavalry in the Civil War*, (Pittsburgh: Art Engraving and Printing Co., 1890); Samuel P. Bates, *History of Pennsylvania Volunteers 1861-1865*, (Harrisburg: D. Singerly, State Printer, 1870), p. 851-897; George S. Westnam, "The Fourteenth Cavalry," *The Pittsburgh Press*, Oct. 8, 1961, p. 4-5.
25. George S. Westnam, op. cit.
26. Samuel P. Bates, op. cit., p. 856; George H. Mowrer, *History of the Organization and Service During the War of the Rebellion of Co. A, 14th Pennsylvania Cavalry,* (N.D.), p.21.
27. Frank S. Reader, op. cit., p. 69; Military Pension Records of Thomas Gibson, (Jr.), National Archives, Washington, D.C.
28. (Lt.) James Abraham, *With the Army of West Virginia 1861-1864: Reminiscences of Lt. James Abraham, Pennsylvania Dragoons, Company A, 1st Regiment (West) Virginia Cavalry*, compiled by Evelyn A. Benson, (Lancaster, Pennsylvania, 1974); Theodore F. Lang, op. cit., p. 163.
29. (Lt.) James Abraham, op. cit., p. 8.
30. Theodore F. Lang, op. cit., p. 194-201; Dorothy Davis, *History of Harrison County, West Virginia* (edit. Elizabeth Sloan), Clarksburg, West Virginia, American Association of University Women, 1970, p. 200; Records of the 3rd West Virginia Cavalry, National Archives, Washington, D.C.
31. (Lt.) James Abraham, op. cit., p. 10; Jasper W. Reece, *Report of the Adjutant General for the State of Illinois, reports for 1861-1865,* (revised by Reece), Springfield Bros., State Printers [etc.], (1900-1902).
32. Military Service Records, Capt. Julius Jaehne, Company C, 16th Illinois Cavalry, National Archives, Washington, D.C.
33. Ohio Roster Commission, *Roster of Ohio Soldiers in the War of the Rebellion 1861-65,* (Cincinnati, Ohio: Valley Publishing - Mfg. Co., 1866), Vol. XI, p. 695-700.
34. Theodore F. Lang, op. cit., p. 275-277 and p. 324-326; H.E. Matheny, *Major General Thomas Maley Harris . . . a member of the Military Commission that tried the President Abraham Lincoln assassination conspirators . . . and Roster of the 10th West Virginia Volunteer Infantry Regiment 1861-1865,* (Parsons, West Virginia: McClain Printing Company, 1963).
35. Theodore F. Lang, op. cit.; H.E. Matheny, op. cit.; Minnie Randall Lowther, *History of Ritchie County*, (Wheeling, West Virginia: Wheeling News Litho. Co. 1911), p. 442-446; Ezra Warner, op. cit., p. 209-210; Patricia L. Faust, op. cit., p. 345.
36. Jacob D. Cox, *Military Reminiscences of the Civil War*, (New York: Charles Scribner's Sons, 1900), p. 110.
37. Papers of Augustus Moor and the 28th Ohio Volunteer Infantry, Ratterman Collection, University of Illinois at Urbana–Champaign, Urbana, Illinois; Jacob D. Cox, op. cit., p. 110.
38. Jacob D. Cox, op. cit., p. 110.
39. Theodore F. Lang, op. cit., p. 106.
40. Theodore F. Lang, op. cit., p. 311-312.
41. Military Pension Records, Capt. John V. Keepers, Battery B, 1st West Virginia Light Artillery, National Archives, Washington, D.C.; Records of Battery B, 1st West Virginia Light Artillery, National Archives, Washington, D.C. (Note: John V. Keepers is the correct spelling of this name,

not Keeper as most historians and writers have employed).
42. Theodore F. Lang, op. cit., p. 317.
43. *Ibid.*, p. 317-318.
44. Frank S. Reader, op. cit., p. 86-87; Military Service Records, Capt. Chatham T. Ewing, Battery G, 1st West Virginia Light Artillery, National Archives, Washington, D.C.
45. Military Pension Records, Capt. Chatham T. Ewing, Battery G, 1st West Virginia Light Artillery, National Archives, Washington, D.C. (Averell Affidavit).
46. George H. Mowrer, op. cit., p. 65.
47. Military Pension Records, Capt. Chatham T. Ewing, op. cit.
48. Military Pension Records, Capt. Catham T. Ewing (Adam Brown Affidavit).
49. *Ibid.*, (Affidavits of Averell and Brown).
50. Noyes Rand, "Reminiscences of the Battle of Dry Creek, August 26-27, 1863," *The Monroe Watchman*, July 8, 1909 (Also appeared in *The Greenbrier Independent,* August 8, 1963); J.G. Stevens, "The Battle of Dry Creek," *The Fayette Tribune*, Jan. 16, 1916.
51. George H. Mowrer, op. cit., p. 65.
52. Frederick Phisterer, *New York in the War of the Rebellion*, Vol. 3, p. 2673.
53. *Official Records*, Series 3, Vol. 4, p. 828.
54. *Official Records*, Vol. XLI, p. 510 (E.A. Denicke to Maj. William J.L. Nicodemus, Nov. 8, 1863).
55. Frederick Phisterer, op. cit., Vol. 3, p. 2676-2677, 2681; Vol. 5, p. 4262-4263.
56. "From the Life of Mark Crayon," West Virginia State Archives, Charleston, West Virginia.
57. Bvt. Maj. Gen. George W. Cullum, *Biographical Register for West Point of the Officers and Graduates of the U.S. Military Academy at West Point, N.Y.*, Vol. II (Boston and New York: Mifflin and Company, The Riverside Press, 1891), p. 866.
58. George H. Mowrer, op. cit., p. 65 (Although Mowrer was describing armament used by Averell's men in the Salem Raid of December 1863, undoubtedly the same equipment was used the prior month at Droop Mountain).
59. George Washington Ordner Diary (2nd West Virginia Mounted Infantry), West Virginia University, Morgantown, West Virginia.
60. *Official Records*, Series 1, Vol. 29, Pt. 1, p. 502 (B.F. Kelley to Brig. Gen Scammon, Oct. 30, 1863).
61. Ezra Warner, op. cit., p. 131-132.
62. Stephen Z. Starr, *The Union Cavalry in the Civil War–Vol. I–From Fort Sumter to Gettysburg 1861-1863*, (Baton Rouge and London: Louisiana State University Press, 1979), p. 96-97.
63. Ezra Warner, op. cit., p. 131-132.
64. "Letter from West Virginia: The Condition of the Department–The Operations of the Army of Occupation–The Rust of Inactivity–Call for a Reformation–The Necessity for the Circulation of the Blood," (Correspondent T.B.F.), *Cincinnati Daily Commercial*, Nov. 26, 1863.
65. Whitelaw Reid, *Ohio In The War*, (Cleveland, Ohio: 1911), p. 221-227; unfinished manuscript regimental history of the 34th Ohio Volunteer Infantry by Bob Cartwright, Dayton, Ohio.
66. "Letter from West Virginia," (T.B.F.), op. cit.
67. Theodore F. Lang, op. cit., p. 177-193.
68. J.J. Sutton, *History of the Second Regiment, West Virginia Cavalry Volunteers, During the War of the Rebellion*, (Portsmouth, Ohio: 1892), p. 50.
69. Theodore F. Lang, op. cit., p. 177-193; J.J. Sutton, op. cit., (summary).
70. Union Soldiers and Sailors Monument Association, *The Union Regiments of Kentucky,* (Louisville, KY: 1897), p. 675-677.
71. "Letter from West Virginia," (Correspondent Q.P.F.), *Cincinnati Daily Commercial*, Dec. 11, 1863.
72. J.E.D. Ward, *12th Ohio Volunteer Infantry*, (Ripley, Ohio: 1864); Whitelaw Reid, op. cit., p. 87-90.
73. Whitelaw Reid, op. cit., p. 87-90.
74. Elmore Ellis Ewing, *The Story of the Ninety-First* [91st O.V.I], (Portsmouth, Ohio: Republican Printing Co., 1868); Elmore Ellis Ewing, *Bugles and Bells; or Stories Told Again, including The Story of the Ninety-First Ohio Volunteer Infantry,* (Cincinnati, Ohio: Press of Curtis and Jennings, 1899).
75. John W. Weed Memoir (91st Ohio Volunteer Infantry), Civil War Miscellaneous Collection, U.S. Army Military History Institute, Carlisle Barracks, Pennsylvania.
76. Whitelaw Reid, op. cit., p. 504-510.

Chapter Three

1. Terry Lowry, *26th Battalion Virginia Infantry*, (Lynchburg, Virginia: H.E. Howard, Inc., 1991), p. 30.
2. Biographical information on Gen. John Echols was assembled primarily from the following sources: (1) Ezra Warner, *Generals In Gray*, (Baton Rouge: Louisiana State University Press, 1964), p. 80; Patricia L. Faust (editor), *Historical Times Illustrated Encyclopedia of the Civil War*, (New York: Harper and Row Publishers, 1986), p. 235-236; Clement A. Evans, general editor, *Confederate Military History* (Virginia volume), (Atlanta: Confederate Publishing Co., 1899), p. 591-593; Gen. John Echols File, V.M.I. Archives, Lexington, Virginia; "Gen. John Echols–Advocate and Defender of the Confederacy," *Confederate Veteran*, Jan. 1896.
3. William C. Davis, *The Battle of New Market*, (Doubleday & Company, Inc., Garden City, New York, 1975), p. 38.
4. *Ibid.*
5. Gen. John Echols, V.M.I. File, op. cit.
6. *Ibid.*
7. George Mathews Edgar Papers, Southern Historical Collection, University of North Carolina, Chapel Hill, North Carolina.
8. Gen. John Echols, V.M.I. File, op. cit.
9. Terry Lowry, *22nd Virginia Infantry*, (Lynchburg, Virginia: H.E. Howard, Inc., 1988 and 1991 revised edition).
10. Virginia Faulkner (compiler), *Dear Annie: A Collection of Letters, 1860-1886*, (Parsons, West Virginia, 1969), p. 28.
11. Terry Lowry, *22nd Virginia Infantry*, op. cit. (quote from dustjacket).
12. Terry Lowry, *22nd Virginia Infantry*, op. cit.
13. Biographical information on Col. George S. Patton was compiled from the various sources employed when I wrote my book on the 22nd Virginia Infantry. Refer to the bibliography of *22nd Virginia Infantry* for details of sources, too numerous to list herein.
14. The bulk of biographical information on Major Robert A. Bailey was extracted from my book on the 22nd Virginia Infantry.
15. John L. Scott, *23rd Battalion Virginia Infantry*, (Lynchburg, Virginia: H.E. Howard, Inc., 1991); Lee A. Wallace, *A Guide to Virginia Military Organizations 1861-65*, (Richmond: Virginia Civil War Commission, 1964) and revised 2nd edition (Lynchburg, Virginia: H.E. Howard, Inc., 1986), p. 106 (of revised edition).
16. United States War Department, *War of the Rebellion: A Compilation of the Official Records of the Union and Confederate Armies*, (Washington, D.C.: Government Printing Office, 1880-1901), Series I, Vol. 29, p. 534 (Col. Geroge S. Patton to Capt. R.H. Catlett, Nov. 19, 1863).
17. Micajah Woods Papers (Jackson's Battery), University of Virginia, Charlottesville, Virginia.
18. John L. Scott, op. cit., p. 49-50.
19. Terry Lowry, *26th Battalion Virginia Infantry*, (Lynchburg, Virginia: H.E. Howard, Inc., 1991); Lee A. Wallace, op. cit., p. 109-110.
20. *Ibid.*, and Micajah Woods Papers, op. cit.
21. George Mathews Edgar Papers, op. cit.; Terry Lowry, op. cit.
22. William C. Davis, op. cit., p. 172.
23. Lee A. Wallace, op. cit., p. 16; Albert S. Johnston, *Captain Beirne Chapman and Chapman's Battery*, (Union, West Virginia: 1905); John L. Scott, *Lowry's, Bryan's and Chapman's Batteries of Virginia Artillery*, (Lynchburg, Virginia: H.E. Howard, Inc., 1988).
24. *Ibid.*, (most direct quotes are from the Albert S. Johnston work on Chapman's Battery).
25. Ezra Warner, op. cit., p. 153-154; Patricia L. Faust (editor), op. cit., p. 392; Clement A. Evans, op. cit. (West Virginia volume), p. 131-133; H.E. Matheny, *Wood County, West Virginia in Civil War Times*, (Parkersburg, West Virginia: Joseph M. Sakach, Jr., Trans-Allegheny Books, Inc., 1987).
26. Jacob William Marshall Papers 1852-1899, West Virginia University, Morgantown, West Virginia.
27.Lee A. Wallace, op. cit., p. 60; Richard L. Armstrong, *19th and 20th Virginia Cavalry*, (Lynchburg, Virginia: H.E. Howard, Inc. 1994).
28. George W. Atkinson and Alvarado F. Gibbens, *Prominent Men of West Virginia*, (Wheeling, West Virginia: W.L. Catlin, 1890), p. 805; William P. Thompson Papers, National Archives, Washington, D.C.; John M. Ashcraft, *31st Virginia Infantry*, (Lynchburg, Virginia: H.E. Howard, Inc., 1988).

29. William P. Thompson Papers, op. cit. (Aug. 15, 1862 letter of J.T. Martin).
30. *Ibid.*, (includes Mar. 31, 1863 letter of Col. William L. Jackson and letter dated Apr. 9, 1863).
31. *Ibid.*, (letter dated Apr. 16, 1863 from Camp Harold, Headquarters, Huntersville Line).
32. Newspaper obituary of William P. Thompson; George W. Atkinson and Alvarado F. Gibbens, op. cit., p. 806-807.
33. William P. Thompson Papers, op. cit., (Mar. 31, 1863 letter of Col. William L. Jackson).
34. *Ibid.*, op. cit.
35. Lee A. Wallace, op. cit.., p. 61; Richard L. Armstrong, op. cit.; David Poe, *Personal Reminiscences of the Civil War by Captain David Poe,* (Buckhannon, West Virginia: Upshur Republican Print, 1911).
36. George W. Atkinson and Alvarado F. Gibbens, op. cit., p. 704; Richard L. Armstrong, op. cit.; "Death Comes Quietly to Col. Arnett," *The Wheeling Intelligencer*, Feb. 20, 1902.
37. Confederate Service Record of Lt. Col. Joseph R. Kessler, National Archives, Washington, D.C.; Lee A. Wallace, op. cit., p. 70; Richard L. Armstrong, op. cit.
38. Lee A. Wallace, op. cit., p. 26; correspondence by author with Robert J. Trout, Myerstown, Pennsylvania.
39. Author's correspondence with Robert J. Trout, Myerstown, Pennsylvania; "Captain Lurty Dead," *Harrisonburg Times*, 1906.
40. Jack L. Dickinson, *16th Virginia Cavalry*, (Lynchburg, Virginia: H.E. Howard, Inc., 1989); Jack L. Dickinson, *Records of the 16th Regiment Virginia Cavalry*, Confederate States Army, (Barboursville, West Virginia: Jack L. Dickinson, 1984).
41. *Ibid.*
42. *Ibid.*
43. Micajah Woods Papers, op. cit.
44. Robert J. Driver, Jr., *14th Virginia Cavalry*, (Lynchburg, Virginia: H.E. Howard, Inc., 1988); Lee A. Wallace, op. cit., p. 55; Micajah Woods Papers, op. cit.
45. Robert J. Driver, Jr., op. cit., p. 111.
46. *Ibid.*, op. cit., p. 119.
47. Lee A. Wallace, op. cit., p. 58; Jack L. Dickinson, op. cit. (both works on the 16th Virginia Cavalry).
48. Jack L. Dickinson, *16th Virginia Cavalry*, op. cit., p. 109; Terry Lowry, *22nd Virginia Infantry*, op. cit., p. 181.
49. Lee A. Wallace, op. cit., p. 23; Micajah Woods Papers, op. cit.
50. Confederate Service Records, Capt. Thomas E. Jackson (Jackson's Battery), National Archives, Washington, D.C.
51. Micajah Woods Papers, op. cit.
52. Micajah Woods Papers, op. cit.; author's correspondence with Gardner D. Beach, Frankfort, Kentucky.
53. Micajah Woods Papers, op. cit.
54. Confederate Service Records, Randolph H. Blain, National Archives, Washington, D.C.; Blain Family Papers 1860-69, Washington & Lee University, Lexington, Virginia.
55. Micajah Woods Papers, op. cit.
56. Jack Dickinson, *16th Virginia Cavalry,* op. cit., p. 33.
57. Robert J. Driver, Jr., *The 1st and 2nd Rockbridge Artillery*, (Lynchburg, Virginia: H.E. Howard, Inc., 1987), p. 72.

Chapter Four

The various troop movements of the Federal and Confederate armies during the Droop Mountain campaign are described in detail in the numerous reports by officers which appear in the *Official Records* (Series I, Vol. 29, Pt. 1), as well as assorted newspapers of the day. Due to space limitations, quotes not acknowledged are generally attributed to these sources.

1. Augustus Moor Papers (28th O.V.I), Ratterman Collection, University of Illinois at Urbana-Champaign, Urbana, Illinois (the Moor Papers contain detailed orders from Gen. Averell and others to Col. Moor for each day of the campaign, specifying marching orders and other military matters. These papers have been used extensively in determining time and order of march for the various commands in Averell's brigade).
2. *Ibid.*

3. *Ibid.*
4. "Lieutenant Colonel John J. Polsley–7th West Virginia Regiment 1861-1865," thesis by Eugene Wise Jones, June 1949, presented to the University of Akron, Akron, Ohio, p. 73; John J. Polsley Papers 1862-1865, University of Akron, Akron, Ohio.
5. Augustus Moor Papers, op. cit.
6. Edgar Blundon Papers (7th West Virginia Cavalry), Boyd Stutler Collection, West Virginia State Archives, Charleston, West Virginia.
7. "Letter from West Virginia: Gen. Averell's Movements–Battle of Droop Mountain," (Correspondent "Union"), *Cincinnati Daily Commercial*, Nov. 27, 1863.
8. *Ibid.*
9. Frank Moore, *The Rebellion Record*, (New York: D. Van Nostrand, Publisher, 1865), Vol. 8 ("A National Account," by Irwin), pg. 156-160.
10. *Ibid.*
11. *Cincinnati Commercial*, Nov. 27, 1863, op. cit.
12. Frank Moore, ("Irwin"), op. cit.
13. *Cincinnati Commercial*, Nov. 27, 1863, op. cit.
14. Frank Moore, ("Irwin"), op. cit.
15. *Ibid.*
16. Barnes Family Papers, 1816-1929, papers of Thomas Rufus Barnes (Co. K, 10th West Virginia Infantry), West Virginia University, Morgantown, West Virginia.
17. George Washington Ordner Diary, (2nd West Virginia Mounted Infantry), West Virginia University, Morgantown, West Virginia.
18. Frank Moore, ("Irwin"), op. cit.
19. James Ireland Diary (Co. A, 12th O.V.I.), Ohio Historical Society, Columbus, Ohio.
20. Augustus Moor Papers, op. cit.
21. G. Douglas McNeill, "Battle of Droop Mountain," *The Marlinton Journal*, June 12, 1958.
22. Frank Moore, ("Irwin"), op. cit.
23. Adeline E. Brown, "Incident of a Boy Confederate," *Confederate Veteran*, Aug. 1912.
24. Andrew Price, *West Virginia Bluebook*, Vol. 13, "Plain Tales of Mountain Trails," Chapter 11, "Battle of Droop Mountain. Tenth West Virginia saved the day," (1928), p. 412.
25. Frank Moore, ("Irwin"), op. cit.
26. Barnes Family Papers, op. cit.
27. George Washington Ordner Diary, op. cit.
28. Augustus Moor Papers, op. cit.
29. Augustus Moor Papers, op. cit.
30. James Ireland Diary, op. cit.
31. "The Late Fight at Droop Mountain–Complete Route of the Rebels," (Correspondent "T.M."), *Wheeling Daily Intelligencer*, Nov. 21, 1863.
32. Frank Moore, ("Irwin"), op. cit.
33. "Particulars of General Averell's Late Campaign," *Wheeling Daily Intelligencer*, (Correspondent "M"), Nov. 25, 1863.
34. Andrew Price, *West Virginia Bluebook* (1928), op. cit., p. 413.
35. "History of Pocahontas County Place Names," paper by Larry Jarvinen, Manister, MI (N.D.)–copy provided to author.
36. George R. Latham Papers–1863 (Josiah Davis, 3rd West Virginia Mounted Infantry–Co. F, 6th West Virginia Cavalry), West Virginia State Archives, Charleston, West Virginia.
37. G. Douglas McNeill, op. cit.
38. Barnes Family Papers, op. cit.
39. George R. Latham Papers (Josiah Davis), op. cit.
40. Area residents often refer to Stamping Creek as Stomping Creek and it shows on modern maps as Mill Creek..
41. Frank S. Reader, *History of the Fifth West Virginia Cavalry, formerly the Second West Virginia Infantry, and of Battery B, 1st West Virginia Light Artillery,* (New Brighton, Pennsylvania: Daily News, 1890), p. 219.
42. *Ibid.*
43. Andrew Price, *West Virginia Bluebook* (1928), op. cit., p. 413.
44. James Ireland Diary, op. cit.
45. Ezra Warner, *Generals In Gray,* (Baton Rouge: Louisiana State University Press, 1959), p. 147; Patricia L. Faust (editor), *Historical Times Illustrated Encyclopedia of the Civil War*, (New York:

Harper and Row Publishers, 1986), p. 378-379; Clement A. Evans, (general editor), *Confederate Military History,* (Atlanta: Confederate Publishing Co., 1899), (Virginia Volume), p. 608-611.
46. Pocahontas County Historical Society, *History of Pocahontas County, West Virginia,* (Dallas, Texas: Taylor Publishing Co., 1981), p. 76.
47. Barnes Family Papers, op. cit.
48. John D. Sutton, "One of Bloodiest Battles of the Civil War Fought at Top of Droop Mountain with 8,000 Troops Engaged in Conflict," *The Charleston Gazette*, Nov. 1, 1925.
49. *Ibid.*
50. Calvin W. Price, "Response to Sutton article on Droop Mountain," *The Pocahontas Times*, Nov. 26, 1925.
51. Augustus Moor Papers, op. cit.
52. "The Battle of Droop Mountain," as related by C.L. Stulting, Droop Mountain Battlefield State Park Files, Hillsboro, West Virginia (C.L. Stulting apparently confused details of the fight at Mill Point with the battle of Droop Mountain. This would not have been unusual for such an impressionable young man).
53. John M. Ashcraft, Jr., *31st Virginia Infantry*, (Lynchburg, Virginia: H.E. Howard, Inc., 1988), p. 126; Richard L. Armstrong, *19th and 20th Virginia Cavalry*, (Lynchburg, Virginia: H.E. Howard, Inc. 1994), p. 118.
54. The Francis S. Reader Papers, "An Average American: The Memoirs of Francis S. Reader," Volume II, completed Nov. 27, 1924, U. S. Army Military History Institute, Carlisle Barracks, Pennsylvania.
55. David Poe, *Personal Reminiscences of the Civil War by Captain David Poe*, (20th Virginia Cavalry), (Buckhannon, West Virginia: Upshur Republican Print, 1911).
56. George R. Latham Papers (Josiah Davis), op. cit.
57. The Francis S. Reader Papers, op. cit.
58. Blain Family Papers 1860-1869 (Jackson's Battery), Washington & Lee University, Lexington, Virginia, (Nov. 8, 1863 letter).
59. L.S. Cochran, "Droop Mountain," *The Pocahontas Times*, Dec. 26, 1935.
60. William F. Bahlmann, "Down In The Ranks," *The Journal of the Greenbrier Historical Society*, Vol. 2, No. 2, Oct. 1970, p. 82.
61. "The Droop Battlefield," *The Pocahontas Times*, Nov. 15, 1895.
62. L.S. Cochran, op. cit.
63. David Poe, op. cit.
64. "The Droop Battlefield," *The Pocahontas Times*, Nov. 22, 1895.
65. Andrew Price, *West Virginia Bluebook,* (1928), op. cit., p. 413.
66. C. L. Stulting Memoir, op. cit.
67. *Ibid.*
68. *Cincinnati Commercial*, Nov. 27, 1863, op. cit.
69. Willaim D. Slease, *The 14th Pennsylvania Cavalry In The Civil War*, (Pittsburgh: Art Engraving and Printing Co., 1890), p. 108 (author could not locate any present-day reference to Little Coal Creek).
70. Frank S. Reader, op. cit., p. 216.
71. George Washington Ordner Diary, op. cit.
72. John D. Sutton, op. cit.
73. James Ireland Diary, op. cit.
74. Micajah Woods Papers, (Jackson's Battery), University of Virginia, Charlottesville, Virginia.
75. George H. Mowrer, *History of the Organization and Service During the War of the Rebellion of Co. A., 14th Pennsylvania Cavalry,* (N.D.), p. 9.
76. George Washington Ordner Diary, op. cit.
77. John D. Sutton, op. cit.
78. Barnes Family Papers, op. cit.
79. "Particulars of General Averell's Late Campaign," *Wheeling Daily Intelligencer*, (Correspondent "M"), Nov. 25, 1863.
80. L.S. Cochran, op. cit.
81. Andrew Price, *West Virginia Bluebook,* Vol. 13, "Plain Tales of Mountain Trails," Chapter 11 ("Battle of Droop Mountain. Tenth West Virginia saved the day"), (1928), p. 407.
82. Frank Moore, ("Irwin"), op. cit.
83. M.A. Dunlap, "From M.A. Dunlap," *The Pocahontas Times*, Dec. 29, 1925.
84. M.A. Dunlap, "From M.A. Dunlap," *The Pocahontas Times,* July 11, 1929.

85. M.A. Dunlap, op. cit.
86. John D. Sutton (and State of West Virginia), *Report of Droop Mountain Battlefield Commission*, (Charleston, West Virginia: 1928), p. 18.
87. M.A. Dunlap, "Battle of Droop Mt.," *The Pocahontas Times*, Sept. 13, 1923.
88. Calvin W. Price, "Response to Sutton article on Droop Mountain," *The Pocahontas Times*, Nov. 26, 1925.
89. Andrew Price, *West Virginia Bluebook*, (1928), op. cit. p. 407.
90. *Ibid.*
91. Pearl Buck, *The Exile*, (John Day Book: Reynal & Hitchcock, New York, 1936), p. 56
92. C.L. Stulting memoir, op. cit.
93. Kermit McKeever, "The First 30 Years," *Outdoor West Virginia*, June 1967.

CHAPTER FIVE

The various troop movements of the Federal and Confederate armies during the Droop Mountain battle are described in detail in the numerous reports by officers which appear in the *Official Records* (Series I, Vol. 29, Pt. 1), as well as assorted newspapers of the day. Due to space limitations, quotes not acknowledged are generally attributed to these sources.

1. Samuel D. Edmonds Memoir (Company B, 22nd Virginia Infantry), Droop Mountain State Park Files, Hillsboro, West Virginia. Courtesy Evelyn McPherson; William F. Bahlmann, "Down In The Ranks," *The Journal of the Greenbrier Historical Society*, Vol. 2, No. 2, Oct. 1970, p. 83; Records of the 23rd Battalion Virginia Infantry, National Archives, Washington, D.C.
2. James H. Mays (editor: Lee Mays), *Four Years for Old Virginia*, (Los Angeles, California: The Swordsman Publishing Company, privately printed, 1970), p. 28; William F. Bahlmann, op. cit.
3. M.A. Dunlap, "Battle of Droop Mountain," *The Pocahontas Times*, Sept. 13, 1923; Micajah Woods Papers (Jackson's Battery), University of Virginia, Charlottesville, Virginia; L.S. Cochran, "Droop Mountain," *The Pocahontas Times*, Dec. 26, 1935; G. Douglas McNeill, "Battle of Droop Mountain," *The Marlinton Journal*, June 19, 1958; *Richmond Whig*, Nov. 14, 1863.
4. William F. Bahlmann, op. cit.; Samuel D. Edmonds Memoir, op. cit.
5. James S. McClung Memoir (Company K, 14th Virginia Cavalry), copy provided to author by John Arbogast, White Sulphur Springs, West Virginia.
6. George H.C. Alderson (Company A, 14th Virginia Cavalry), "From the *Greenbrier Independent*, letter concerning the battle of Droop Mountain and death of Major Bailey," *Nicholas Chronicle*, Feb. 17, 1927; E.E. Bouldin Papers (14th Virginia Cavalry), University of Virginia, Charlottesville, Virginia.
7. M.A. Dunlap, "From Mr. Dunlap," *The Pocahontas Times*, July 11, 1929.
8. Augustus Moor Papers (28th O.V.I.), Ratterman Collection, University of Illinois at Urbana-Champaign, Urbana, Illinois; Frank S. Reader, *History of the Fifth West Virginia Cavalry, formerly the Second Virginia Infantry, and of Battery G, First West Virginia Light Artillery*, (New Brighton, Pennsylvania: Daily News, 1890), p. 216; "Particulars of General Averell's Late Campaign," (Correspondent "M"), *Wheeling Daily Intelligencer*, Nov. 26, 1863.
9. M.A. Dunlap, *The Pocahontas Times*, Nov. 14, 1935; M.A. Dunlap, "From M.A. Dunlap," *The Pocahontas Times*, Dec. 29, 1925.
10. "Letter from West Virginia: Gen. Averell's Movements – Battle of Droop Mountain," (Correspondent "Union"), *Cincinnati Daily Commercial*, Nov. 27, 1863; *Wheeling Daily Intelligencer*, (Correspondent "M"), op. cit.
11. Calvin W. Price, *The Pocahontas Times*, Oct. 31, 1935; Letter in *West Virginia Hillbilly*, concerns Collins Family, author's collection [n.d.].
12. John D. Sutton (and State of West Virginia), *Report of Droop Mountain Battlefield Commission*, (Charleston, West Virginia, 1928), p. 21; Micajah Woods Papers, op. cit.
13. Blain Family Papers 1860-1869 (Jackson's Battery), Washington & Lee University, Lexington, Virginia; Micajah Woods Papers, op. cit.
14. William F. Bahlmann, "Battle of Droop," *The Pocahontas Times*, Nov. 29, 1923; Samuel D. Edmonds Memoir, op. cit.
15. William F. Bahlmann, "Down In The Ranks," op. cit.
16. Thomas Algernon Roberts, "Droop Mountain Battle," *The Pocahontas Times*, Sept. 27, 1928.
17. Droop Mountain Battlefield State Park Files, Hillsboro, West Virginia (notes on Capt. James M. McNeill).

18. James Z. McChesney Letters, H.E. Matheny Collection, West Virginia University, Morgantown, West Virginia (letter dated Nov. 24, 1863).
19. G. Douglas McNeill, op. cit.; "Brilliant Exploits in Western Virginia," (Correspondent "C"), *The Ironton Register,* Dec. 3, 1863; Micajah Woods Papers, op. cit.
20. Calvin W. Price, *The Pocahontas Times*, Nov. 26, 1925; G. Douglas McNeill, op. cit.; George H. Mowrer, *History of the Organization and Service During the War of the Rebellion of Co. A, 14th Pennsylvania Cavalry*, N.D. (author's collection), p. 9-10; John D. Sutton (and State of West Virginia), op. cit., p. 15.
21. M.A. Dunlap, "Battle of Droop Mountain," Sept. 13, 1923; Calvin W. Price, *The Pocahontas Times*, Nov. 26, 1925; M.A. Dunlap, "From M.A. Dunlap," *The Pocahontas Times,* Dec. 29, 1925.
22. William D. Slease, *The 14th Pennsylvania Cavalry In The Civil War*, (Pittsburgh: Art Engraving and Printing Co., 1890), p. 252.
23. John D. Sutton (and State of West Virginia), op. cit., p. 21; "From the 14th Pennsylvania Cavalry," *The Pittsburgh Gazette,* Nov. 21, 1863; Notes on George W. McCoy in the Droop Mountain State Park Files, Hillsboro, West Virginia; James S. McClung Memoir, op. cit.; Notes on Timothy McCune, Droop Mountain State Park Files, Hillsboro, West Virginia; Micajah Woods Papers, op. cit.; Blain Family Papers, op. cit.; James Z. McChesney Letters, op. cit.
24. M.A. Dunlap, "Battle of Droop Mt.," *The Pocahontas Times*, Sept. 13, 1923; James Z. McChesney Letters, op. cit.
25. L.S. Cochran, op. cit.; Larry Jarvinen, "History of Pocahontas County Place Names," (N.D.), copy provided to author by Mr. Jarvinen Oct. 13, 1983; Andrew Price, *West Virginia Bluebook*, Vol. 13, 1928, p. 408; Micajah Woods Papers, op. cit.; Pearl Buck, *The Exile,* (John Day Book: Reynal & Hitchcock, New York, 1936).
26. Micajah Woods Papers, op. cit.; M.A. Dunlap, "From Mr. Dunlap," *The Pocahontas Times,* July 11, 1929.
27. M.A. Dunlap, "From M.A. Dunlap," *The Pocahontas Times*, Dec. 29, 1925; L.S. Cochran, op. cit.
28. Calvin W. Price, *The Pocahontas Times*, Nov. 26, 1925.
29. *Cincinnati Daily Gazette,* Nov. 27, 1863 (account by Capt. Francis Mathers).
30. Frank S. Reader, op. cit., p. 216; *Cincinnati Daily Gazette*, Nov. 27, 1863; Micajah Woods Papers, op. cit.; Samuel D. Edmonds Memoir, op. cit.
31. *The Ironton Register*, Nov. 26, 1863; M.A. Dunlap, "Battle of Droop Mt.," *The Pocahontas Times*, Sept. 13, 1923; *The Ironton Register*, op. cit.; James Z. McChesney Letters, op. cit.; Micajah Woods Papers, op. cit.; *The Ironton Register*, Dec. 10, 1863; Military Pension Records for Capt. John V. Keepers, National Archives, Washington, D.C.
32. Micajah Woods Papers, op. cit.
33. L.S. Cochran, op. cit.
34. "Lieutenant Colonel John J. Polsley –7th West Virginia Regiment 1861-1865," thesis by Eugene Wise Jones, June 1949, presented to the University of Akron, Akron, Ohio, p. 74; *Cincinnati Daily Gazette*, Nov. 27, 1863.
35. Droop Mountain Battlefield State Park Files, Hillsboro, West Virginia.
36. Samuel D. Edmonds Memoir, op. cit.
37. M.A. Dunlap, "From Mr. Dunlap," *The Pocahontas Times*, July 11, 1929; John D. Sutton (and State of West Virginia), op. cit., p. 18; Droop Mountain Battlefield State Park Files, op. cit.; E.R. Howery (Captain, 398th Infantry), *The Battle of Droop Mountain November 6, 1863,* [n.d.]–probably 1935–p. 6 (statement of George H.C. Alderson, 14th Virginia Cavalry).
38. M.A. Dunlap, "From M.A. Dunlap," *The Pocahontas Times*, Dec. 29, 1925; Blain Family Papers, op. cit.
39. M.A. Dunlap, "From M.A. Dunlap," *The Pocahontas Times*, Dec. 29, 1925; Droop Mountain Battlefield State Park Files, op. cit.; L. S. Cochran, op. cit.
40. Andrew Price, "Battle of Droop Mountain," *The Pocahontas Times,* [n.d.] - ca. 1920's, clipping in Droop Mountain State Battlefield Park Files; John D. Sutton (and State of West Virginia), op. cit., p. 31 (Sutton) and p. 17 (Bender).
41. E.R. Howery, op. cit., p. 6-7.
42. Andrew Price, *West Virginia Bluebook*, op. cit.; p. 415; "The Droop Battlefield," *The Pocahontas Times,* Nov. 22, 1895; L.S. Cochran, op. cit.
43. "Letter from West Virginia: Gen. Averell's Movements–Battle of Droop Mountain," (Correspondent "Union"), *Cincinnati Daily Commercial,* Nov. 27, 1863; M.A. Dunlap, "From Mr. Dunlap," *The Pocahontas Times*, July 11, 1929; *Cincinnati Daily Commercial*, Nov. 27, 1863, op. cit.

44. David Poe, *Personal Reminiscences of the Civil War by Captain David Poe,* (20th Virginia Cavalry), (Buckhannon, West Virginia: Upshur Republican Print, 1911).
45. Robert Driver, Jr., *14th Virginia Cavalry,* (Lynchburg, Virginia: H.E. Howard, Inc. 1988), p. 19.
46. Frank S. Reader, op. cit., p. 88.
47. Frank S. Reader, op. cit., p. 58-60 and p. 217.
48. William M. Dearing Information, (Co. F, 3rd West Virginia Mounted Infantry), letters to author from Brian Kesterson, Washington, West Virginia.
49. Civil War Miscellaneous Papers 1859-1937, (includes letters of B.F. Hughes, Co. F, 3rd West Virginia Mounted Infantry), Item #1021, West Virginia University, Morgantown, West Virginia.
50. Barnes Family Papers, op. cit.; David Poe, op. cit.; John D. Sutton (and State of West Virginia), op. cit., p. 17; "The Late Fight at Droop Mountain–Complete Rout of the Rebels," (Correspondent "T.M."), *Wheeling Daily Intelligencer*, Nov. 21, 1863; John D. Sutton (and State of West Virginia), op. cit., p. 20.
51. John D. Sutton (and State of West Virginia), op. cit., p. 25.
52. John D. Sutton (and State of West Virginia), op. cit. p. 20 (Clothier) and p. 15 (Bender).
53. George A. Otis (Assistant Surgeon U.S. Army), Prepared under the direction of Joseph K. Barnes (Surgeon General, U.S. Army), *The Medical and Surgical History of the War of the Rebellion, Part II, Volume II, Surgical History,* First Issue, (Washington: Government Printing Office, 1876), p. 82.
54. "The Late Fight at Droop Mountain–Complete Rout of the Rebels," (Correspondent "T.M."), *Wheeling Daily Intelligencer*, Nov. 21, 1863.
55. John D. Sutton (and State of West Virginia), op. cit. p. 23.
56. Henry Bender Papers, (Co. F, 10th West Virginia Infantry), Droop Mountain Battlefield State Park Files; W.F. Morrison, "Henry Bender, 90, Veteran of Civil War, Dead," *The Braxton Central*, March 27, 1931.
57. John D. Sutton (and State of West Virginia), op. cit., p. 18, 23-24; A.P. Lockard, *The Pocahontas Times*, Jan. 24, 1929; "The Late Fight at Droop Mountain–Complete Rout of the Rebels," (Correspondent "T.M."), *Wheeling Daily Intelligencer*, Nov. 21, 1863; "Letter from West Virginia: Gen. Averell's Movements–Battle of Droop Mountain," (Correspondent "Union"), *Cincinnati Daily Commercial*, Nov. 27, 1863.
58. L.S. Cochran, op. cit.
59. Frank S. Reader, op. cit., p. 104-105 and 217.
60. Military Service Records, John Y. Bassell, National Archives, Washington, D.C.; Jacob S. Hall Letters, (Co. E, 19th Virginia Cavalry), original in possession of Homer C. Cooper, Athens, Georgia. Courtesy Droop Mountain Battlefield State Park Files; Correspondence with Brian Kesterson, Washington, West Virginia concerning Abram Lowers.
61. Samuel T. Wiley, *History of Monongalia County, West Virginia*, (Kingwood, West Virginia: Preston Publishing Co., 1883), p. 530; William F. Bahlmann, "Down In The Ranks," op. cit., p. 86.
62. David Poe, op. cit.
63. George R. Latham Papers–1863 (Josiah Davis, 3rd West Virginia Mounted Infantry), West Virginia State Archives, Charleston, West Virginia.
64. George H.D. Alderson, op. cit.; James Z. McChesney Letters, op. cit.
65. Calvin W. Price, *The Pocahontas Times*, Nov. 26, 1925; Robert J. Driver, Jr., op. cit., p. 122, 156, 176; Ann Monson (editor), "The Battle of Droop Mountain" and letter of Samuel H. Lucass, Co. F, 14th Virginia Cavalry, *Grape and Canister Newsletter of the Civil War Roundtable of Wilmington*, Feb. 1988.
66. Frank S. Reader, op. cit., p. 217; The Francis S. Reader Papers, "An Average American: The Memoirs of Francis S. Reader–Volume II," completed Nov. 17, 1924, U.S. Army Military History Institute, Carlisle Barracks, Pennsylvania; Frank S. Reader, op. cit., p. 217; John Henry Cammack, *Personal Recollections of Private John Henry Cammack, a Soldier of the Confederacy 1861-1865*, (Huntington, West Virginia, Paragon Ptg. & Pub. Co., 1920), p. 66-67.
67. James Siburt, "The Confederates' Last Stand in West Virginia," *Civil War Times Illustrated*, Mar. 1989, p. 26; Terry Lowry, *22nd Virginia Infantry* (Lynchburg, Virginia: H.E. Howard, Inc., 1988), p. 60; Frank S. Reader, op. cit., p. 217; Droop Mountain Battlefield State Park Files, op. cit., (letter concerning James L. Workman); "Letter from Thomas Swinburn," *The Pocahontas Times*, Apr. 11, 1901; "Brilliant Exploits in Western Virginia," (Correspondent "C"), *The Ironton Register*, Dec. 3, 1863.
68. Terry Lowry, op. cit., p. 201.
69. Roy Bird Cook, "The Battle of Droop Mountain," *The West Virginia Review*, Oct. 1928, p. 15;

Confederate Service Records, Company A, 22nd Virginia Infantry, National Archives, Washington, D.C.
70. Andrew R. Barbee Letters, Lt. Col. 22nd Virginia Infantry, Misc. 1864, Letters Received 1863-1865, Dept. of West Virginia (Grant), Entry 5691 (Inventory Vol. I), Record Group 393, U.S. Army Continental Commands 1821-1920 (located in Stack Area 10wz, Row 11, Bottom Shelf), National Archives, Washington, D.C.; Andrew Price, *West Virginia Bluebook*, op. cit., p. 408; Andrew R. Barbee, op. cit.
71. Terry Lowry, op. cit.; Andrew R. Barbee Letters, op. cit.
72. Terry Lowry, op. cit.; "The Late Fight at Droop Mountain–Complete Rout of the Rebels," (Correspondent "T.M."), *Wheeling Daily Intelligencer*, Nov. 21, 1863.
73. Frank Moore, *The Rebellion Record*, (New York: D. Van Nostrand, Publisher, 1865), Vol. 8, (Correspondent "Irwin"), p. 156-160; Samuel D. Edmonds Memoir, op. cit.; Frank Moore, op. cit., ("Irwin").
74. William F. Bahlmann, "Down In The Ranks," op. cit., p. 83-84.
75. George H.C. Alderson, op. cit.; Terry Lowry, op. cit.; William F. Bahlmann, "Down In The Ranks," op. cit., p. 85; Andrew R. Barbee Letters, op. cit.; Joseph Alleine Brown, op. cit., p. 29; Micajah Woods Papers, op. cit.
76. James H. Mays, op. cit. p. 29.
77. James S. McClung Memoir, op. cit.; John D. Sutton (and State of West Virginia), op. cit., p. 21, 23.
78. Captain James W. Barnes, (23rd Battalion Virginia Infantry), unknown newspaper clipping of letter describing the Droop Mountain battle, Droop Mountain Battlefield State Park Files.
79. Frank Moore, *The Rebellion Record*, ("Irwin"), op. cit.; Samuel D. Edmonds Memoir, op. cit.; Letters of William Ludwig, 34th Ohio Volunteer Infantry, West Virginia University, Morgantown, West Virginia; Augustus Moor Papers, op. cit.; Micajah Woods Papers, op. cit.
80. Micajah Woods Papers, op. cit.
81. Blain Family Papers, op. cit.; Micajah Woods Papers, op. cit.
82. Micajah Woods Papers, op. cit.
83. Samuel D. Edmonds Memoir, op. cit.
84. James Z. McChesney Letters, op. cit.; Blain Family Papers, op. cit.; Robert J. Driver, Jr., op. cit., p. 30; *The Richmond Examiner*, Nov. 9, 1863.
85. John D. Sutton (and State of West Virginia), op. cit., p. 24; Frank Moore, *The Rebellion Record*, ("Irwin"), op. cit.
86. Pocahontas County Historical Society, *History of Pocahontas County, West Virginia*, (Dallas, Texas: Taylor Publishing Co., 1981), p. 317.
87. Samuel D. Edmonds Memoir, op. cit.; George H.C. Alderson, op. cit.; James H. Mays, op. cit., p. 29; Micajah Woods Papers, op. cit.; Terry Lowry, op. cit., p. 52.
88. Samuel D. Edmonds Memoir, op. cit.; James H. Mays, op. cit., p. 29; Andrew Price, *West Virginia Bluebook*, op. cit., p. 408.
89. Calvin W. Price, *The Pocahontas Times*, Nov. 26, 1925.
90. Andrew Price, *West Virginia Bluebook*, op. cit., p. 408; L.F. Fortney, "Droop Mt. Battle," *The Pocahontas Times*, Aug. 1, 1929.
91. C.R. Williams (compiler), *Southern Sympathizers: Wood County Confederate Soldiers & A Sketch of the Nighthawk Rangers of Wood, Jackson, Wirt and Roane Counties in West Virginia*, (Parkersburg, West Virginia: Inland River Books, 1979), p. 22.
92. Roy Bird Cook, op. cit., p. 15; George H.C. Alderson, op. cit.; Micajah Woods Papers, op. cit.; Lanty F. McNeal Collection, Hillsboro, West Virginia.
93. David Poe, op. cit.
94. George H. Mowrer, op. cit., p. 10; M.A. Dunlap, "Battle of Droop Mt.," *The Pocahontas Times*, Sept. 13, 1923.
95. Frank Moore, *The Rebellion Record*, ("Irwin"), op. cit.; Barnes Family Papers, op. cit.; Micajah Woods Papers, op. cit.; Joseph Alleine Brown, op. cit., p. 29.
96. Edward K. Eckert and Nicholas J. Amato (editors), *Ten Years in the Saddle: The Memoirs of William Woods Averell*, (San Rafael, California: Presidio Press, 1978), p. 391; Frank Moore, *The Rebellion Record*, ("Irwin"), op. cit.
97. L.S. Cochran, op. cit.; George H.C. Alderson, op. cit.; Frank S. Reader, op. cit., p. 218.
98. Micajah Woods Papers, op. cit.
99. L.S. Cochran, op. cit.; M.A. Dunlap, "Battle of Droop Mt.," *The Pocahontas Times*, Sept. 13, 1923.

100. Blain Family Papers, op. cit.; Droop Mountain Battlefield State Park Files, op. cit.
101. M.A. Dunlap, "Battle of Droop Mt.," *The Pocahontas Times*, Sept. 13, 1923; J.W. Benjamin, "The Battle of Droop Mountain," *The Charleston Gazette*, Jan. 12, 1958.
102. Droop Mountain Battlefield State Park Files, op. cit.
103. Andrew Price, *West Virginia Bluebook*, op. cit., p. 414.
104. Augustus Moor Papers, op. cit.; Frank Moore, *The Rebellion Record*, ("Irwin"), op. cit.; "The Late Fight at Droop Mountain–Complete Rout of the Rebels," (Correspondent "T.M."), *Wheeling Daily Intelligencer*, Nov. 21, 1863.
105. Augustus Moor Papers, op. cit.
106. "Brilliant Exploits in Western Virginia," (Correspondent "C"), *The Ironton Register*, Dec. 3, 1863; Frank Moore, *The Rebellion Record*, ("Irwin"), op. cit.
107. "Brilliant Exploits in Western Virginia," op. cit.; John D. Sutton (and State of West Virginia), op. cit., p. 18; Calvin W. Price, *The Pocahontas Times*, Nov. 26, 1925; James Z. McChesney Letters, op. cit.
108. Samuel D. Edmonds Memoir, op. cit.; George H.C. Alderson, op. cit.; L.S. Cochran, op. cit.
109. J.J. Sutton, *History of the Second Regiment, West Virginia Cavalry Volunteers, During the War of the Rebellion*, (Portsmouth, Ohio: 1892), p. 109.
110. James Ireland Diary (Co. A, 12th O.V.I.), Ohio Historical Center, Columbus, Ohio.
111. "The Battle of Droop Mountain," as related by C.L. Stulting, Droop Mountain Battlefield State Park Files, Hillsboro, West Virginia; L.S. Cochran, op. cit.; Ann Griffith, "Preserving the Pioneer Spirit," *Charleston Daily Mail*, Aug. 18, 1982; "Letter from Thomas Swinburn," *The Pocahontas Times*, Apr. 11, 1901.
112. William F. Bahlmann, "Down In The Ranks," op. cit., p. 84-85.
113. *Ibid.*, p. 85.
114. *Ibid.*, p. 85-86.
115. C.L. Stulting Memoir, op. cit.; "Brilliant Exploits in Western Virginia," (Correspondent "C"), *The Ironton Register*, Dec. 3, 1863.
116. John D. Sutton (and State of West Virginia), op. cit., p. 24.
117. *Ibid.*; *Staunton Vindicator*, Nov. 13, 1863 and Nov. 27, 1863; George H.C. Alderson, op. cit.; Droop Mountain Battlefield State Park Files, op. cit.
118. "Casualties in the 10th [West] Virginia Volunteers at the Battle of Droop Mountain," *Wheeling Daily Intelligencer*, Nov. 30, 1863; "The Late Fight at Droop Mountain–Complete Rout of the Rebels," (Correspondent "T.M."), *Wheeling Daily Intelligencer*, Nov. 21, 1863.
119. A vast amount of sources were used in compiling Federal casualties at Droop Mountain, particularly the *Official Records*, period newspaper accounts, the Augustus Moor Papers, and original regimental casualty lists at the National Archives, Washington, D.C.; "From the 14th Pennsylvania Cavalry," *The Pittsburgh Gazette*, Nov. 21, 1863.
120. "Casualties in the 10th [West] Virginia Volunteers at the Battle of Droop Mountain," op. cit.
121. Peterson Family Papers 1858-1913 (Lt. David T. Peterson, Co. B, 10th West Virginia Inf.), West Virginia University, Morgantown, West Virginia.
122. Confederate casualties at Droop Mountain have been compiled from a number of sources, particularly, the *Official Records*, period newspapers, Confederate service records, and county histories; William F. Bahlmann, "Bahlmann Gives History of Men Serving in '61," *CharlestonGazette*, Sept. 17, 1922; James H. Mays, op. cit., p. 29-30.

Chapter Six

1. James Ireland Diary (12th O.V.I), Ohio Historical Center, Columbus, Ohio.
2. Micajah Woods Papers (Jackson's Battery), University of Virginia, Charlottesville, Virginia.
3. Joseph Alleine Brown, *The Memoirs of a Confederate Soldier, As Told to His Grandson, Samuel Hunter Austin*, (Sam Austin-Forum Press: Abingdon, Virginia, 1940), p. 29.
4. Micajah Woods Papers, op. cit.
5. Blain Family Papers 1860-1869 (Jackson's Battery), Washington & Lee University, Lexington, Virginia.
6. Droop Mountain Battlefield State Park Files, Hillsboro, West Virginia.
7. Micajah Woods Papers, op. cit.
8. Droop Mountain Battlefield State Park Files, op. cit.
9. George H.C. Alderson, (Co. A, 14th Virginia Cavalry), "From the *Greenbrier Independent*–letter concerning the battle of Droop Mountain and death of Major Bailey," *Nicholas Chronicle*, Feb.

17, 1927.
10. Samuel D. Edmonds Memoir (Co. B, 22nd Virginia Infantry), Droop Mountain State Park Files, Hillsboro, West Virginia.
11. Pearl Buck, *The Exile*, (John Day Book: Reynal & Hitchcock, New York, 1936), p. 56.
12. Augustus Moor Papers (28th O.V.I.), Ratterman Collection, University of Illinois at Urbana-Champaign, Urbana, Illinois.
13. Frank Moore, *The Rebellion Record*, (New York: D. Van Nostrand, Publisher, 1865), Vol. 8 ("A National Account," by Irwin), p. 156-160.
14. Barnes Family Papers, 1816-1929, papers of Thomas Rufus Barnes (Co. K, 10th West Virginia Infantry), West Virginia University, Morgantown, West Virginia.
15. James Ireland Diary, op. cit.
16. Droop Mountain Battlefield State Park Files, op. cit.
17. George H.C. Alderson, op. cit.
18. Dunlap Family Papers 1863-64, courtesy Helen Steele Ellison, Droop Mountain State Park Files, Hillsboro, West Virginia.
19. James Z. McChesney Letters (14th Virginia Cavalry), H.E. Matheny Collection, West Virginia University, Morgantown, West Virginia.
20. Letters of William Ludwig, 34th Ohio Volunteer Infantry, West Virginia University, Morgantown, West Virginia.
21. Micajah Woods Papers, op. cit.
22. *Ibid.*
23. James Ireland Diary, op. cit.
24. "Brilliant Exploits in Western Virginia," *The Ironton Register* (Correspondent "C"), Dec. 3,1863.
25. James Z. McChesney Letters, op. cit.
26. Frank Moore, *The Rebellion Record,* ("Irwin"), op. cit.
27. Milton Humphreys Papers (Bryan's Battery), University of Virginia, Charlottesville, Virginia.
28. Oren F. Morton, *A History of Monroe County, West Virginia,* (Staunton, Virginia: The McClure Co., Inc., 1918)—contains the Civil War diary of Rev. S.R. Houston of Union, West Virginia, p. 177.
29. Blain Family Papers, op. cit.
30. Augustus Moor Papers, op. cit.
31. Frank Moore, *The Rebellion Record,* ("Irwin"), op. cit.
32. James Ireland Diary, op. cit.
33. Oren F. Morton (S.R. Houston diary), op. cit.
34. James Ireland Diary, op. cit.
35. Clarence Shirley Donnelly, *David S. Creigh—The Greenbrier Martyr*, (limited edition), (Oak Hill, West Virginia, 1950); "The David S. Creigh Execution," *The Journal of the Greenbrier Historical Society*, Vol. II, No. 2, Oct. 1970; Pvt. Benjamin F. Zeller, Co. L, 8th O.V.C., "Over the Mountains and Down the Valleys of Virginia: Thirty Days in the Saddle," original manuscript at the Garst Museum, Greenville, Ohio.
36. Barnes Family Papers, op. cit.
37. Micajah Woods Papers, op. cit.
38. Frank S. Reader, *History of the Fifth West Virginia Cavalry, formerly the Second Virginia Infantry, and of Battery G, First West Virginia Light Artillery,* (New Brighton, Pennsylvania: *Daily News*, 1890), p. 218.
39. Andrew Price, *West Virginia Bluebook,* (1926), p. 408.
40. George Washington Ordner Diary, (2nd West Virginia Mounted Infantry), West Virginia University, Morgantown, West Virginia.
41. Civil War Diary of Amos A. Vandervort, 14th West Virginia Infantry, courtesy William D. Wintz, St. Albans, West Virginia.
42. Micajah Woods Papers, op. cit.
43. Blain Family Papers, op. cit.
44. *The Ironton Register,* op. cit.
45. Frank M. Imboden, "War Diary," Co. H, 18th Virginia Cavalry, C.S.A (Typescript copy), West Virginia State Archives, Charleston, West Virginia.
46. George Washington Ordner Diary, op. cit.
47. Letters of William Ludwig, op. cit.
48. Anthony H. Windsor, *History of the 91st Regiment, O.V.I.,* (Cincinnati: Gazette Steam Printing House, 1865).

49. J.E.D. Ward, *12th Ohio Volunteer Infantry*, (Ripley, Ohio: 1864), p. 69.
50. James Ireland Diary, op. cit.
51. Frank M. Imboden, op. cit.
52. Frank Moore, *The Rebellion Record*, ("Irwin"), op. cit.
53. James Ireland Diary, op. cit.
54. Blain Family Papers, op. cit.
55. James Z. McChesney Letters, op. cit.
56. Micajah Woods Papers, op. cit.
57. James Z. McChesney Letters, op. cit.
58. Samuel D. Edmonds Memoir, op. cit.
59. Frank Moore, *The Rebellion Record*, ("Irwin"), op. cit.
60. Frank Reader, op. cit., p. 219.
61. Barnes Family Papers, op. cit.
62. Frank M. Imboden, op. cit.
63. Micajah Woods Papers, op. cit.
64. George Washington Ordner Diary, op. cit.
65. Frank Moore, *The Rebellion Record*, ("Irwin"), op. cit.
66. James Ireland Diary, op. cit.
67. Civil War Diary of Amos A. Vandervort, op. cit.
68. Frank Moore, *The Rebellion Record*, ("Irwin"), op. cit.
69. Ralph Haas, (edited by Philip Ensley), *Dear Esther: The Civil War Letters of Private Aungier Dobbs, Centerville, Pennsylvania, Company "A," the Ringgold Cavalry Company, 22nd Pennsylvania Cavalry, June 29, 1861 to October 31, 1865,* (Apollo, Pennsylvania: Clossen Press, 1991), p. 185.
70. Frank Moore, *The Rebellion Record*, ("Irwin"), op. cit.
71. Micajah Woods Papers, op. cit.
72. Frank Moore, *The Rebellion Record*, ("Irwin"), op. cit.; George Washington Ordner Diary, op. cit.; Barnes Family Papers, op. cit.
73. "Lieutenant Colonel John J. Polsley–7th West Virginia Regiment 1861-1865," thesis by Eugene Wise Jones, June 1949, presented to the University of Akron, Akron, Ohio; John J. Polsley Papers 1862-1865 (8th West Virginia Mounted Infantry), Letters 108 and 109 (Nov. 14 and 20, 1863), University of Akron, Akron, Ohio.
74. Barnes Family Papers, op. cit.
75. Frank M. Imboden, op. cit.
76. Shaver Family Papers 1861-67, Civil War Collection, West Virginia State Archives, Charleston, West Virginia.
77. Samuel D. Edmonds Memoir, op. cit.
78. Micajah Woods Papers, op. cit.
79. Letters of William Ludwig, op. cit.
80. Anthony H. Windsor, op. cit.
81. Civil War Diary of Amos A. Vandervort, op. cit.
82. Frank M. Imboden, op. cit.
83. Stephen Cresswell (editor), *We Will Know What War Is: The Civil War Diary of Sirene Bunten,* (Parsons, West Virginia: McClain Printing, 1993), p. 57.
84. Micajah Woods Papers, op. cit.
85. *Ibid.*
86. Civil War Diary of Amos A. Vandervort, op. cit.
87. *Ibid.*
88. Dunlap Family Papers, op. cit.
89. In addition to incorporating the various reports in the *Official Records* on the Nov. 16, 1863 attack on the Federal supply train near Burlington, West Virginia, the following sources were utilized: Roger U. Delauter, Jr., *McNeill's Rangers,* (Lynchburg, Virginia: H.E. Howard, Inc., 1986); Roger U. Delauter, Jr., *62nd Virginia Infantry,* (Lynchburg, Virginia: H.E. Howard, Inc., 1988); Samuel Clarke Farrar (Co. C, 22nd P.V.C.), *Twenty-Second Pennsylvania Cavalry and Ringgold Battalion 1861-1865*, (Akron, Ohio and Pittsburgh: The New Werner Co., 1911); Ralph Haas (edited by Philip Ensley), *Dear Esther: The Civil War Letters of Private Aungier Dobbs, Centerville, Pennsylvania Cavalry, Company A , The Ringgold Cavalry Company, 22nd Pennsylvania Cavalry, June 29, 1861 to October 31, 1865,* (Apollo, Pennsylvania: Clossen Press, 1991); Maryland General Assembly, *History and Roster of Maryland Volunteers, War of 1861-1865,* (Baltimore, Maryland: Press of Guggenheimer, Weil & Co., 1898); Charles F. Miller, "Capture of Wagon Train By McNeill's

Men," *Confederate Veteran,* #21, Jan. 1913; Civil War Diary of Amos A. Vandervort, courtesy William D. Wintz, St. Albans, West Virginia.
90. Frank Moore, *The Rebellion Record*, ("Irwin"), op. cit.
91. Frank Moore, *The Rebellion Record*,("Irwin"), op. cit.
92. Frank Reader, op. cit., p. 219.
93. Frank Moore, *The Rebellion Record*, ("Irwin"), op. cit.
94. Stephen Cresswell (editor), op. cit., p. 57.
95. The majority of quotes in the "Epilogue" section are from the following sources: Stephen Z. Starr, *The Union Cavalry in the Civil War–Vol. I–From Fort Sumter to Gettysburg 1861-1863,* (Baton Rouge and London: Louisiana State University Press, 1979); Stephen Z. Starr, *The Union Cavalry in the Civil War–Vol. II–The War in the East–From Gettyburg to Appomattox 1863-1865,* (Baton Rouge and London: Louisiana State University Press, 1981); John Alexander Williams, *West Virginia–A Bicentennial History (The States and the Nation Series)*, W.W. Norton & Co., Inc., N.Y., N.Y.; Dallas B. Shaffer, *The Battle at Droop Mountain,* (West Virginia Department of Natural Resources, 1966).

Chapter Seven

1. The bulk of biographical material on Gen. William W. Averell was culled from the following sources: Edward K. Eckert and Nicholas J. Amato (editors), *Ten Years in the Saddle: The Memoirs of William Woods Averell,* (San Rafael, California: Presidio Press, 1978); Ezra Warner, *Generals In Blue,* (Baton Rouge: Louisiana State University Press, 1964) p. 12-13; Patricia L. Faust (editor), *Historical Times Illustrated Encyclopedia of the Civil War,* (New York: Harper and Row Publishers, 1986); William W. Averell Papers 1836-1910, New York State Library Cultural Education Center, Albany, New York.
2. Biographical information on Gen. John Echols was assembled primarily from the following sources: (1) Ezra Warner, *Generals In Gray*, (Baton Rouge: Louisiana State University Press, 1964), p. 80; Patricia L. Faust (editor), *Historical Times Illustrated Encyclopedia of the Civil War,* (New York: Harper and Row Publishers, 1986), p. 235-236; Clement A. Evans, general editor, *Confederate Military History* (Virginia volume), (Atlanta: Confederate Publishing Co., 1899), p. 591-593; Gen. John Echols File, V.M.I. Archives, Lexington, Virginia; "Gen. John Echols–Advocate and Defender of the Confederacy," *Confederate Veteran*, Jan. 1896.
3. Biographical information on Lt. Col. Alexander Scott was culled from the following sources: Frank S. Reader, *History of the Fifth West Virginia Cavalry, formerly the Second Virginia Infantry, and of Battery G, First West Virginia Light Artillery*, (New Brighton, Pennsylvania: Daily News, 1890), p. 33; Theodore F. Lang, *Loyal West Virginia from 1861 to 1865,* (Baltimore: Deutch Publishers, 1895) p. 207-213; Records of the 5th West Virginia Cavalry; National Archives, Washington, D.C.
4. Samuel T. Wiley, *History of Monongalia County, West Virginia,* (Kingwood, West Virginia: Preston Publishing Co., 1883), p. 529-531; newspaper obituary for Francis W. Thompson.
5. Biographical information on John H. Oley was culled from the following sources: Rick Baumgartner, *First Families of Huntington,* (Huntington Publishing, April 1977), p. 37-38; obituary for John H. Oley, *The Huntington Advertiser*, Mar. 17, 1888; Roger D. Hunt and Jack R. Brown, *Brevet Brigadier Generals in Blue*, (Olde Soldier Books, 1990), p. 454.
6. William D. Slease, *The 14th Pennsylvania Cavalry in the Civil War,* (Pittsburgh: Art Engraving and Printing Co., 1890); Samuel P. Bates, *History of Pennsylvania Volunteers 1861-1865,* (Harrisburg: D. Singerly, State Printer, 1870), p. 851-897; George SWestnam, "The Fourteenth Cavalry," *The Pittsburgh Press*, Oct. 8, 1961, p. 4-5. *Pittsburgh of Today,* p. 966-968 (no other details available).
7. Military Pension Records of Thomas Gibson, (Jr.), National Archives, Washington, D.C.
8. Military Service Records, Capt. Julius Jaehne, Company C, 16th Illinois Cavalry, National Archives, Washington, D.C.
9. H.E. Matheny, *Major General Thomas Maley Harris . . . a member of the Military Commission that tried the President Abraham Lincoln assassination conspirators . . . and Roster of the 10th West Virginia Volunteer Infantry Regiment 1861-1865,* (Parsons, West Virginia: McClain Printing Company, 1963). Minnie Randall Lowther, *History of Ritchie County*, (Wheeling, West Virginia: Wheeling News Litho. Co. 1911), p. 442-446; Ezra Warner, op. cit., p. 209-210; Patricia L. Faust, op. cit., p. 345.
10. Papers of Augustus Moor and the 28th Ohio Volunteer Infantry, Ratterman Collection, University of Illinois at Urbana - Champaign, Urbana, Illinois; Jacob D. Cox, op. cit., p. 110.

11. Military Pension Records, Capt. John V. Keepers, Battery B, 1st West Virginia Light Artillery, National Archives, Washington, D.C.; Records of Battery B, 1st West Virginia Light Artillery, National Archives, Washington, D.C. (Note: John V. Keepers is the correct spelling of this name, not Keeper as most historians and writers have employed).
12. Military Pension Records, Capt. Chatham T. Ewing, Battery G, 1st West Virginia Light Artillery, National Archives, Washington, D.C. (Averell Affidavit).
13. Frank S. Reader, *History of the Fifth West Virginia Cavalry, formerly the Second Virginia Infantry, and of Battery G, First West Virginia Light Artillery*, (New Brighton, Pennsylvania: Daily News, 1890).
14. Bvt. Maj. Gen. George W. Cullum, *Biographical Register for West Point of the Officers and Graduates of the U. S. Military Academy at West Point, N. Y.*, Vol. II (Boston and New York: Mifflin and Company, The Riverside Press, 1891), p. 866. Jeffry D. Wert, *From Winchester to Cedar Creek: The Shenandoah Campaign of 1864*, (Carlisle, Pennsylvania: South Mountain Press, Inc., Publishers, 1987), p. 145.
15. Ezra Warner, op. cit., p. 131-132. Stephen Z. Starr, *The Union Cavalry in the Civil War—Vol. I—From Fort Sumter to Gettysburg 1861-1863,* (Baton Rouge and London: Louisiana State University Press, 1979), p. 96-97.
16. Whitelaw Reid, *Ohio In The War*, (Cleveland, Ohio: 1911), p. 87-90.
17. Terry Lowry, *22nd Virginia Infantry,* (Lynchburg, Virginia: H.E. Howard, Inc., 1988 and 1991 revised edition), p. 110-112.
18. *Ibid.*, p. 201.
19. *Ibid.*, p. 125.
20. *Ibid.*, p. 145.
21. *Ibid.*, p. 173.
22. *Ibid.*, p. 147.
23. John L. Scott, *23rd Battalion Virginia Infantry,* (Lynchburg, Virginia: H.E. Howard, Inc., 1991), p. 49-50; 1870 Census of Smyth County, Virginia.
24. Terry Lowry, *26th Battalion Virginia Infantry*, (Lynchburg, Virginia: H.E. Howard, Inc., 1991), p. 116. George Mathews Edgar Papers, Southern Historical Collection, University of North Carolina, Chapel Hill, North Carolina.
25. Albert S. Johnston, *Captain Beirne Chapman and Chapman's Battery,* (Union, West Virginia: 1905); John L. Scott, *Lowry's, Bryon's and Chapman's Batteries of Virginia Artillery,* (Lynchburg, Virginia: H.E. Howard, Inc., 1988).
26. Ezra Warner, op. cit., p. 153-154; Patricia L. Faust (editor), op. cit., p. 392; Clement A. Evans, op. cit. (West Virginia volume), p. 131-133; H.E. Matheny, *Wood County, West Virginia in Civil War Times*, (Parkersburg, West Virginia: Joseph M. Sakach, Jr., Trans-Allegheny Books, Inc., 1987).
27. George W. Atkinson and Alvarado F. Gibbens, *Prominent Men of West Virginia*, (Wheeling, West Virginia: W.L. Catlin, 1890), p. 805; William P. Thompson Papers, National Archives, Washington, D.C.; John M. Ashcraft, *31st Virginia Infantry*, (Lynchburg, Virginia: H.E. Howard, Inc., 1988).
28. George W. Atkinson and Alvarado F. Gibbens, op. cit., p. 704; "Death Comes Quietly to Col. Arnett," *The Wheeling Intelligencer*, Feb. 20, 1902.
29. Author's correspondence with Robert J. Trout, Myerstown, Pennsylvania; "Captain Lurty Dead," *Harrisonburg Times,* 1906.
30. Jack L. Dickinson, *16th Virginia Cavalry*, (Lynchburg, Virginia: H.E. Howard, Inc., 1989); Jack L. Dickinson, *Records of the 16th Regiment Virginia Cavalry,* Confederate States Army, (Barboursville, West Virginia: Jack L. Dickinson, 1984).
31. Jack L. Dickinson, *16th Virginia Cavalry*, op. cit., p. 109; Terry Lowry, *22nd Virginia Infantry*, op. cit., p. 181.
32. Robert J. Driver, Jr., *14th Virginia Cavalry*, (Lynchburg, Virginia: H.E. Howard, Inc., 1988); p. 111.
33. *Ibid.*, p. 126.
34. *Ibid.*, p. 119.
35. *Ibid.*, p. 103.
36. Confederate Military Service Records, Capt. Thomas E. Jackson (Jackson's Battery); author's correspondence with Gardner D. Beach, Frankfort, Kentucky.
37. Blain Family Papers 1860-69, Washington & Lee University, Lexington, Virginia.
38. Robert J. Driver, Jr., *The 1st and 2nd Rockbridge Artillery,* (Lynchburg, Virginia: H.E. Howard, Inc., 1987), p. 72.

Ezra Warner, op. cit. (Confederate), p. 147; Clement A. Evans, op. cit. (Virginia volume), p. 608-611; Patricia L. Faust (editor), op. cit., p. 378-379.

Chapter Eight

The majority of this chapter was written by Michael Smith, Superintendent, Droop Mountain Battlefield State Park, and originally was published in the book *Where People and Nature Meet: A History of the West Virginia State Parks,* (Charleston, West Virginia: Pictorial Histories Publishing Company, 1988). Additions and revisions to the original article have been made by Terry Lowry, utilizing the Droop Mountain Battlefield State Park Files and various issues of *The Pocahontas Times.*

Chapter Nine

The information presented in this chapter has been selected from the various papers and clippings in the Droop Mountain Battlefield State Park files. For these I am indebted to park superintendent Michael Smith who generously made these documents available to me for research. The bulk of the material is from the following: (1) Nancy Adams, "Ghosts of Droop Mountain," *The Charleston Gazette,* (Oct. 30,1985); (2) Anna Atkins (contributor), *History of Pocahontas County*, (1981), "Do You Know About The Ghosts Of Droop Mountain?" (p. 33) and "Some Other Tales," (p. 159); (3) Henry J. Johnson, "At Droop Mountain–CCC Workers Haunted By Ghosts Of Battle," *Wonderful West Virginia*, (Nov. 1982); (4) Skip Johnson, "Droop Mountain Battlefield Ghosts," *The Charleston Sunday Gazette-Mail*, (Sept. 18, 1977); (5) Skip Johnson, "The Ghosts Of Droop Mountain," *Wonderful West Virginia,* (Nov. 1982); (6) James Gay Jones, *A Wayfaring Sin-Eater And Other Tales of Appalachia,* (McClain Printing Co., Parsons, West Virginia, 1983), "Restless Spirits On Droop Mountain," (p. 36-39): (7) Johnie Keen, "Ghosts Of Droop Mountain," unpublished manuscript, Droop Mountain Battlefield State Park Files: (8) Chuck Poliafico, "Local Man Baffled by Ghostly Vision," *The Parkersburg News*, (Dec. 23, 1990).

Bibliography

Books and Pamphlets

Abraham, (Lt.) James, *With The Army of West Virginia 1861-64: Reminiscences of Lt. James Abraham, Pennsylvania Dragoons, Company A, 1st Regiment (West) Virginia Cavalry*, compiled by Evelyn A. Benson, (Lancaster, Pennsylvania, 1974).

Armstrong, Joan Tracy, *Ante-Bellum Years Through the Civil War; History of Smyth County, Virginia,* Vol. II, (Marion, Virginia: 1986).

Armstrong, Richard L., *West Virginian vs. West Virginian: The Battle of Bulltown, W. Va.,* (R. L. Armstrong, Hot Springs, Va., 1994).

Armstrong, Richard L., *19th and 20th Virginia Cavalry*, (Lynchburg, Virginia: H.E. Howard, Inc., 1994).

Ashcraft, John M., Jr., *31st Virginia Infantry*, (Lynchburg, Virginia: H.E. Howard, Inc., 1988).

Atkinson, George W. and Alvarado F. Gibbens, *Prominent Men of West Virginia*, (Wheeling, West Virginia: W.L. Catlin, 1890).

Bates, Samuel P., *History of Pennsylvania Volunteers 1861-5,* (Harrisburg: D. Singerly, State Printer, 1870).

Baumgartner, Rick, *First Families of Huntington*, (Huntington Publishing, April 1977).

Brown, Joseph Alleine, *Memoirs of a Confederate Soldier, Joseph Alleine Brown, As Told to his Grandson, Samuel Hunter Austin*, (Abingdon, Virginia: Sam Austin-Forum Press, 1940).

Buck, Pearl, *The Exile*, (New York: John Day Book: Reynal & Hitchcock, 1936).

Address of Capt. Jas. Bumgardner, Jr. Before R.E. Lee Camp, Confederate Veterans, Friday, May 8, 1903, Presenting Portrait of General John Echols, (Staunton, Va.: Daily News Printing Company, 1903).

Cammack, John Henry, *Personal Recollections of Private John Henry Cammack, a Soldier of the Confederacy 1861-1865*, (Huntington, West Virginia: Paragon Ptg. & Pub. Co., 1920).

Cohen, Stan, *The Civil War In West Virginia: A Pictorial History,* (Charleston, West Virginia: Pictorial Histories Publishing Company, 1976).

______, *A Pictorial Guide to West Virginia's Civil War Sites and Related Information,* (Charleston, West Virginia: Pictorial Histories Publishing Company, 1990).

Cole, J.R., *History of Greenbrier County*, (Lewisburg, West Virginia: 1917—reprinted for the West Virginia Genealogical Society, Waynesville, NC: Dan Mills, Inc., 1995).

Comstock, Jim and Peter Wallace, *West Virginia Picture Book—West Virginia Heritage Encyclopedia,* Vol. 51, (Richwood, West Virginia: 1978). Articles "Averell's Army In Major Victory At Droop Mountain" and "Moor Back in Beverly With Wounded Prisoners" p. 153.

Cox, Jacob D., *Military Reminiscences of the Civil War,* (New York: Charles Scribner's Sons, 1900).

Cresswell, Stephen (editor), *We Will Know What War Is: The Civil War Diary of Sirene Bunten,* (Parsons, West Virginia: McClain Printing, 1993). (original diary in rare books collection at West Virginia Wesleyan College and reprinted in the *West Virginia Hillbilly* newspaper).

Cullum, Bvt. Maj. Gen. George W., *Biographical Register for West Point of the Officiers and Graduates of the U. S. Military Academy at West Point, N.Y.*, Vol. II (Boston and New York: Mifflin and Company, The Riverside Press, 1891).

Davis, Dorothy, *History of Harrison County, West Virginia* (edit. Elizabeth Sloan), (Clarksburg, West Virginia: American Association of University Women, 1970).

Davis, William C., *The Battle of New Market*, (Garden City, New York: Doubleday & Company, Inc., 1975).

Delauter, Roger U., Jr., *18th Virginia Cavalry,* (Lynchburg, Virginia: H.E. Howard, Inc., 1985).

______, *McNeill's Rangers*, (Lynchburg, Virginia: H.E. Howard, Inc., 1986).

______, *62nd Virginia Infantry*, (Lynchburg, Virginia: H.E. Howard, Inc., 1988).
Dickinson, Jack, *16th Virginia Cavalry*, (Lynchburg, Virginia: H.E. Howard, Inc., 1989).
______, *Records of the 16th Regiment Virginia Cavalry, Confederate States Army,* (Barboursville, West Virginia: Jack L. Dickinson, 1984).
Donnelly, Clarence Shirley, *David S. Creigh—the Greenbrier Martyr*, (limited edition), (Oak Hill, West Virginia: 1950).
Driver, Robert J., Jr., *The Staunton Artillery—McClanahan's Battery*, (Lynchburg, Virginia: H.E. Howard, Inc., 1988).
______, *14th Virginia Cavalry*, (Lynchburg, Virginia: H.E. Howard, Inc., 1988).
______, *The 1st and 2nd Rockbridge Artillery*, (Lynchburg, Virginia: H.E. Howard, Inc., 1987).
Dyer, Frederick M., *Compendium of the Rebellion*,(Dayton, Ohio: Morningside Bookshop, 1978—original printing 1908).
Eckert, Edward K. and Nicholas J. Amato (editors), *Ten Years in the Saddle: The Memoirs of William Woods Averell*, (San Rafael, California: Presidio Press, 1978).
Edgar, Betsy Jordan, *Our House*, (Parsons, West Virginia: McClain Printing Co., 1965).
Evans, Clement A., general editor, *Confederate Military History* (13 volumes), (Atlanta: Confederate Publishing Co., 1899)—particular emphasis on the Virginia and West Virginia volumes.
Ewing, Elmore Ellis, *The Story of the Ninety-First [91st O.V.I.],* (Portsmouth, Ohio: Republican Printing Co., 1868).
______ *Bugles and Bells; or Stories Told Again, including The Story of the Ninety-First Ohio Volunteer Infantry*, (Cincinnati, Ohio: Press of Curtis and Jennings, 1899).
Farrar, Samuel Clarke (Co. C, 22nd P.V.C.), *Twenty Second Pennsylvania Cavalry and Ringgold Battalion 1861-1865*, (Akron, Ohio and Pittsburgh: The New Werner Co., 1911).
Faulknier, Virginia (compiler), *Dear Annie: A Collection of Letters, 1860-1886*, (Parsons, West Virginia: 1969).
Faust, Patricia L. (editor), *Historical Times Illustrated Encyclopedia of the Civil War,* (New York: Harper and Row Publishers, 1986).
Geiger, Joe, Jr., *Civil War in Cabell County, West Virginia 1861-1865,* (Charleston, West Virginia: Pictorial Histories Publishing Co., 1991).
Gordon, Armistead C., *An Address on the Occassion of the Presentation of Judge William McLaughlin's Portrait to the Trustees of W & Lee University*, (June 17, 1903).
Haas, Ralph, (edited by Philip Ensley), *Dear Esther: The Civil War Letters of Private Aungier Dobbs, Centerville, Pennsylvania, Company "A," the Ringgold Cavalry Company, 22nd Pennsylvania Cavalry, June 29, 1861 to October 31, 1865*, (Apollo, Pennsylvania: Clossen Press, 1991).
Howery, E. R. (Captain, 398th Infantry), *The Battle of Droop Mountain November 6, 1863* [n.d.]—probably 1935.
Hunt, Roger D. and Jack R. Brown, *Brevet Brigadier Generals In Blue*, (Olde Soldier Books, 1990).
Johnson, Patricia Givens, *The United States Army Invades The New River Valley May 1864,* (Christiansburg, Virginia: Walpa Publishing, 1986).
Johnston, Albert S., *Captain Beirne Chapman and Chapman's Battery,* (Union, West Virginia: 1905).
Johnston, Jane Echols and Brenda Lynn Williams, *Hard Times 1861-1865,* (A Collection of Confederate Letters, Court Minutes, Soldiers Records and Local Lore from Craig Court, Virginia), Vol. II, (Craig County Historical Society, 1990).
Jones, James Gay, *A Wayfaring Sin-Eater and Other Tales of Appalachia*, (Parsons, West Virginia: McClain Printing Co., 1983).
Lang, Theodore F., *Loyal West Virginia from 1861 to 1865,* (Baltimore: Deutch Publishers, 1895).
Lowry, Terry, *The Battle of Scary Creek: Military Operations in the Kanawha Valley: April-July 1861,* (Charleston, West Virginia: Pictorial Histories Publishing, 1982).
______, *September Blood: The Battle of Carnifex Ferry,* (Charleston, West Virginia: Pictorial

Histories Publishing, 1985).
______, *22nd Virginia Infantry,* (Lynchburg, Virginia: H.E. Howard, Inc., 1988 and 1991 revised edition).
______, *26th Battalion Virginia Infantry,* (Lynchburg, Virginia: H.E. Howard, Inc., 1991).
Lowther, Minnie Randall, *History of Ritchie County*, (Wheeling, West Virginia: Wheeling News Litho. Co., 1911).
Ludlum, John Charles, . . . *The Geology of Watoga and Droop Mountain Battlefield State Park, West Virginia*, (Morgantown, West Virginia: Geological and Economic Survey, 1954).
Marvin Chapel History Committee, *A History of Marvin Chapel and Community—Mill Point, West Virginia 1953-1954*, (bound paper at West Virginia State Archives, Charleston, West Virginia).
Maryland General Assembly, *History and Roster of Maryland Volunteers, War of 1861-65*, (Baltimore, Maryland: Press of Guggenheimer, Weil & Co., 1898).
Matheny, H.E., *Major General Thomas Maley Harris . . . a member of the Military Commission that tried the President Abraham Lincoln assassination conspirators . . . and Roster of The 10th West Virginia Volunteer Infantry Regiment 1861-1865*, (Parsons, West Virginia: McClain Printing Company, 1963).
______, *Wood County, West Virginia In Civil War Times,* (Parkersburg, West Virginia: Joseph M. Sakach, Jr., Trans-Allegheny Books, Inc., 1987).
Mays, James H. (editor: Lee Mays), *Four Years for Old Virginia*, (Los Angeles, California: The Swordsman Publishing Company, privately printed, 1970).
McKinney, Tim, *Robert E. Lee and the 35th Star,* (Charleston, West Virginia: Pictorial Histories Publishing Company, 1993).
Memorial of Captain Jacob W. Marshall of Mingo Flats, Randolph County, W. Va. (1830-1899), Stoneburner & Ptufer, Steam Printers, Staunton, Va. (Courtesy Lanty McNeel, Hillsboro, West Virginia).
Monongalia Historical Society, *Morgantown: A Bicentennial History*, (Morgantown, West Virginia: Pioneer Press of West Virginia, Inc., Terra Alta, WV, 1985).
Moore, Frank, *The Rebellion Record* (12 Volumes), (New York: D. Van Nostrand, Publisher, 1865).
Morton, Oren F., *A History of Monroe County, West Virginia*, (Staunton, Virginia: The McClure Co., Inc., 1918—contains the Civil War diary of Rev. S.R. Houston of Union, West Virginia).
Mowrer, George H., *History of the Organization and Service During the War of the Rebellion of Co. A, 14th Pennsylvania Cavalry,* N.D. (rare book printed for veterans of the regiment. Presentation copy signed by Mowrer with a presentation date of Aug. 29, 1892 in author's collection).
New York (State) Adjutant General's Office, *Annual Report . . . for the year 1901*, Serial #27, (Albany: 1902).
History of Cincinnati and Hamilton County, Ohio, (Cincinnati, Ohio: S.B. Nelson & Co. Publishers, 1894).
Ohio Roster Commission, *Roster of Ohio Soldiers in the War of the Rebellion 1861-65*, (Cincinnati, Ohio: Valley Publishing-Mfg. Co., 1866).
Otis, George A. (Assistant Surgeon U.S. Army), Prepared under the direction of Joseph K. Barnes (Surgeon General, U.S. Army), *The Medical and Surgical History of the War of the Rebellion, Part II, Volume II, Surgical History,* First Issue, (Washington: Government Printing Office, 1876). [p. 82]
Payne, Dale and Bob Beckelheimer (compilers), *Tales and Trails—From the Fayette Tribune*, (Fayetteville, West Virginia: L.W. Printing, 1991).
Phisterer, Frederick , *New York in the War of the Rebellion.*
Pocahontas County Historical Society, *History of Pocahontas County, West Virginia*, (Dallas, Texas: Taylor Publishing Co., 1981).
Poe, David, *Personal Reminiscences of the Civil War by Captain David Poe*, (20th Virginia

Cavalry), (Buckhannon, West Virginia: Upshur Republican Print, 1911).
Pollard, Edward A., *Southern History of the War*, (Fairfax Press), (New York: C.B. Richardson, 1866).
Price, Andrew, *West Virginia Bluebook*, Volume 11, "Incidents Concerning General Averell," (1926).
______, *West Virginia Bluebook*, Volume 13, "Plain Tales of Mountain Trails," Chapter 11 (Battle of Droop Mountain. Tenth West Virginia saved the day). (1928).
Price, William Thomas, *Historical Sketches of Pocahontas County*, West Virginia, (Marlinton, West Virginia: Price Brothers, 1901).
Reader, Frank S., *History of the Fifth West Virginia Cavalry, formerly the Second Virginia Infantry, and of Battery G, First West Virginia Light Artillery*, (New Brighton, Pennsylvania: Daily News, 1890).
Reece, Jasper W. (Brig. Gen.), *Report of the Adjutant General for the State of Illinois*, reports for 1861-1865, (revised by Reece), (Springfield bros., State Printers [etc.], 1900-02).
Reid, Whitelaw, *Ohio In The War*, (Cincinnati, New York: Moore, Wilstach and Baldwin, 1868).
Rice, Otis K., *A History of Greenbrier County,* (Parsons, West Virginia: McClain Printing Company, 1986).
Ryan, Daniel J., *The Civil War Literature of Ohio*, (Cleveland, Ohio: 1911).
Serrano, Domenick A., *Still More Confederate Faces*, (Bayside, New York: Metropolitan Company, 1992).
Scott, John L., *Lowry's, Bryan's and Chapman's Batteries of Virginia Artillery,* (Lynchburg, Virginia: H.E. Howard, Inc., 1988).
______, *23rd Battalion Virginia Infantry,* (Lynchburg, Virginia: H.E. Howard, Inc., 1991).
Shaffer, Dallas B., *The Battle at Droop Mountain*, (West Virginia Department of Natural Resources, 1966).
Shetler, Charles, *West Virginia Civil War Literature,* (Morgantown, West Virginia: University Press, 1963).
Slease, William D., *The 14th Pennsylvania Cavalry In The Civil War*, (Pittsburgh: Art Engraving and Printing Co., 1890).
Starr, Stephen Z., *The Union Cavalry in the Civil War—Vol. I—From Fort Sumter to Gettysburg 1861-1863,* (Baton Rouge and London: Louisiana State University Press, 1979).
______, *The Union Cavalry in the Civil War—Vol. II—The War in the East—From Gettysburg to Appomattox 1863-1865*, (Baton Rouge and London: Louisiana State University Press, 1981).
Stutler, Boyd B., *West Virginia in the Civil War*, (Charleston, West Virginia: Education Foundation, Inc., 1963).
Sutton, J.J., *History of the Second Regiment, West Virginia Cavalry Volunteers, During the War of the Rebellion*, (Portsmouth, Ohio: 1892).
Sutton, John D. (and State of West Virginia), *Report of Droop Mountain Battlefield Commission*, (Charleston, West Virginia: 1928).
Sutton, John Davison, *History of Braxton County and Central West Virginia*, (Sutton, West Virginia: 1919).
Turner, Ronald R., *7th West Virginia Cavalry,* (Manassas, Virginia: 1989).
Union Soldiers and Sailors Monument Association, *The Union Regiments of Kentucky*, (Louisville, Kentucky: 1897).
United States War Department, *War of the Rebellion: A Compilation of the Official Records of the Union and Confederate Armies*, (Washington, D.C.: Government Printing Office, 1880-1901).
Wallace, Lee A., *A Guide to Virginia Military Organizations 1861-65*, (Richmond: Virginia Civil War Commission, 1964), and revised edition (Lynchburg, Virginia: H.E. Howard, Inc., 1986).
Walls, R. Hal, *A History of the White Sulphur Rifles (Company G) and the Scouts and Guides (Company E) of Edgar's 26th Battalion of Patton's Brigade*, (Lewisburg, West Virginia: Roadrunner Press, 1989).

Ward, J.E.D., *12th Ohio Volunteer Infantry*, (Ripley, Ohio: 1864).
Warner, Ezra, *Generals in Gray*, (Baton Rouge: Louisiana State University Press, 1959).
______, *Generals in Blue*, (Baton Rouge: Louisiana State University Press, 1964).
Wert, Jeffry D., *From Winchester to Cedar Creek: The Shenandoah Campaign of 1864*, (Carlisle, Pennsylvania: South Mountain Press, Inc., Publishers, 1987).
West Virginia State Park History Committee, *Where People and Nature Meet: A History of the West Virginia State Parks,* (Charleston, West Virginia: Pictorial Histories Publishing Company, 1988).
Wiley, Samuel T., *History of Monongalia County West Virginia*, (Kingwood, West Virginia: Preston Publishing Co., 1883).
Williams, C.R. (Compiler), *Southern Sympathizers: Wood County Confederate Soldiers & A Sketch of the Nighthawk Rangers of Wood, Jackson, Wirt and Roane Counties in West Virginia*, (Parkersburg, West Virginia: Inland River Books, 1979).
Williams, John Alexander, *West Virginia—A Bicentennial History (The States and the Nation Series),* (W.W. Norton & Co., Inc., N.Y., N.Y.), (contains Droop Mountain on pg. 56).
Windsor, Anthony H., *History of the 91st Regiment, O.V.I.*, (Cincinnatti: Gazette Stern Printing House, 1865).

Magazines and Journals

Bahlmann, William F., "Down In The Ranks," *The Journal of the Greenbrier Historical Society*, Vol. 2, No. 2, Oct. 1970.
Brown, Adeline E., "Incident of a Boy Confederate," *Confederate Veteran*, Aug. 1912.
Cook, Roy Bird, "The Battle of Droop Mountain," *Confederate Veteran*, #36, Sept. 1928.
______, "The Battle of Droop Mountain," *The West Virginia Review*, Oct. 1928.
"The David S. Creigh Execution," *The Journal of the Greenbrier Historical Society*, Vol. II, No. 2, Oct. 1970.
Crockett, Maureen, "Re-Enacting The Battle of Droop Mountain," *Wonderful West Virginia*, Vol. 56, #7, Sept. 1992.
"Gen. John Echols—Advocate and Defender of the Confederacy," *Confederate Veteran*, Jan. 1896.
Johnson, Henry J., "At Droop Mountain—CCC Workers Haunted By Ghosts Of Battle," *Wonderful West Virginia*, Nov. 1982.
Johnson, Skip, "The Ghosts of Droop Mountain," *Wonderful West Virginia*, Nov. 1982.
McKeever, Kermit, "The First 30 Years," *Outdoor West Virginia*, June 1967.
McNeil, George Douglas, "The Battle of Droop Mountain," *Davis and Elkins Historical Magazine*, #6, Apr. 1953.
______, "Tales of Pocahontas County," *Davis and Elkins Historical Magazine*, #19.
McNeill, Louise, "Droop Mountain," (poem), *Confederate Veteran*, Nov. 1928.
McNeill, Louise Pease, "The Prison Notebook of Captain James M. McNeill, C.S.A." *West Virginia History*, #3, April 1970.
Miller, Charles F., "Capture of Wagon Train By McNeill's Men," *Confederate Veteran*, #21, Jan. 1913.
Monson, Ann (edit.), "The Battle of Droop Mountain" and letter of Samuel H. Lucass, Co. F, 14th Virginia Cavalry, *Grape and Canister Newsletter of the Civil War Round Table of Wilmington*, Feb. 1988.
Siburt, James, "The Confederates' Last Stand in West Virginia," *Civil War Times Illustrated*, Mar. 1989.

Newspapers and Newspaper Articles

Adams, Nancy, "Ghosts of Droop Mountain," *The Charleston Gazette*, Oct. 30, 1985.
Alderson, George H.C. (Co. A, 14th Va. Cav.), "From the *Greenbrier Independent*—Letter concerning the battle of Droop Mountain and death of Major Bailey," *Nicholas Chronicle*, Feb. 17, 1927.
"Death Comes Quietly to Col. Arnett," *The Wheeling Intelligencer*, Feb 20, 1902.
"Gen. Averell," *The Pittsburgh Gazette*, Dec. 30, 1863.
Bahlmann, Captain William F. (Co. K, 22nd Va. Inf.), "Bahlmann Gives History of Men Serving in '61," *Charleston Gazette*, Sept. 17, 1922
______, "Battle of Droop," *The Pocahontas Times*, Nov. 29, 1923 (Originally in *Fayette Tribune*).
Barnes, Captain James W. (23rd Battalion Virginia Infantry), unknown newspaper clipping of letter describing the Droop Mountain battle, Courtesy Droop Mountain Battlefield State Park Files.
"Battle Anniversary," *The Pocahontas Times*, Oct. 24, 1935.
"The Battle of Droop Mountain," *The Pocahontas Times*, Nov. 8, 1895.
"Battlefield To Be Marked," *Clarksburg Express*, Nov. 5, 1935.
Benjamin, J.W., "The Battle of Droop Mountain," *The Charleston Gazette*, Jan. 12, 1958.
Bowman, Mary K., "Civil War in Wyoming County," *West Virginia Hillbilly*, Mar. 14, 1964.
"Brilliant Exploits in Western Virginia," (Correspondent "C," probably with the 2nd West Virginia Mounted Infantry), *The Ironton Register*, Dec. 3, 1863.
The Cannonball, newsletter of Droop Mountain Battlefield C.C.C. Camp Price, Company 2598, July 31, 1936 (Courtesy of Anna Atkins, Chesterfield, Virginia).
"Camp Price," *The Pocahontas Times*, Apr. 9, 1936.
"Captain James M. McNeill," unknown clipping.
"Casualties In the 10th [West] Virginia Volunteers at the Battle of Droop Mountain," *Wheeling Intelligencer*, Nov. 30, 1863.
Cochran, Lincoln S., "Droop Mountain," *The Pocahontas Times*, Dec. 26, 1935.
Comstock, Jim, "Outnumbered Rebs Licked at Droop," *West Virginia Hillbilly*, date unknown.
Untitled article on dedication of the Confederate Monument at Droop Mountain State Park, *The Pocahontas Times*, July 16, 1931.
Information from Corp. Daniels concerning Keepers' Battery and the 2nd West Virginia Mounted Infantry at Droop Mountain, *The Ironton Register*, Nov. 26, 1863.
Donnelly, Shirley, "Droop Mountain Battle Recalled," *The Beckley-Post Herald*, Nov. 4, 1967.
______, "History of the Battle of Droop Mountain," *The Beckley-Post Herald*, Mar. 20, 1957.
______, "Maj. 'Gus' Bailey, Hero of Droop Battle," *The Beckley Post-Herald*, Apr. 4, 1957.
______, "94th Anniversary of Droop Battle Today," *The Beckley Post-Herald*, Nov. 6, 1957.
"The Droop Battlefield," *The Pocahontas Times*, Nov. 15, 1895.
"Droop Mountain," *West Virginia Hillbilly*, Nov. 16, 1963.
"Droop Mountain," *The Pocahontas Times*, Oct. 17, 1974.
"Droop Mountain Battle Soldier," (David Van Buren—U.S. Soldier) *West Virginia Hillbilly*, Apr. 6, 1989.
"Droop State Park," *The Pocahontas Times*, Jan. 26, 1928.
Dunlap, M.A., "Battle of Droop Mt.," *The Pocahontas Times*, Sept. 13, 1923.
______, "From M.A. Dunlap," *The Pocahontas Times*, Dec. 29, 1925.
______, "From M.A. Dunlap," *The Pocahontas Times*, July 11, 1929.
______, untitled article on Droop Mountain battle, *The Pocahontas Times*, Nov. 14, 1935.
F.H.V., "Why Not Droop?," *The Pocahontas Times*, November 27, 1930.
"The Fight In Greenbrier and Retreat of Gen. Echols," (from the *Richmond Whig* of Nov. 14, 1863) *Wheeling Daily Intelligencer*, Nov. 20, 1863.
"Fourth of July," *The Pocahontas Times*, May 15, 1930.

Fortney, L.F., "Droop Mt. Battle," (letter concerning 3rd West Virginia Mounted Infantry), *The Pocahontas Times*, August 1, 1929.
"Gen. Averell's Work," *Point Pleasant Weekly Register*, Dec. 3, 1863.
"Major Gibson," *The Pittsburgh Post*, Dec. 2, 1863.
Griffith, Ann, "Preserving the Pioneer Spirit," *Charleston Daily Mail*, Aug. 18, 1982.
"Intercepted Letter," to *The Amelia Bulletin Monitor*, April 23, 1987 (letter from Water Lee Painter mentioning Thomas E. Wood, 19th Virginia Cavalry, at Droop Mountain).
Jackson, Col. William L., "Droop Mountain Battle," (report of Nov. 9, 1863), unknown clipping.
Johnson, Skip, "Droop Mountain Battlefield Ghosts," *Charleston Gazette-Mail*, Sept. 18, 1977.
"The Late Fight at Droop Mountain–Complete Rout of the Rebels," (Correspondent "T.M.") *Wheeling Daily Intelligencer,* Nov. 21, 1863.
"Letter" dated Nov. 28, 1863, written from Beverly, West Virginia (written by Hiram Depew concerning Keepers' Battery in the Droop Mountain battle), *The Ironton Register*, Dec. 10, 1863.
"Letter to Editor," from Pvt. Clark Kellison, Co. C, 3rd West Virginia Infantry, *The Pocahontas Times*, Nov. 15, 1895.
"Letter from Thomas Swinburn," *The Pocahontas Times*, Apr. 11, 1901.
"Letter from West Virginia," (Correspondent Q.P.F.), *Cincinnati Daily Commercial*, Dec. 11, 1863.
"Letter from West Virginia: The Condition of the Department–The Operations of the Army of Occupation–The Rust of Inactivity–Call for a Reformation–The Necessity for the Circulation of the Blood," (Correspondent T.B.F.), *Cincinnati Daily Commercial*, Nov. 26, 1863.
"Letter from West Virginia: Gen. Averell's Movements–Battle of Droop Mountain," (Correspondent "Union"), *Cincinnati Daily Commercial,* Nov. 27, 1863.
"Letter from West Virginia: Impressions of Charleston–Details of the Battle of Lewisburg by an Eyewitness," (Correspondent J.P.F.), *Cincinnati Daily Commerical*, Nov. 20 1863.
"Letter from West Virginia: Traveling on the Kanawha–The Broken Doctor–Southern Internal Improvements–Congratulatory Order on the Victory of Lewisburg," (Correspondent J.P.F.), *Cincinnati Daily Commercial*, Nov. 21, 1863.
Lockard, A.P., (letter concerning 10th West Virginia Infantry), *The Pocahontas Times*, January 24, 1929.
"Lone Veteran of the Grey Accepts Park," unknown clipping, Nov. 7, 1935.
"Captain Lurty Dead," *Harrisonburg Times*, 1906.
"Memorial Day," *The Pocahontas Times*, May 29, 1930.
"Memorial on Droop," *The Pocahontas Times*, May 15, 1930.
McNeill, G. Douglas, "Battle of Droop Mountain," *The Marlinton Journal*, June 12, 1958.
______, "Battle of Droop Mountain," *The Marlinton Journal*, June 19, 1958.
McNeill, Louise, "Droop," *The Pocahontas Times*, August 16, 1928.
Moor, Augustus, (28th O.V.I.), "Droop Mountain Battle," *The Pocahontas Times*, June 27, 1929.
Morris, Dorothea, "Droop Mountain Soldier," *West Virginia Hillbilly*, April 6, 1989.
Morrison, W.F., "Henry Bender, 90, Veteran of Civil War, Dead," *The Braxton Central*, March 27, 1931.
Muscari, Geraldine, "Key Roles Played by Area Men in Crucial Civil War Battle," *Parkersburg News*, Nov. 8, 1969.
N.R.P., "Droop Mountain," *The Pocahontas Times*, July 10, 1930.
Obituary for John H. Oley, *The Huntington Advertiser*, Mar. 17, 1888.
"Obituary," (for Wilson Rider), *The Pocahontas Times*, July 10, 1930.
"Official Reports of the West Virginia Success," *Gallipolis Journal*, Nov. 19, 1863.
Paragraph concerning rebels robbing a store at Scary Creek and their "recent repulse at

Lewisburg," *Gallipolis Journal*, Nov. 19, 1863.
"Particulars of General Averell's Late Campaign," *Wheeling Daily Intelligencer,* (Correspondent "M"), Nov. 25, 1863.
"From the 14th Pennsylvania Cavalry," *The Pittsburgh Gazette*, Nov. 21, 1863.
The Pittsburgh Gazette, Nov. 10, 1863 (news item about Averell and Duffié at Droop Mountain).
The Pittsburgh Gazette, Nov. 11, 1863 (news item about Averell and Duffié at Droop Mountain).
Poliafico, Chuck, "Local Man Baffled by Ghostly Vision," *The Parkersburg News*, Dec. 23, 1990.
Price, Andrew, "Battle of Droop Mountain," *The Pocahontas Times*, 1920's.
______, "Droop Mountain Battlefield State Park, Site of Most Bitter Civil War Engagement, will be Presented to State Wednesday," *The Charleston Gazette*, July 1, 1928.
Price, Calvin W., "Response to Col. W.H. Waldron," *The Pocahontas Times*, Oct. 31, 1935.
______, "Response to Sutton article on Droop Mountain," *The Pocahontas Times*, Nov. 26, 1925.
______, untitled article on battle anniversary, *The Pocahontas Times*, Nov. 14, 1935.
Price, Norman R., "On Droop Mountain," *The Pocahontas Times*, August 28, 1930.
"Program," *The Pocahontas Times*, June 28, 1928.
Rand, Noyes, "Reminiscences of the Battle of Dry Creek, August 26-27, 1863," *The Monroe Watchman*, July 8, 1909.
Roberts, Thomas Algernon, (22nd Virginia Infantry), "Droop Mountain Battle," *The Pocahontas Times*, September 27, 1928.
Siburt, James, "Battle of Droop Mountain," *West Virginia Hillbilly*, July 6, 1989.
"State Park To Be Dedicated July 4th," *The Pocahontas Times*, June 21, 1928.
"State Park Dedicated," *The Pocahontas Times*, July 12, 1928.
Staunton Vindicator, Nov. 13, 1863 and Nov. 27, 1863 (information on the Droop Mountain battle).
Stephens, J.G., "The Battle of Dry Creek," *The Fayette Tribune*, Jan. 6, 1916.
Sutton, J.D., (10th WV Inf.), "Droop Mountain," *The Pocahontas Times*, Dec. 1, 1927.
______, "One of Bloodiest Battles of the Civil War Fought at Top of Droop Mountain with 8,000 Troops Engaged in Conflict," *The Charleston Gazette*, Nov. 1, 1925.
Sutton, John D., "Droop Mountain," *The Pocahontas Times*, June 19, 1930.
Swetnam, George, "The Fourteenth Cavalry," *The Pittsburgh Press*, Oct. 8, 1961.
Waldron, Colonel William S., "Flashes of American History: Battle of Droop Mountain," *The Pocahontas Times*, Nov. 19, 1936.
"120 Years Later, What's the True Story?," *The Beckley Post-Herald*, July-Aug. 1985.

Manuscript Collections

William W. Averell Papers 1836-1910, New York State Library Cultural Education Center, Albany, New York.
"Autobiography of William F. Bahlmann," (22nd Virginia Infantry), Unpublished handwritten account. Copy in author's collection provided by Larry Legge, Barboursville, West Virginia.
Andrew R. Barbee Letters, Lt. Col. 22nd Virginia Infantry, Mics. 1864, Letters Received 1863-1865, Dept. of West Va. (Grant), Entry 5691 (Inventory Vol. 1), Record Group 393, U.S. Army Continental Commands 1821-1920 (Located in Stack Area 10wz, Row 11, Bottom Shelf), National Archives, Washington, D.C.
Barnes Family Papers, 1816-1929, papers of Thomas Rufus Barnes (Co. K, 10th West Virginia Infantry), West Virginia University, Morgantown, West Virginia.
John D. Baxter Letters (Co. F, 10th West Virginia Infantry), Droop Mountain Battlefield

State Park Files.
"Henry Bender–1840-1931," by Michael A. Smith, March 6, 1986, Droop Mountain Battlefield State Park Files.
Henry Bender Papers, (Co. F, 10th West Virginia Infantry), Droop Mountain Battlefield State Park Files.
Blain Family Papers 1860-1869 (Jackson's Battery), Washington & Lee University, Lexington, Virginia.
Edgar Blundon Papers (7th West Virginia Cavalry), Boyd Stutler Collection, West Virginia State Archives, Charleston, West Virginia.
E.E. Bouldin Papers (14th Virginia Cavalry), University of Virginia, Charlottesville, Virginia.
"From the Life of Mark Crayon," West Virginia State Archives, Charleston, West Virginia.
William M. Dearing Information, (Co. F, 3rd West Virginia Mounted Infantry), letters to author from Brian Kesterson, Washington, West Virginia.
Dunlap Family Papers 1863-64, courtesy Helen Steele Ellison. Droop Mountain State Park Files, Hillsboro, West Virginia.
John Echols, Alumni File, Virginia Military Institute, Lexington, Virginia.
George Mathews Edgar Papers, Southern Historical Collection, University of North Carolina, Chapel Hill, North Carolina.
Samuel D. Edmonds Memoir (Company B, 22nd Virginia Infantry), "The Battle of Droop Mountain," Courtesy Mike Smith, Superintendent, Droop Mountain Battlefield State Park, Hillsboro, West Virginia and Evelyn McPherson.
Jacob S. Hall Letters, (Company E, 19th Virginia Cavalry), original in possession of Homer C. Cooper, Athens, Georgia. Courtesy Droop Mountain Battlefield State Park Files.
Thomas M. Harris Papers 1861-1892 (10th West Virginia Infantry), West Virginia University, Morgantown, West Virginia.
Civil War Miscellaneous Papers 1859-1937, (includes letters of Sgt. B.F. Hughes, Company F, Third West Virginia Mounted Infantry–later Company F, Sixth West Virginia Cavalry), Item #1021, West Virginia University, Morgantown, West Virginia.
Milton Humphreys' Papers (Bryan's Battery), University of Virginia, Charlottesville, Virginia.
Frank M. Imboden (Co. H, 18th Virginia Cavalry, C.S.A.), "War Diary," (Typescript copy), West Virginia State Archives, Charleston, West Virginia.
James Ireland Diary (Co. A, 12th O.V.I.), Ohio Historical Center, Columbus, Ohio.
Larry Jarvinen, "History of Pocahontas County Place Names," paper prepared by Larry Jarvinen, Manister, MI, (N.D.), copy provided to author.
George R. Latham Papers–1863, (Josiah Davis, 3rd West Virginia Mounted Infantry), West Virginia State Archives, Charleston, West Virginia.
Diary of Thomas H. B. Lemley (copy), 1st West Virginia Cavalry, West Virginia State Archives, Charleston, West Virginia.
Letters of William Ludwig, 34th Ohio Volunteer Infantry, West Virginia University, Morgantown, West Virginia.
Jacob Williamson Marshall Papers 1852-1899, West Virginia University, Morgantown, West Virginia.
H.E. Matheny Collection (14th Virginia Cavalry, C.S.A.), Letters, West Virginia University, Morgantown, West Virginia.
James Z. McChesney Letters (14th Virginia Cavalry), H.E. Matheny Collection, West Virginia University, Morgantown, West Virginia.
James S. McClung Memoir (Co. K, 14th Virginia Cavalry), Copy provided to author by John Arbogast, White Sulphur Springs, West Virginia.
William McLaughlin Papers, Washington & Lee University, Lexington, Virginia.
Lanty F. McNeel Collection, Hillsboro, West Virginia.
"Nicholas Blues," poem by Capt. James M. McNeil (Co. D, 22nd Virginia Infantry), Courtesy John Gregory, Covington, Virginia.

Augustus Moor Papers (28th O.V.I.), Ratterman Collection, University of Illinois at Urbana-Champaign, Urbana, Illinois.
Company books of the Ohio Volunteer Infantry, Co. D, 28th Regiment in the Civil War, Cincinnati Historical Society, Cincinnati, Ohio.
George Washington Ordner Diary, (2nd West Virginia Mounted Infantry), West Virginia University, Morgantown, West Virginia.
Ronn Palm Photo Collection, (14th Pennsylvania Cavalry), United States Army Military History Institute, Carlisle Barracks, Pennsylvania.
Peterson Family Papers 1858-1913, (Lt. David T. Peterson, Co. B, 10th WV Inf.), West Virginia University, Morgantown, West Virginia.
"Lieutenant Colonel John J. Polsley—7th West Virginia Regiment 1861-1865," thesis by Eugene Wise Jones, June 1949, presented to the University of Akron, Akron, Ohio.
John J. Polsley Papers 1862-1865 (8th West Virginia Mounted Infantry), Letters 108 and 109 (Nov. 14 and 20, 1863), University of Akron, Akron, Ohio.
The Francis S. Reader Papers, "An Average American: The Memoirs of Francis S. Reader—Volume II, " completed Nov. 17, 1924, U. S. Army Military History Institute, Carlisle Barracks, Pennsylvania. (2nd West Virginia Mounted Infantry).
Renick Farm History, Droop Mountain State Park Files, Hillsboro, West Virginia.
Shaver Family Papers 1861-67, Civil War Collection, West Virginia State Archives, Charleston, West Virginia, (includes letters of George T. Shaver, 36th Virginia Infantry).
"The Battle of Droop Mountain," as related by C.L. Stulting, Droop Mountain Battlefield State Park Files, Hillsboro, West Virginia.
Boyd Stutler Collection, West Virginia State Archives, Charleston, West Virginia.
William P. Thompson Papers (19th Virginia Cavalry), National Archives, Washington, D.C., courtesy of John Ashcraft.
Civil War Diary of Amos A. Vandervort (14th West Virginia Infantry), courtesy William D. Wintz, St. Albans, West Virginia.
John W. Weed Memoir (91st Ohio Volunteer Infantry), Civil War Miscellaneous Collection, U.S. Army Military History Institute, Carlisle Barracks, Pennsylvania.
Micajah Woods Papers (Jackson's Battery), University of Virginia, Charlottesville, Virginia.
James L. Workman Information (Co. B, 7th West Virginia Cavalry), Droop Mountain State Park Files, Hillsboro, West Virginia.
Denver Yoho Collection (contains material on the 7th West Virginia Cavalry, the 1st and 3rd West Virginia Cavalry, and various other West Virginia material both North and South), copies sent to author from Denver Yoho, Gallipolis, Ohio.
Pvt. Benjamin F. Zeller (Co. L, 8th O.V.C.), "Over the Mountains and Down the Valleys of Virginia: Thirty Days in the Saddle," original manuscript at the Garst Museum, Greenville, Ohio (Thanks to Bob Cartwright, Dayton, Ohio, for bringing this valuable resouce to my attention).

About the Author

PHOTO COURTESY TOM WILLS, CHARLESTON, WEST VIRGINIA

Terrance (Terry) David Lowry was born November 18, 1949 in Charleston, West Virginia. A 1967 graduate of South Charleston High School where he worked on the school newspaper. Graduated in 1974 with a B.A. in History from West Virginia State College. Studied Civil War history at Marshall University Graduate School. Professional musician for twenty-six years. Contributing music editor for *The Charleston Gazette* 1970-75. Music editor for *The Charleston Gazette* 1977-78. Two years with the Circulation Department of *The Atlanta Journal*. Published his first book, *The Battle of Scary Creek: Military Operations in the Kanawha Valley, April-July 1861* in July of 1982. Has also published *September Blood: The Battle of Carnifex Ferry* (1985); and two volumes in the *Virginia Regimental Histories Series*, 22nd Virginia Infantry (1988) and 26th (Edgar's) Battalion Virginia Infantry (1991). Mr. Lowry has also had Civil War articles published in *North South Trader, Wonderful West Virginia,* and *Confederate Veteran* magazines, as well as the *West Virginia Hillbilly* newspaper. A contributor to the *Time-Life* series of books on the American Civil War and an avid collector of Civil War memorabilia. Currently employed by the Circulation Department of Charleston Newspapers, Inc., Terry can be reached at 237 Kenna Drive, South Charleston, W.Va. 25309

About the Cover Artist

Tim Decker grew up in Columbus, Ohio where he attended the Columbus College of Art and Design. He received his B.A. from Appalachian Bible College in 1980, and his M.A. from Marshall University in Fine Art in 1982, as well as certification to teach Art for grades K-12. Since that time he has taught art to students from pre-school age to the college level. He has worked as an instructor at the Huntington Museum of Art and has supervised their summer Art Camp program. He does free lance art work for various businesses and has illustrated books for Aegina Press and University Editions publishers.

Presently he is an instructor of art at Russel Middle School in Kentucky and a part-time instructor at Marshall University teaching undergraduate classes. He is in his final year of completeing his doctorate degree in education administration from the University of Kentucky. Tim resides in Barboursville, WV with his wife and son. Besides drawing and painting, Tim does Civil War re-enacting in his free time. He began re-enacting in 1986 and belongs to the 91st Ohio Volunteers, Company B.

Index

Droop Mountain is not indexed.

Made in the USA
Monee, IL
22 June 2023

36651277R00177